CATACLYSMS

CATACLYSMS

An Environmental History of Humanity

LAURENT TESTOT

Translated by Katharine Throssell

The University of Chicago Press
Chicago and London

The University of Chicago Press, Chicago 60637
The University of Chicago Press, Ltd., London
© 2020 by The University of Chicago
All rights reserved. No part of this book may be used or reproduced in any manner whatsoever without written permission, except in the case of brief quotations in critical articles and reviews. For more information, contact the University of Chicago Press, 1427 E. 60th St., Chicago, IL 60637.
Published 2020
Printed in the United States of America

29 28 27 26 25 24 23 22 21 20   1 2 3 4 5

ISBN-13: 978-0-226-60912-6 (cloth)
ISBN-13: 978-0-226-60926-3 (e-book)
DOI: https://doi.org/10.7208/chicago/9780226609263.001.0001

Originally published as *Cataclysmes: Une histoire environnementale de l'humanité*.
© 2017, Editions Payot & Rivages

Library of Congress Cataloging-in-Publication Data

Names: Testot, Laurent, author. | Throssell, Katharine, translator.
Title: Cataclysms : an environmental history of humanity / Laurent Testot ; translated by Katharine Throssell.
Other titles: Cataclysmes. English (Throssell)
Description: Chicago : University of Chicago Press, 2020. | Includes bibliographical references and index.
Identifiers: LCCN 2020018400 | ISBN 9780226609126 (cloth) | ISBN 9780226609263 (ebook)
Subjects: LCSH: Human ecology—History. | Human beings—Effect of environment on—History. | Nature—Effect of human beings on—History.
Classification: LCC GF13 .T47 2020 | DDC 304.2/09—dc23
LC record available at https://lccn.loc.gov/2020018400

♾ This paper meets the requirements of ANSI/NISO Z39.48-1992 (Permanence of Paper).

To Philippe Norel (1954–2014), for having led me into global history

Many thanks to Geneviève Darles, for all her help in preparing the manuscript

CONTENTS

Introduction   ix

PART I: MONKEY CONQUERS THE WORLD
1. We Are the Children of the Climate   3
2. The End of the Elephants   25
3. The Wheat Deal   50
4. Collapse   79

PART II: MONKEY DOMINATES NATURE
5. When Gods Guide the Way   95
6. All Empires Will Fall   116
7. After Summer Comes Winter   139
8. Biological Hazards   164
9. Demographic Hazards   186

PART III: MONKEY TRANSFORMS THE EARTH
10. The Promises of Quicksilver   207
11. Cold, Cold Earth   226
12. Dying for the Forest   241
13. Unlimited Energy   259
14. The Cold Chill of Catastrophe   285
15. A Time of Excess   308
16. The Blind Flock   332
17. Tomorrow's World   348

Conclusion   379

Epilogue to the English Edition: Two and a Half Years after the French Edition . . . 387

Appendix A: Glossary   395

Appendix B: Chronology   399

Notes   407

Bibliography   425

Index   445

INTRODUCTION

This book first began to take shape as I was sitting on the edge of a hot volcanic pool near Yamanouchi, a village deep in the Japanese Alps. At first sight the place seems idyllic if you overlook its theme park–like name, Jigokudani, meaning "Hell Valley." Perhaps you already know of this park and the Japanese snow monkeys who live there, now immortalized in numerous documentaries and photographs. The pool is where the monkeys bathe. Once, bathing might have been a perfectly spontaneous event for them, but these simian ablutions have become a boon for tourism. So now the snow monkeys are gently encouraged to take a dip.

I arrive early in the afternoon. Some young monkeys are playing. They dive into the water, swimming and squabbling. The biggest one delights in dunking the smaller ones under the watchful but sporadic gaze of a few adults until the game goes too far and an older female intervenes with a growl and a smack. It could almost be a human* kindergarten. The monkeys hold the visitors' gaze, their eyes heavy with all the emotions we normally think of as reserved for our own kind.

Tourist photos of this place are ubiquitous but misleading. They are generally taken in winter, in the snow, with the monkeys huddled together in the hot water while a tempest rages. They hint at a place lost in time, inaccessible, in the depths of a lost valley. It seems so "natural."

In reality, the snow falls on concrete. The pool was artificially built in an easy-to-access location—easy, that is, if the *gaijin* (foreigner) has mastered the subtle dance of Japanese driving. It is only a ten-minute walk from the parking lot up to the house of the park's guards, where a small fee will grant you entry to the gorge that leads to the pool.

Two hundred monkeys live here. A peaceful tribe. The afternoon stretches out, marked only by the cavorting of the young ones. At the end of the day, it becomes clear why they stay by the pool. Two employees appear carrying a large crate of apples. The macaques converge on them, organizing themselves in concentric circles. A few punches are thrown. A large male moves forward, insistent, toward the humans.

He will be the first to be fed, but not without also being served a reminder that he is inferior to his feeders. The two employees reinforce the group hierarchy and impose themselves as superior while also ensuring that no one is forgotten. They throw the apples violently, like baseballs, smashing them on the rocks and on the concrete. The monkeys run in all directions. Some jump into the water. The dominants gobble down the fruit while the subordinates fight for the scraps.

The sun is setting. The monkeys are also going, climbing up the cliffs. This is nature, Japanese style. There is no overt trace of human intervention, yet it is totally artificial, anthropized,* shaped entirely by human hands. It is a striking analogy for our planet today.

*The Saga of Monkey**

This book is like a film. It relates how humans have progressively transformed the planet, creating peaceful places and urban hells. It also recounts how nature, distorted, has retaliated: in return for the metamorphoses it has been subjected to, it has reshaped humans' bodies and minds.

It is blockbuster material. The narrative covers three million years, conservatively speaking. Of course, given just a few hundred pages, we will be staging key scenes and focusing on pivotal stories. And we have cast some actors to bring this planetary drama to life.

The main character is Monkey, because of all the animals, he is the closest to us. We are, after all, "naked apes."[1] The figure of Monkey provides a condensed vision of humanity* as a whole. He is also a major mythological character in both China and India, two of the most historically important cultures on the planet.

In China, Monkey, known as Sun Wukong, is the protagonist in *Journey to the West*,[2] a picaresque sixteenth-century novel that is more popular in China than its Western equivalents—*Pantagruel, Gargantua, Gulliver's Travels*—are in Europe. *Journey to the West* has two parts. The first puts Monkey center stage. He is a peasant among supernatural beings, destined to embody the underdog, a rube who must live in the shadows, a stable boy to the gods. But Monkey has a cunning mind. He tricks his way into learning sorcery and steals a magic sword from the Dragon King. Something like a *Star Wars* light saber, this 6-meter (20-foot) iron bar can be shrunk to the size of an embroidery needle. Monkey breaks into the Heavenly Peach Garden, whose peaches bestow everlasting life, and eats them all. Furious that the secret of immortality has been lost, the gods send their most powerful armies to punish the thief. But to no avail. Monkey cannot be captured; the heavenly peaches have given him astronomical power, and he gives a good beating to any immortal who comes near.

Only the intervention of Buddha puts an end to Monkey's antics. As punishment for his wanton ways, Buddha orders him to be the bodyguard for a young monk who is traveling into the West (to India) to revive the sacred word of Buddhism at its source. Overcome with remorse, Monkey accepts. This pilgrimage constitutes the second part of the book, which is just as rich in social satire and fantastical battles as the first. At the service of pious humanity, Monkey and his companions strike down all the chimerical forces that nature throws in their path.

In India, Monkey takes the form of Hanuman, King of the Monkeys. He has enormous strength and can lift mountains and leap as far as Sri Lanka in a single bound. In the epic poem *Rāmāyaṇa*, Hanuman helps the god Rama rescue his wife Sita, who has been abducted by the demon Ravana. This Monkey-god is extremely popular because he symbolizes the wisdom of the people, defends peasants, and incarnates the generosity of those who have nothing other than their word. The monkey weeps not for himself but for others, holds an old Indian proverb.

These two Monkey figures provide a perfect metaphor for

humankind, a hyperpredator who has become the unlawful king of the earth. Yet we also owe our special status to our acute sense of empathy that enhances cooperation between humans. Monkey is an animal whose vitality has been boosted by culture.* It is through collaboration that humanity can move mountains, alter the vegetation of continents, and fly through the skies from London to Japan.

Moreover, using Monkey as a metaphor for humanity helps us remember a fundamental premise: humans are animals. We are animals who consider ourselves exceptional, and yet today we struggle to define just what sets us apart. We have culture. But other animals demonstrate culture. Tools? Cognition? We are not alone in these either. Humanity is above all characterized by the scale on which these qualities have been applied; no other species can alter nature to the same extent.

Our story will therefore be that of Monkey, a concentrated essence of humankind. We must keep in mind that Monkey is always a trickster—like Loki, the mischievous Scandinavian god of fire, or Prometheus, the polytechnic Titan who gave humans fire and tricked the gods out of the tastiest morsels of sacrificial meat. In punishment for these crimes, Zeus chained Prometheus to a mountaintop where every day a giant eagle would devour his liver and every night his liver would grow back again.

Prometheus is often held up as a tutelary deity personifying our technical age, marked by the industrial revolution of fire. He is the reflection of a humanity that must pay for the liberation of the terrestrial forces of coal and oil in suffering that sometimes gnaws at its organs like some endocrine-disrupting eagle.

Monkey's saga is made up of seven revolutions (detailed below), each of which is the object of one of our chapters. These seven revolutions are capitalized because they are major evolutionary processes predated by long periods of adaptation.[3] The succession of these revolutions has progressively become faster and faster as the cumulative effects of human culture have made themselves felt. It took five to seven million years to amass the effects of the Physiological Revolution that transformed a frugivorous, quadrupedal primate into an omnivorous, bipedal, tool-using human. Hundreds of thousands of years then paved the way for the Cognitive Revolution, while tens of thousands of years (and a global heat wave) provided the prerequisites for the Agricultural Revolution. The Moral Revolution began over a few thousand years, and the Energy Revolution emerged in a few hundred. The Digital Revolution that followed took only a few decades. The next, the Evolutive Revolution, will take only a few years. In fact, it is already here.

Monkey has initiated an extraordinary acceleration of time itself.

The scene is set: the whole planet and its different environments. Monkey, the lead actor, has signed on without hesitation. The screenwriter is yours truly, professional journalist, lecturer, and teacher in world history submerged in this discipline for more than a decade. But there can be no film without a script. How can we trace the history of the world over three million years? We need a method—global history—and a field—world environmental history.

Toward a Global History

Global history can be defined as a method that allows us to explore the field of world history—all the different pasts of humanity— from its tentative beginnings in Africa three million years ago to the globalization we see today.[4] It is the living tool that allows us to produce this world history, and it is brought to life by four strands of DNA. Global history is (1) transdisciplinary; it brings together other disciplines in equal measure, including economics, demographics, archaeology, geography, anthropology, philosophy, social sciences, and evolutionary biology. It (2) analyzes the past over the long term. It (3) encompasses a broad space. And it (4) plays out on different levels, both temporal and spatial. It produces a narrative that opens the door wide to humanity's varied pasts, emphasizing a biographical anecdote, for example, before looking at its global implications. Could the lost harvest of a peasant in 1307 be attributed to a global cold snap? And what might that cold period tell us about global warming today?

I have written an in-depth review of the Anglophone studies in world history, soon to be published as a book, combining different historiographic approaches, and this increased my awareness of the importance of the natural environment in human history.⁵ If Monkey is an actor in his own story, the environment is its stage and determines its possibilities.

Toward an Environmental Narrative

Environmental history was officially born in the United States in the 1970s, although it is possible to trace its origins much farther back, first to Montesquieu and then to Aristotle and his Chinese contemporaries. American authors also emphasize the fundamental role of Anglophone pioneers, such as George Perkins Marsh. In *Man and Nature* (1864), this linguist documented the impact of human action on the lands of the ancient Mediterranean civilizations and deduced that deforestation was the systematic prelude to desertification. By way of conclusion, he called (even then) for the restoration of ecosystems, forests, soils, and rivers. And he prayed for the advent of a humanity that would collaborate with nature rather than destroy it. In 1915, the geographer Ellsworth Huntington diagnosed the aridification of Asia in *Civilization and Climate*. He also noted that in the past, variations in climate have led to the destruction of civilizations.

In the period after the Second World War, the geographer William M. Thomas edited the book *Man's Role in Changing the Face of the Earth* (1956), which documents the extent of the environmental change produced by humans from prehistory to today. A little later on, Roderick F. Nash set about demonstrating the social evolution of the perception of nature in America in his book *Wilderness and the American Mind* (1967). In the same year, the geographer Clarence J. Glacken published his landmark work *Traces on the Rhodian Shore*, a monumental history of human attitudes toward nature in the West from antiquity to the eighteenth century. Environmental history was officially baptized in 1972 by the historian Alfred W. Crosby Jr. with his book *The Columbian Exchange* (see chap. 9). By a happy coincidence, the same year saw Nash establish the first chair of environmental history at the University of California, Santa Barbara. The benefit of continuing in this intellectual direction was confirmed in 1976 by the historian William H.

McNeill with *Plagues and Peoples*, a masterful analysis of microbes as a driving force in history (see chap. 8).

Since then, publications in this area have abounded. In addition to work in North America, certain European historians—especially British; sometimes Swiss, German, Dutch, and Italian; and more recently French[6]—are also involved in this movement. South Africa, India, and Australia have also established solid traditions in this field, but the environmental histories of China, Japan, Russia, and the Islamic world* still remain largely the domain of American historians.

Schematically speaking, environmental history can take three main forms: one that aims to bring nature into history, to historicize it; one that studies the impact of humankind on the environment, which is particularly in demand today as societies fight environmental damage; and, finally, one that looks at the impact of the environment on humanity—for example, in terms of health or the trajectories of societies. The discipline is by nature eclectic. It incorporates social sciences and geography as well as physical and biological sciences. But it sometimes struggles to reconcile these different forms and is often accused of overreaching. This book, for example, will look at wars, religions, political ideologies, and economics because these products of human societies are not only subjects for the social sciences but also ways of interacting with our surroundings. Religions and political ideologies dictate the ways in which we engage with the environment. The economy exploits natural resources. And war leaves biotopes battered and scarred.

A Film on Human-Nature Relations

Clearly, a book like this cannot exhaustively cover the three-million-year history of the whole world. Choices had to be made. Certain scenes illustrate global processes. Chapter 12, for example, will focus on forests in the modern era, but at other points in the book they will be mentioned only in passing even though their evolution has always been crucial for humanity. Elephants will often be in the spotlight while salmon will not, yet both of these animals have things to teach us about humans' relationships with nature.

Africa will be mentioned only rarely, because the environmental historiography provides us with few sources on it. China, India, and Europe, the decisive spaces of global history as it is written today, provide our regular backdrops.

Before we go any further, let us state the obvious. Like any animal, a human organism has three obsessions: (1) finding food, to ensure short-term survival; (2) sleeping, to ensure medium-term survival; and (3) reproduction, to ensure long-term survival.

I am going to spoil the suspense right away and reveal the thesis that underpins this book. As with all animals, evolution pushes us to have as many descendants as possible regardless of their quality of life. We live in societies of incomparable wealth and comfort, yet we are not programmed to make rational choices in terms of food nor to force ourselves to exercise. If we were, obesity would run less rampant. Nature, seen through the magnifying glass of evolutionism, is laughing at us individuals. All that matters to it is the perpetuation of the species, its expansion. Individuals matter for their multiplication, not for their qualities. In view of obsession number three, human history reads like Monkey's success story, with the expansion of the population to a genuinely incredible scale. But what if the trickster has tricked us? What if we have signed a pact with the devil? Will there not be a price to pay at the end of the story?

Monkey has achieved an unprecedented feat. We have transformed our surroundings in a way that was previously unimaginable. But although we can radically alter our environment, we can never be free of its influence. Like Prometheus, we have usurped the power of the gods—in the form of energy—only to discover that it is destroying us from the inside. Monkey has overcome epidemics. We now live longer and better lives. But we pay for it in cancers, diabetes, and heart disease, much of which is caused by the invisible modifications we have inflicted on the environment.

All books must be selective, and I do not think that there is a right way to explore history, particularly when working on very large temporal, disciplinary, and spatial scales. Much as there is no neutral journalism, there is no historian presenting "real history." All history is written out of the subjective experience of its author. I have therefore tried to avoid the pitfalls of "tunnel history," denounced by the geographer James M. Blaut, in which we use the present to explain why—in light of the past—we could not possibly be elsewhere than where we are. If history were that deterministic, mathematicians would have long since had the absolute monopoly on the production of historical knowledge. History is malleable. At any moment it could have led to other trajectories. It is important to understand that. The realm of possibilities remains open as far as the environment is concerned. The state of the world may have been quite different if in 1048 the embankments of the Yellow River had been reinforced enough to resist the devastating floods that carried away the Song dynasty (see chap. 7). If in 2009 US president Barack Obama had chosen, as Iceland did, to consider the banks responsible for compensation after the financial crisis, our present may have been very different.[7] The point here is not to produce counterfactual history but to bear in mind that we can always shape our future.[8] I simply hope that by presenting certain key elements from our long, shared history with mother nature, we will be able to think more clearly about the future that we desire in the hope that we can make the vital decisions that are needed to achieve it.

The trailers are over, the lights have gone down. The film opens with the African savanna, where our story begins . . .

The Seven Revolutions

1. The Physiological Revolution (also called anatomical, around three million years ago): emergence of the *Homo* genus and of tools, bipedalism, running, throwing objects, omnivorous feeding, global expansion. Monkey becomes **human** (chap. 1).
2. The Cognitive Revolution (also called symbolic, between 500,000 and 100,000 BCE): fire, art and language, domination of the environment and extinction of all the *Homo* species except *sapiens*. Monkey becomes a **hunter** (chap. 2).
3. The Agricultural Revolution (also called the Neolithic, begins nearly twelve thousand years ago): leads to the domestication of nature and a demographic boom. Monkey becomes a **farmer** (chap. 3).

4. The Moral Revolution (also called axial, 2,500 years ago): societies become connected over long distances, generating collective groups—empires and religions—that aspire to universality, collaborating more effectively to exploit their surroundings, and inventing money to boost their interactions. Monkey finds **religion** (chap. 5).
5. The Energy Revolution (also called industrial, around the year 1800): the choice to burn fossil fuels for energy pushes humanity onto a new trajectory. Like the preceding ones, this revolution is multifaceted. Depending on the discipline and on which component is emphasized, it can be read as scientific, military, economic, or demographic. What is important is its effect: the unification of the world under European hegemony followed by the profound modification of the global environment and the beginning of the Anthropocene.* Monkey becomes a **worker** (chap. 13).
6. The Digital Revolution (also called the media revolution, around the year 2000): communication technologies enable intricate connections over the whole planet in real time. Monkey becomes a **communicator** (chap. 16).
7. The Evolutive Revolution (also called demiurgic, over the course of the twenty-first century). Two main trends coexist: (1) the "great convergence" of NBIC technologies—nanotechnology, biotechnology, information technology, and cognitive science—leads to the emergence of new entities (augmented humans, cyborgs, artificial intelligence, and so forth) who will replace or coexist with humanity; (2) the inability of humanity to change its behavior will alter the planet's environment to the point where humans will involuntarily be transformed into "mutants" adapted to the new ecological situation of the Anthropocene. Monkey will become either a **god** or a **mutant** (chap. 17). The future, by definition unpredictable, should fall somewhere between these two extremes. Or perhaps it will combine them? It is easy to imagine superrich elites able to indefinitely prolong their precious existence with exorbitantly expensive technology while common mortals suffer the burden of increasing environmental degradation.

PART I

Monkey Conquers the World

1

WE ARE THE CHILDREN OF THE CLIMATE

Three million years ago, Africa began to dry out. One primate, the winner of the evolutionary lottery, shot up to the rank of Lord of All Creation. Monkey set out to conquer the world.

Three men walk in the desert, eyes down. It is the early morning and already the sand shimmers in the heat. We are in southern Africa, in the Kalahari Desert, which is arid and scattered with thorn bushes. The men wear shorts and sneakers. They are bare chested, thin and muscular. They are built to run.

The Race to Death

Hoofprints in the sand, traces of greater kudu, antelopes with magnificent spiraled horns. For our three hunters, an adult kudu means a feast that would last several days in their community. But they have to manage to kill one. Without guns. They are from the San people in the Kalahari, one of the last human communities in the world to still practice our oldest hunting technique: the race to exhaustion.

They approach in silence, downwind to ensure their smell and the noise of their footsteps do not alert their prey. They know that these animals are faster than them and that if the herd remains together, their prints will be impossible to track. The first task is to isolate an individual. They jog, then sprint, yelling to scatter the antelope, then slow down . . . One of the men throws a stick at a powerful adult male kudu who has broken away from the group. The challenge is on.

Everything will now play out between the bipedal runner, the only hunter who will follow the tracks, and his quadrupedal quarry. The hunter's only equipment is a large knife, a flask, and a spear.

The sun rises, its heat beating down on the hunter's head and on the thick skin of his prey. For the human, the sun is an ally. His curly hair traps cool air close to his head while his body, covered in pores, expels excess heat as sweat. His body encases a powerful machine, a heart able to endure prolonged physical effort. Being bipedal means his lungs do not suffer from the pounding of forefeet on the ground, and his internal organs are firmly attached to his hips by core abdominal muscles. His legs are endowed with long, lean muscles providing him powerful propulsion. Humans have evolved to excel at endurance. If he loses sight of his prey, he will follow its tracks until he eventually, inevitably, finds it.

For the antelope, however, the sun is an enemy. Under its fur, it does not sweat. It cannot expel the excess heat that progressively engulfs it except by desperate panting. Over short distances it can easily outrun the presumptuous hunter, but it will soon run out of breath and, gasping, be obliged to stop. Moments later, a cry from the hunter will scare it into flight again.

It is now midday by the sun. For the hunter, this is the time when his body leaves no shadow on the ground. For the antelope, it is the last chance, after five hours of unrelenting pursuit. It darts into a forest of thorny acacias that will provide some shade and—could it know this?—a cover of vegetation that may conceal its tracks.

The runner enters the thicket. No sign of his prey. He pauses in the shade of a tall tree, rests a moment, takes two or three swigs from his flask, and, dusty and overheated, tips some water over head and shoulders. He performs a few propitiatory gestures and tries to think like the animal: if he were a kudu, where would he go? Over there, of course. A dozen or so paces on, the race begins again.

It goes on for another three hours until the antelope finally collapses, asphyxiated by the effort. The man moves purposefully toward his prey, despite his own burning muscles. For an instant he watches the dying animal and then, with a skillful thrust, he spears it in the chest, a symbolic movement that coincides with the kudu's last breath.

Kneeling before the carcass, the hunter gathers his thoughts. He thanks his quarry for its valiant struggle, for its strength and endurance. He sprinkles dust over the animal so that its spirit may leave its body in peace; this body that will provide a moment of abundance to a group of humans who are among the poorest on our so opulent planet.

Where Do We Come From?

The biologist and long-distance runner Bernd Heinrich summarized evolution like this: "Every morning in Africa, an antelope wakes up. It knows it must outrun the fastest lion, or it will be killed. Every morning in Africa, a lion wakes up. It knows that it must run faster than the fastest antelope, or it will starve. It doesn't matter whether you're a lion or an antelope—when the sun comes up, you'd better be running."[1] Movement is life.

For animals, movement is internal. It comes from muscles that burn the energy obtained from eating plants or other animals whose food (at the bottom of the food chain) is necessarily derived from plants.

For plants, movement is external. In growing, they use the nutritional elements of minerals to lead tendrils, leaves, and branches toward the light. Plants store solar energy, and their only movement is growth. But their seeds travel far, sometimes very far, on the wind or water, via pollenating insects, or as stowaways in birds' stomachs or on mammals' fur. More and more often they are also carried by human vehicles (boats, trucks, planes, or shoes).

Originally, humans were like any other animals. Our energy came exclusively from the plants and animals we could eat. But today, like plants, we use other sources of energy drawing on animals, wind, and water to move more efficiently. We have learned to burn ever more energy-rich substances, such as wood, coal, oil, and gas. Even the atom is used. This special relationship with energy is part of what makes our species an exception in the animal world. But this relatively recent triumph should not mask the fact that for millions of years the only way humans could move was with their feet.

Where did we learn to walk and run like this? How did we become bipedal? In short, where do we come from?

Paleoanthropologists (archaeologists who study the evolution of ancient humans and prehumans) agree that humans are a member of the great ape family known as Hominidae (its members being known as hominids*). Our closest cousins are chimpanzees and bonobos; gorillas are slightly more removed. According to geneticists, the common ancestor of humans, chimpanzees, and bonobos diverged from the ancestors of gorillas around nine million years ago. The gorilla-human last common ancestor is referred to by its acronym, GHLCA.

Humans and chimpanzees diverged from each other approximately six million years ago (although it seems that some genetic exchanges continued periodically between the two before they became completely distinct), and bonobos broke away from chimpanzees around one million years ago. Most of the traces of these species have been found in Africa, so we assume that is where these evolutions took place. The scenario sketched out by geneticists is consolidated by two fossils—Toumaï and Orrorin—at least for the moment, because if there is one discipline that is constantly challenging its own recent findings, it is paleoanthropology.

Let us take a closer look at the first of these fossils: Toumaï, or *Sahelanthropus tchadensis*. In Greek *Sahelanthropus* means "Sahel man." The name of the species, *tchadensis*, refers to the fact that it was first found in Chad. This seven-million-year-old fossil was found in July 2001 by Ahounta Djimdoumalbaye, a member of the paleoanthropological team led by Frenchman Michel Brunet. This first fossil of the *Sahelanthropus* genus's nickname "Toumaï" means "hope for life," a name suggested by the then president of the Republic of Chad, Idriss Déby. This nickname is often used in the Gorani language for children born just before the dry season who therefore face a higher risk of mortality during their early months. In fact, this name was above all a reference to the memory of one of Déby's brothers in arms who was killed in their struggle for power.

Fossils are clearly a vehicle for political and identity issues—

particularly when they are exceptional. Toumaï was, because near-complete skulls are exceedingly rare, and we can learn many things from a skull. From a quick glance at this one, we can see that its owner was male, had a small brain (360 cubic centimeters [22 cubic inches], roughly the size of a chimp's), weighed about 35 kilograms (77 pounds), and was about a meter (3.3 feet) tall. The position of the hole in the base of the skull, the foramen magnum, seems to suggest that this animal could have been a bipedal ape. To understand the function of this hole, we have to imagine the bony human skull as a kind of armor to protect the most important organ in our body: the brain. The foramen magnum is where the cables come out, connecting the nerves between the brain and the rest of the body. When this hole is positioned closer to the middle of the skull, this suggests that the head is balanced on the spinal cord, as it is for humans. When it is more toward the back of the skull, the animal generally walks on all fours, like the great apes and our remote ancestors.

Toumaï competes with Orrorin for the title of the oldest homininan*—the group that excludes chimps, gorillas, and bonobos but includes our now-extinct partially bipedal kin of which we are the last representatives.

We have found only fragments of Orrorin: three femurs, two jaw bones, a phalanx from a finger, six teeth, and a few other bits of bone, all belonging to five different individuals. They were found in Kenya by Ezra Kiptalam Cheboi and identified as belonging to a prehuman bipedal ape by paleoanthropologists Brigitte Senut and Martin Pickford at the end of the year 2000, only a few months before Toumaï was discovered. This earned the fossils the nickname "thousand-year man." The name Orrorin refers to a mythical Kenyan figure and means "original man," which is somewhat biblical and befitting of his age. Orrorin is nearly six million years old, the oldest homininan known to us until the discovery of Toumaï, now presumed to be the oldest.

These two near-simultaneous discoveries dethroned a usurper, *Ardipithecus*, from the top of our family tree. Initially identified in Ethiopia in 1992 by Tim D. White, Gen Suwa, and Berhane Asfaw, this fossil's name comes from *ardi*, which means "on the ground" in the Afar language, and *pithecus*, derived from the Greek word for ape. Physically, *Ardipithecus* seems to be closer to a chimpanzee than to a human, which now leads us to believe that it is probably their ancestor rather than ours. Initially, however, paleoanthropologists considered it a homininan because they did not have any comparable fossils from the same period (four million years ago).

The discoveries of Toumaï, Orrorin, and *Ardipithecus* teach us three things:

1. Evidence of the distant past is limited. It is open to interpretation but also to the risk of reading into it confirmation of what we want to find. A few fragments of bone might enable us to build a theory or they might not, and any conclusions we do reach are liable to be challenged with each new discovery.
2. The more distant the past, the more tenuous the evidence, and the more difficult it is to date. For something that is six million years old, the different dating methods have a margin of error of more or less two hundred thousand to three hundred thousand years.
3. We believe ourselves to be heirs of a continual process of physical improvement. The typical image of five or six silhouettes beginning with a chimpanzee and progressively evolving into the upright position of humans (before the invention of desks and computers) does not reflect the reality of human evolution. Our family tree is a bush whose branches are twisted, broken, and intertwined. In the past, several species of hominids coexisted with our ancestors, and genetics now provides ample evidence that these species clearly hybridized. Evolution led certain branches to dead ends, which means that those whose fragments we dig up might not necessarily have any descendants.

We are the only survivors of a family massacre. We have always had brothers and cousins. Two million years ago, and even one hundred thousand years ago, there were still somewhere between six and twenty first cousins, some of whom we had direct genetic exchanges with. They all died out. The only thing that remains of them now is a few genes in us. In the next chapter we will investigate what happened to them, but for now, let us try to understand what makes us human.

The Human Exception

This raises the inevitable question of what sets humans apart. Language? Many animals vocalize and exchange information. Making and using tools? We have no monopoly on that. Empathy? No monopoly on that either. Societies? Ant and bee colonies are better organized than we are. Cultures? They exist in several animal societies. What makes humans exceptional is not a group of characteristics or abilities but rather the momentum that our species has been able to create in exploiting our environment and improving our skills. This is what has led us to transform the entire planet. Our uniqueness cannot be reduced to a single trait; it is a complex phenomenon. Here, we will look at the major trends in this development and emphasize the environmental processes that influenced them.

What makes us unique in the living world is momentum; the ability to bring about continual change is specific to humans. Human culture is constantly evolving under the combined pressure of societal choices and environmental transformations. And this culture has, in turn, long modified its environment, creating continual feedback between culture and nature, of which our bodies are the product. Humans escaped natural determinism the day we were able to use culture to make a lasting impact on our surroundings. If we could identify some point in the past at which our species first altered its environment, and if this impact could be correlated to the development of our skills (cooperation, shaping and using tools, etc.), then we would have a good starting point to begin tracing the interactions between humans and the environment. We just have to work out when this might have taken place.

Let's go back in time. Our ancestors probably had to come down out of the trees before they learned to walk on two feet. Geneticists think this probably happened around ten million years ago. They can deduce this thanks to a shared gene that exists in humans, gorillas, bonobos, and chimpanzees. This gene allows us to metabolize ethanol, or in other words, digest alcohol. Our closest surviving relatives are thus fellow drinkers—which should make them all the more likable! It seems our common ancestor was a frugivorous primate, forced to come down out of the trees (probably by selective pressure from its environment—the decline of tropical forests, perhaps?). Used to eating ripe fruit in the trees, our ancestors now found themselves obliged to eat the overripe and partially fermented fruit that had fallen on the ground. Our genes therefore adapted to this new diet.

We must remember that humans did not descend from monkeys—we are one kind of monkey among the many others who evolved alongside us, chimps, bonobos, and gorillas. Observing these animals in the wild (which constantly leads to new discoveries) shows that chimpanzees know how to resolve conflicts and plan murders, how to share food, deduce intentions in others, laugh, work together, demonstrate empathy, vary sexual behavior, transmit knowledge, perpetuate technical culture, and use combinations of tools. Our common ancestors probably had the same abilities six million years ago (the alternative, less likely, hypothesis is that humans and chimpanzees developed these skills in parallel, independent from one another).

Chimpanzees, gorillas, and humans all live on the ground. The first two live in the tropical jungle, while humans have spread into other areas. Chimpanzees still spend at least some of their time in trees, moving through the branches with infinitely more ease than we could. They also have trouble walking for long periods whereas we can cover distances on foot that would be unimaginable for our cousins. Conversely, the loss of prehensile feet has left us stranded on the ground; when we do venture to the treetops, we are ridiculously ungainly by comparison.

The light that paleontologists are able to cast on this evolutionary past reveals a multitude of prehuman primates descended from the same branch as us. For every fossilized Toumaï, Orrorin, or *Ardipithecus* found in Africa, how many others remain unknown? These species apparently had little ability to make a lasting impact on their environment. But everything seems to have changed sometime between 3.5 and 2.5 million years ago.

In the Beginning Was the Great Interchange

Up until that point, our ancestors had enjoyed a warm, comfortable climate that had existed on Earth for tens of millions of years. The weather was still variable, however, and we have found evidence of rapid upheavals in climate. In their tropical zone, prehumans lived in various environments, from dry savannas, to seasonal marshlands, on riverbanks, and in lush, dense forests. We will probably only ever deduce a fraction of the many adaptations their bodies developed in response to these changing environments. If a forest replaced the savanna, some prehumans might have gone back to all fours. Or perhaps the savanna persisted? We can tell that our hardy bipedal cousins *Paranthropus*, who lived with us in Africa between three and one million years ago, had powerful molars for grinding vegetation that had become tough and fibrous because of lack of water.

Back then, the climate was warmer than what we have experienced over the last ten millennia. It began to cool down, very slowly, the world over, around thirty-four million years ago. Antarctica was quietly covered with ice and eventually completely frozen over around fourteen million years ago. Then this cooling began to speed up, first around six million years ago and then again around three million years ago.

What was to blame? The movements of the tectonic plates that make up the crust of the earth and the continents with them. Around sixty million years ago, the South American plate broke away from the African plate and eventually ended its drift across the Atlantic when it collided with the North American plate. North and South America were now joined. But the less fragmented the continents are, the less the ocean currents circulate freely, and these are what regulate the temperature of the earth, moderating the fluctuations in the thermal exchanges around the world.

The formation of the Isthmus of Panama began very slowly 5.5 million years ago and led to a first period of global cooling because it progressively restricted water flows between the Pacific and the Caribbean, changing the way heat circulated and leading to aridification in southern and eastern Africa. We can briefly sketch the evolution of our ancestors in light of this upheaval in the climate. Living in the forested zones of southern and eastern Africa, they progressively improved their ability to walk on two feet while still conserving the characteristics that allowed them to take refuge in the trees if necessary. Their legs were shorter than ours are, their arms were more developed, and their prehensile feet still allowed them to grip and climb. These primates are grouped together under the genus *Australopithecus* (from the Latin *australo* [southern] and the Greek *pithekos* [ape]), which covers various species of hominians. Lucy—a young woman who lived 3.3 million years ago and whose partial skeleton was exhumed in Ethiopia in 1974—is, of course, the most famous of them. As for the ancestors of chimpanzees, bonobos, and gorillas, they were well adapted to life in the jungle and remained in the western half of Africa, which preserved its dense vegetation.

It was then, when the two Americas were beginning to join together some 3.5 million years ago, that the climate shifted into a new phase. The ocean currents, which had up until then circulated unimpeded around the globe, spreading their warmth from the tropics to the poles, became trapped. They were forced to turn in circles around the Pacific and the Atlantic as the Isthmus of Panama progressively joined the two continents. When this process was complete, roughly 2.8 million years ago, the earth entered a series of ice ages punctuated by more temperate periods. The North Pole, Greenland, Canada, Scandinavia, and Siberia became covered in ice. Much of Africa became arid. The colder it got, the further the Sahara spread. At the coldest point, the desert took up the whole of the northern half of Africa; during the most temperate periods it shrank to a thin band of sand some 500 kilometers (about 311 miles) wide.

The contact between the two American continents led to two major transformations of their biotopes.

American historians call the first of these biological consequences the Great American Interchange. The land animals of the two continents came into contact with each other for the first time. North America had been dominated by placental mammals (where the females carry their young to full term, a group that includes humans, dogs, elephants, beavers, and many other animals with fur and teats). Up until then, South America had been the kingdom of marsupial mammals (in which the females give birth to premature young who finish their development in pouches). There were also a few placental mammals who had previously immigrated from Africa (rodents, monkeys), and there were large birds. Placental mammals have a decided advantage in an environment rich in natural resources. On average they use more energy to feed themselves, reproduce, and move around than marsupials do[2]—but they are better at competing for ecological niches. Thus, with a few exceptions, the placental mammals from the north progressively replaced the marsupials from the south. This Great American Interchange constituted the most substantial biological upheaval the earth had experienced for over thirty million years. But this world record will have lasted for only a brief geological moment, because humans are in the process of ending it as, "unbeknownst to ourselves," we bring about the sixth extinction.*

The second consequence of the contact between the two continents was that certain species died out because of the aridification of the eastern half of Africa. Mastodons and hipparions (ancestors of horses) went extinct there but survived in North America, for example. Others—such as elephants, *Phacochoerus* (primitive warthogs), and rhinoceroses—adapted by developing more powerful teeth that were able to grind tough, dry vegetation. *Australopithecus* was driven out of the forest, which progressively gave way to savanna. They learned to feed on roots rather than the fruit they could no longer find, and they proved themselves more or less bipedal. The next homininans, probably their children, seem to have developed two kinds of adaptive responses. *Paranthropus*, like elephants, developed excessively enlarged teeth—capable of extracting what they needed to survive from vegetation—along with a large sturdy frame to dissuade most predators in search of easy prey. By contrast, our ancestors, *Homo*, developed more fragile teeth suited to an omnivorous diet that probably included meat. Had we already become hunters? Or did we content ourselves with insects and small game? Or perhaps we were scavengers, feeding off the remains of the prey of the great predators? Paleoanthropologists are still debating this.

Let us return to the importance of this shift in diet for *Australopithecus*. Fruit makes up three quarters of a chimpanzee's diet, but these fruits are far removed from the bananas that human farmers genetically select to be soft and sweet. Chimpanzees eat figs, grapes, or palm fruit; wild fruit that is fibrous, sour or slightly bitter, and often encased in a thick skin. These cousins of ours dedicate half their lives to feeding themselves. The fruit is so poor in nutritional value that they have to stuff down a kilogram in an hour and then spend the next hour digesting it, along with a few extra treats (insects, larvae, etc., which make up 20% of their diet), and then start eating again. If fruit is scarce, chimps eat leaves, which are even less nutritious, tougher, and more fibrous. Sometimes they manage to improve their daily rations with meat they scavenge somewhere, such as antelope or a small monkey. When they do get some meat, they have so much trouble chewing and digesting it that it takes them a whole day to eat 1 or 2 kilograms (2.2 or 4.5 pounds). Only 5% of their diet is carnivorous, though that can potentially increase to 12% in situations of nutritional stress.

It was therefore *Australopithecus*, driven out of the forests by the desertification of eastern Africa, who took the first step toward the comfort we have today. On average we spend two hours a day preparing and eating (generally ready-made) food, while our chimp cousins spend half their lives eating and digesting. The transformations of *Australopithecus*'s face and teeth show that they made a decisive nutritional shift. They adapted to life on the savanna, where fruit trees are rare and seasonal, unlike in the jungle where fruit is available all year round. They made chimpanzee survival foods—roots, bulbs, herbs, seeds, insects, eggs, and scavenged meat—their everyday fare. And that caused their teeth to evolve, too. *Australopithecus*'s teeth are a third bigger than chimpanzees' teeth, and their enamel is twice as thick. As for *Paranthropus*, their teeth are twice as large as those of chimpanzees, and their enamel is three times as thick. All this led to a decisive evolution in the anatomy of these prehumans. Footprints left in volcanic ash in Laetoli (Kenya) 3.6 million years ago suggest that at least some of these *Australo-*

pithecus were perfectly bipedal. This is logical, of course. Chimps can find all the food they need just by wandering 2 or 3 kilometers (1.25 to 1.9 miles) in a forest rich in potential resources. But in arid zones, where resources are few and far between, evolution had to favor ways of moving that required less energy and made it possible to cover long distances. Being bipedal must have made all the difference for the *Australopithecus*, the primates who survived, who were probably the ancestors of both *Paranthropus* and *Homo*. They may have even made the first tools, carved out of wood, for digging for roots.

Another tectonic phenomenon that seems to have contributed to the desertification of eastern Africa is the volcanic valley known as the Great Rift, which has run between the Red Sea and South Africa for roughly the last ten million years. According to the east side story theory popularized by paleoanthropologist Yves Coppens, this immense and active fault line, which pushed up a mountainous barrier between eastern and western Africa, changed rainfall patterns, radically transformed vegetation, and hampered the circulation of animals.

These two phenomena—the connection of the American plates and the Great Rift dividing Africa in two—have clearly contributed to the aridification of eastern Africa. This in turn drove our ancestors out of the dwindling forests and made them bipedal. They adapted to life on the savanna freed from the burden of having to survive solely on fruit.

Then Arose the Naked Ape

The first of the seven revolutions that made us the curious animals we are thus occurred at the same time as a major climatic shift. The Physiological Revolution transformed one prehuman primate among others into a human. Monkey (capitalized because of his obsessive egocentrism) became the proverbial "naked ape" when the planet began to cool down. *Homo* appeared just as eastern Africa was becoming a desert in the wake of the global cooling that resulted from the forming of the Isthmus of Panama.

The development of the savanna probably accentuated the environmental pressure encouraging bipedalism, which had already been adopted by *Australopithecus*. Although *Homo* was apparently not the first to stand, they seem to have been the first to be permanently bipedal. From a structural perspective, the upright position provoked several further adaptations. The foramen magnum, where the spinal cord joins the head, moved toward the center of the skull to stabilize the head. The spinal cord became sinuous with four precise curves that act as shock absorbers. The pelvis became smaller and broader both to improve balance (walking on two feet is less stable than on four) and to accommodate the increase in the size of the head (the opening in the female pelvis determines the maximum size of a newborn's head, which will have to get through it safely to survive). Knees became more centered, thicker, and more solid, losing some of their flexibility but providing better balance. The gluteus maximus muscle (in the buttocks), the largest muscle in the human body, developed to provide power and endurance during running. Feet became longer, flatter, and more rigid, losing their ability to grasp things in exchange for providing greater propulsive force. Finally, the overdevelopment of the Achilles tendon also boosted our ability to run. This ligament works like a spring. As we run, it stores a third of the force produced in each stride and then releases it with the next step. In humans this tendon is ten times as long as it is in a chimpanzee. Evolution made us runners where our cousins are ideally suited to alternating between the trees and land.

The climate has coached us well; we are athletes built for endurance.

Prehensile feet are useless when there are no trees. But standing on two feet—combined with the excellent trichromatic and three-dimensional vision shared by all the Old World primates of Afro-Eurasia*—allows us to see far into the distance, which is very useful for spotting predators as well as food. It also ensures a better regulation of body temperature. Standing upright means the body suffers less from the heat of the sun while still maximizing the surface cooled by the wind. It is likely that our loss of body hair coincided with this to allow the network of veins under the skin to evacuate excess heat. Of course, we did not really lose our hair (there are as many on a person as on a chimpanzee), but it became softer and finer, to the point of becoming nearly invisible. Often only our head, armpits, and genitals still have "fur," a natural protection against the sun's rays. Our African ancestor undoubtedly had dark skin and curly hair, as the latter traps a maximum amount of cool air around the head. We also developed more sweat glands to improve the regulation of body temperature through perspiration. Prehumans straightened up and became *Homo*.

Monkey, the naked ape, child of the climate, now walks upright in this world.

His skull straightened out and his face became flatter. As a result, his nose decreased in both size and skill. Freed from the weight of the nose, his mouth was able to develop more, and he began to vocalize broader ranges of sounds, which eventually enabled articulate language. His hearing improved, because the inner ear was important for balance. Improved hearing enabled him to hear more varied sounds, which helped communication. Modern humans have particularly good hearing between 1 and 6 kilohertz, a range that includes most spoken language. Most vowels are situated between 1 and 2 kilohertz, and most consonants are between 2 and 6 kilohertz. In contrast, chimpanzees, like most primates, have trouble differentiating sounds between 1 and 4 kilohertz. Recent studies suggest that two million years ago *Paranthropus* and *Australopithecus* had hearing more like that of chimpanzees, which would imply that they did not use articulate language. For the moment, such studies are not possible for the *Homo* who lived then because we have not found enough skulls in good condition.

Finally, over the next two or three million years, the brain of this naked ape grew. It developed new zones, leading to the progressive enhancement of cognition and the development of a broader forehead. Increased intelligence and communicative ability were associated with an ability to adapt to all kinds of biotopes. Our canine teeth became smaller, probably owing to the change to an omnivorous diet; being able to eat everything was a significant adaptive advantage. Incidentally, walking on two feet also meant using less energy to get around. Over the same distance, humans use 25% less energy than a chimpanzee of equivalent size walking on all fours. Over thousands of years, our upright posture also facilitated the development of an omnivorous or carnivorous diet. Herbivores, who have to digest tough vegetation, need much larger intestines to break down what they consume. Carnivores have smaller, shorter digestive systems because processing meat proteins does not require as much effort. Becoming bipedal demanded a smaller stomach and intestines, which in turn meant adopting an omnivorous, protein-rich diet, as our ancestors did. If their intestines were too big, *Homo* would not have been able to catch their prey and would have become easy prey themselves.

Having smaller digestive organs is an anatomical requirement for running. In other primates, digestive organs are attached to the pelvis. In humans, they are independent, held in close to the body by the core abdominal muscles. This is the physique of a runner. It protects the organs while preventing them from floating around when we run. There are other anatomical characteristics that set us apart from our cousins the chimpanzees. We have a nose that sticks out from our face to improve aspiration and a larynx in the throat to optimize airflow in the lungs, and we can use our muscles to control our breath. In other words, in our respiratory system, evolution encouraged the development of the essential characteristics for endurance.

We now have the identikit portrait of the first *Homo* regardless of the species. Numerous, intrinsically connected characteristics set *Homo* apart from our primate cousins, including an inflexible foot and the ability to stand for long periods; longer leg bones; a supple, flexible spinal cord; smaller teeth; a genuinely omnivorous diet; smaller intestines; less hair; and more sweat glands. These ancestors already had lungs built for endurance, a flat face, and a large brain (around 900 cubic centimeters [55 cubic inches], three times that of *Australopithecus*, *Paranthropus*, and chimpanzees). These features were already ensuring that we were the only mammal to be able to run a marathon in the midday sun. *Homo* also tended to have smaller size differences between males and females, which possibly indicates a tendency toward inegalitarian monogamy.

Our *Australopithecus* ancestors had significant levels of sexual dimorphism: the males were twice the size of the females, a characteristic shared today by chimpanzees and gorillas. This is generally the sign of a society organized in "harems" in which one (or sometimes several) dominant male(s) monopolize access to the females, pushing the other males to the margins of the group. It is associated with strategies for alliances and fights for dominance.[3] For other primates, where monogamy is the rule, both sexes are generally around the same size.

We are somewhere in between. Human males on average tend to have around 10%–15% greater body mass than females. Is this a throwback to a distant past? It is more likely cultural. All around the planet, males have monopolized access to the best food, particularly meat, for tens of thousands of years (at least). And this remains the case for most of humankind. This systematic monopolization by males may well be imprinted into our genes, possibly resulting in a slight but significant size difference between males and females.

An Artisan Primate, New Lord of Creation?

If you ask paleoanthropologists what is particular about humanity, they will reply without hesitation: tools. You can interrupt all you like and object that we know now that tools are used by other species. You can argue that chimpanzees carve sticks to flush out lizards or use combinations of tools—one rock as a hammer, another as a chisel—to break nuts in a technique we still use today. You can protest that these animals even have cultures, because they have transmitted these techniques over centuries through imitation. Moreover, because they know what to look for, ethologists observe all kinds of animals using tools in the wild: other mammals, birds, even fish.

So what makes us different from them? A matter of degree. Chimps can carve sticks. If shown how, they can also carve stone, and they will probably be more dexterous at it than an average *Homo sapiens* today. But they will not be able to make a shovel. If we let our paleoanthropologists finish, the answer is tools, yes, but manufactured tools. In other words, a stone broken with another stone whose fragments suggest to the informed eye that long ago someone carved a lithic tool (from the Greek *lithos*, meaning "stone") useful for its sharp cutting edge. Most probably our ancestors also used other more perishable materials (wood, bone, leather, horns, etc.) for their tools, but they are lost to us now. If lithic tools are the key to paleoanthropology, it is because they are by far the most visible trace our ancestors have left us of their presence.

Today, the oldest stone tools we have found are around 3.3 million years old. They were discovered in Kenya in 2011 by a team led by French archaeologist Sonia Harmand. Before this, the oldest cutting tools we had found were between 2.6 and 2.9 million years old. However, butchering marks on 3.3-million-year-old bovid bones found in 2013 suggest that there must be older tools still to be found. This idea throws doubt on the anthropocentric theory that our genus *Homo* was the first to use tools, because the oldest *Homo* found so far is only 2.8 million years old. For the moment we have no idea who carved these first tools: early *Homo*, *Paranthropus*, *Australopithecus*, or even *Kenyanthropus*? The latter is a homininan found in Kenya in 1991 whose skull combines prehuman (very long teeth) and human (flat face) characteristics. This leads paleoanthropologists to concurrently argue that *Kenyanthropus* is a separate genus, a species of *Australopithecus*, or even a prehuman. In any case, this unknown artisan was omnivorous because he or she scraped bones to tear off meat some 3.3 million years ago. Amusingly, some of the traces left by these tools suggest that whoever carved them had great skill, creating much longer blades than those of the tools that would follow them half a million years later.

Contrary to what has long been believed, it now seems that neither the development of lithic tools nor prehuman evolution is linear. There is no direct progression here from simple to more complex but rather an evolution that branched off in different directions, some leading down radically different paths, others dying out completely. Other unknown artisans await us in India or China, where archaeologists say they have found stone tools carved 2.6 million years ago. These tools therefore predate, by some six hundred thousand years, the generally accepted date at which early humans are thought to have left Africa. In other words, we have absolutely no idea who could have carved them unless we considerably revise the date of human expansion out of Africa.

The most probable scenario of early tools is presented by Daniel E. Lieberman.[4] Back in Africa, the first prehumans to change their diets were the *Australopithecus*, because they were forced to venture farther out into the savanna that gradually replaced the forest. Their teeth attest to the fact that they had to fall back on survival foods such as roots and tubers. To get at them, they had to dig. Our cousin the chimpanzee knows how to use sticks to flush out termites and lizards. *Australopithecus*'s hands had evolved to be strong and thick, good for scratching at the dirt, but they probably also developed cultures around the digging stick. This tool is still widely used in the hunter-gatherer peoples that survive today, but, being made of wood, it has left no visible archaeological traces.

The Importance of Fat for a Monkey with a Fat Head

So what does this first revolution consist of, this Physiological Revolution that will make Monkey, the naked ape, an accomplished biped and master of tools? Using stone tools implies synergy, a cascade of prerequisites and consequences. First, it requires an opposable thumb, a throwback to the tree-climbing ancestors we share with many primates. But our thumb, consolidated by three million years of toolmaking, is now much stronger and makes us better craftsmen. This precious prehensile hand is essential for gripping and hitting as well as shaping and then using tools. Strong shoulders help with wielding and throwing—especially when they are as well developed as our shoulders are. Indeed, humans are infinitely better at throwing than our cousins, the chimpanzees, bonobos, and gorillas; we are both more precise and more powerful. Our arms, which swing while we run, help our lungs in their work, ensure balance, and optimize the distribution of kinetic energy. To make this possible, evolution freed up the articulation of the shoulder. The great apes have arms that cannot rotate as fully as ours and do not have the same reflexes. Our monkey-like physique has been optimized not only for running but also for throwing.

The upright position also facilitates the growth of the brain. An animal walking on all fours, leaning forward, would not be able to support a heavier head; the neck and back muscles would have to be impossibly strong. A big head is possible, however, when one's weight is evenly distributed from the top of the spinal "column." The growth in the size of the brain was also made possible by an increase in protein consumption, which meant access to meat either through hunting or scavenging. Omnivores are able to adapt to a broad variety of ecological contexts. Tools also primarily serve to make it easier to absorb food. For the last three million years, whether they are hand axes or industrial blenders, tools have been used to cut up meat and vegetables to make them more digestible and quicker to eat and to increase access to the available calories and nutrients.

The path of human evolution began when Monkey took his first bite of steak tartare around 3.3 million years ago. As soon as they were invented, tools were primarily used to satisfy the main obsession of all animals: eating. What do chimpanzees do with their tools? They dig out termites, flush out lizards, and break nuts.

Obtaining more nutritious food contributed to the success of the species. But we needed to be able to store it. Our brain had to be fed constantly, and to satisfy this need our body became good at storing fat. When food is short, the body burns fat to feed the brain. We are fat naked apes. In adult primates, fat generally makes up about 6% of body mass; for infants this is only around 1% because they use all their body's resources for growing. Hunter-gatherer babies, by comparison, have a plethora of fat at their disposal— around 15% of their body mass. For hunter-gatherer children, the figure rises to about 25% before dropping in adulthood to around 10% for men and 20% for women (who need to store more for pregnancy and breastfeeding). The current trend toward obesity seems to be the result of three million years of evolution favoring those of us who carry the genes for storing fat.

When food is guaranteed, our second obsession is rest, a vital need. The Olduvai Gorge in Tanzania has provided traces of what might be the first identifiable construction project, a stone windbreak 1.9 million years old. Bits of broken bone found on the ground nearby suggest that those who built it probably ate meat. But the site gives rise to speculation: both prey and stones were brought there, but was it a dwelling or simply an occasional shelter? Was it possible for its inhabitants to be so immobile while formidable megapredators—colossal hyenas, bears, and lions—prowled around?

Our third obsession, reproduction, has clearly been unprecedentedly successful for our species. In just a few hundred million years, the dynamic demographics of the *Homo* genus enabled us to colonize the ancient world as well as the continental block made up of Africa, Asia, and Europe; every generation pushed exploration farther than the previous one.

In 2013 the Swedish paleontologist Lars Werdelin put forward an interesting hypothesis arguing that humans have always been hyperpredators. Apparently, Monkey has been a killer since his conception. Werdelin says there was a spectacular collapse of diversity in the ecological niches of the great carnivores wherever and whenever *Homo* appeared. In eastern Africa, the "amount of niche spaces filled by species in the community," what he calls "functional richness,"[5] declined from 100% 3.3 million years ago to 1% around 1.5 million years ago. Nearly all carnivores weighing over 20 kilograms (44 pounds) appear to have died out in numerous biotopes. And this moment coincides with the rise of human evolution and the changes in our ancestors' diets as they began eating more meat and demonstrating their new abilities to colonize various ecosystems. As far as we know today, *Homo erectus* seems to have been the first hominid to venture out of Africa, no later than two million years ago.

Given that they apparently had no mastery of fire, however, how did humans kill or scare off more powerful predators 2.5 million years ago? The use of running, throwing weapons, and collective hunting probably gave *Homo* a decisive advantage from the very beginning. As a sociable animal, humans were perhaps able to avoid direct confrontation. Working together to steal prey from other predators, for example, they may have kept them at bay by yelling and throwing stones, much as a lion tamer dominates a lion by sheer force of will and the ridiculous threat of the whip.

And perhaps these first *Homo* made the most of their advantage by overhunting. Carnivores have been known to adopt this technique, which consists of eradicating all competition in an ecological niche. Wolves, for example, will kill baby foxes without eating them to ensure that they will not become competitors in their search for food.

Do you want to know what a hyperpredator looks like? No need to watch *Jaws*, just take a look in the mirror.

We are left with three puzzles, which we will try to resolve in the next chapter. First, how did we manage to adapt to the whole planet, from intertropical deserts to the polar ice of the great north? Second, why are we now the only humans left on Earth when our ancestors coexisted with several other human species? And third, did our expansion necessarily occur at the expense of global biodiversity?

2

THE END OF THE ELEPHANTS

Between 500,000 and 100,000 BCE, Monkey used his language skills to develop a new superpower: mass cooperation. From Australia to the Americas, a wave of extinctions swept the world in step with his progress.

Paris, January 21, 1796, or rather, the first day of Pluviose in Year 4 according to the republican calendar of the French revolutionary period. A young naturalist by the name of Jean Leopold Nicolas Frederic Cuvier, only twenty-seven years old, was holding a conference.[1] His speech would contribute to the emergence of modern biology. He was to present on the subject of giant teeth, including a few enormous molars of unknown origin that the revolutionaries had confiscated among the personal collections of the late King Louis XVI. The speaker is known today simply as Georges Cuvier (1769–1832), one of France's founding fathers of science. And the teeth in question have a long and enigmatic history that began in 1739.

A Mastodon—How Revolutionary

That year, Charles III Le Moyne (1687–1755), second Baron of Longueuil, was exploring the future state of Kentucky at the head of an expedition of four hundred men. Escorted by their Algonquin and Iroquois allies, a handful of French adventurers were with them. The idea was to put down the members of the Chickasaw tribe to facilitate circulation between the French territories of Nouvelle France, which comprised Canada, Acadia, and Louisiana. It was a risky venture. Supplies were short, and the men were forced to grind and cook acorns to eat. From their base camp near what is now Cincinnati, the French explorers and their American Indian allies discovered a stinking, sulfurous marshland—a kind of hell. Attracted by salt deposits, hundreds of bison had drowned there, and their corpses had decomposed alongside other older remains. The explorers returned to their base camp with a femur over 1 meter (3.3 feet) long, a gigantic tusk, and a few enormous teeth, the heaviest of which weighed nearly 11 royal pounds, later converted into 5 kilograms during the French Revolution.

Longueuil was fascinated by this discovery to the point where he was ready to take any risks necessary to ensure safe passage for his treasures. The expedition suffered illness, military defeat at the hands of native tribes, and near annihilation before retreating toward Mississippi. However, there the survivors were able to connect with another French expedition. The precious bones were saved. They made the journey across the Atlantic without incident and were added to King Louis XV's royal cabinet of curiosities.

The teeth intrigued the experts of the time. Although the tusks and the femur could have been those of an elephant, the molars lacked the ridges found on pachyderms' teeth. They could almost have been human with their bumpy surface, but those of a gargantuan human. For want of a better explanation, these French scientists concluded the teeth must be the remains of a giant hippopotamus.

The story picks up again five years later in 1744. The French and the English were competing to colonize the territories that would become the United States. This time it was Robert Smith, an American Indian trader who, having heard the local legends about this place, brought back long bones, tusks, and teeth identical to the French royal collection to be offered to George II. The doctors of the royal British court considered (contrary to their French counterparts) that these must be the remains of an unknown gigantic animal that they presumed to be carnivorous. They called it the *incognitum*.

Then Georges Louis Leclerc, comte de Buffon (1707–1788), revered among French naturalists, joined the debate. For him, these teeth were indeed those of an unknown animal. And the fact that this mysterious creature could no longer be found meant that animals could die out. This conclusion profoundly unsettled the scientific certainties of the period, at a time when the world was considered unchanging, created by God with a set number of animal species. Thomas Jefferson (1743–1826) added to the diagnosis of the English scientists: this American *incognitum* would have to have been the largest terrestrial animal, five or six times as big as an elephant. He saw this as a powerful contradiction of the prejudice of European thinkers who considered the animal species of the New World to be "degenerate" forms of Old World fauna and therefore necessarily physically smaller. When he was elected president of the United States in 1804, Jefferson launched an expedition in pursuit of these *incognita*, assuming they could not have died out.

In 1796 Cuvier's blue eyes and inspired charisma galvanized his audience. As a professor at the National Museum of Natural History (created as a result of the nationalization of the royal collections), he was able to compare the different bones placed under his responsibility. He began with the elephants. Asian elephants were docile with humans; African elephants had remained wild. After examining the jawbones of each of the specimens, bones from Ceylon and from South Africa, he pronounced his verdict: these two pachyderms were distinct species rather than a single one as had commonly been believed. The difference lay in their teeth: curved ridges for Asian elephants and diamond-shaped ones for African elephants. Other teeth, sent from Russia, were clearly from a third species, today classified in a different genus: the mammoth. This meant that Cuvier could conclude that the teeth brought back from America by Baron de Longueuil and the bones and tusks found with them all came from a single animal, a new kind of elephant. In 1806 he named it the mastodon.

In an hour-long speech about the bones of proboscideans—the family of mammals with prehensile trunks—Cuvier demonstrated the existence of four elephant species before announcing that two of them—the Russian and the American—had died out. In so doing he contributed to the birth of what would become modern biology and proved the existence of animal extinction.[2] However, Cuvier remained convinced that although extinction was possible, the number of species on Earth remained invariable. He assumed that animals were "recreated" after each wave of extinction to replace those that were lost, and he vigorously and disingenuously challenged all those who developed other ideas.

The Empathetic Killer

As we saw at the end of the previous chapter, for the paleontologist Lars Werdelin, our greatest adaptive advantage as a species was being omnivorous. Being able to eat anything meant being able to adapt to unexpected environmental changes, colonize a large number of environmental niches, and survive where rivals with more selective diets perished.

For the historian Yuval Noah Harari,[3] it was our imagination, our ability to represent the past and anticipate the future, that were decisive in this. Perhaps we survived because of our ability to narrate our existence, to constantly fictionalize ourselves.

Other authors consider that it is our capacity for collaboration that ensured our survival. This last hypothesis deserves further examination, although these three attributes—being omnivorous, imaginative, and cooperative—probably had combined effects.

With no protective fur, minuscule canines, and laughable little claws, Monkey is a runt. In a match of physical strength, any chimpanzee of equal weight would leave even the most athletic human wrestler in tatters. Our primate cousins have muscles that develop four times the strength ours have, their teeth are larger than those of a big dog, they can rip up phone books with their bare hands, and their acrobatic abilities put the best human gymnasts to shame. Clearly, we are little more than a prematurely born primate with a hyperdeveloped brain. Over a period of some two or three million years our brain progressively quadrupled in size, which meant that the learning phase of human children also increased. We are the only medium-sized animal species that allows the luxury of having dependent offspring for the first twelve years of life. Baby antelopes can run alongside the herd hours after birth, and kittens are quite independent by six months old. To ensure the survival of the group, humans had to collectivize child rearing. This undoubtedly contributed to enriching language structures as well as improving the success of the hunt. Predators are always more efficient as a coordinated group, and they are always more intelligent than their prey. If they were not, they would die of starvation.

Our ability to work together now allows us to throw our weight around on a planetary level. We flatten mountains, wipe out forests and marshland, and force rivers to flow in straight lines. How did we come to have such a gift? British anthropologist Claude Dunbar argues that the size of primate groups is dependent on the ability to conduct affective exchanges. A primate can identify (and therefore correctly interact with) at most one or two hundred other individuals. Group members look after each other; they turn to others for cleaning and lice picking as well as for licking, cuddling, and maybe sex. This preserves social harmony but automatically limits the size of the group to fewer than two hundred individuals.

Humans have broken down these barriers thanks to language. Its collaborative potential, boosted by syntax, is unique in the animal kingdom. Complex language enables us to transmit messages more clearly and manipulate others more efficiently. It allows us to remember the past, project ourselves into the future, and anticipate things that need doing. It means we can categorize, mythologize, and symbolize. We are Monkey, the storyteller, the chatterbox. Among primates, the dominant male is the object of all the others' attentions; he is the one to be groomed by the others. Among humans, the leader is the one who speaks first. And when it came to imagining a supreme leader, what form did it take? The Word.

Our words foster empathy. We use them to help those close to us, those we identify as being "like us." But empathy has a dark side. It means we develop the idea of belonging, of "us" and "them." And also the idea of "us" *against* "them." Monkey may be cooperative, but he is also racist, nationalist, and speciesist (ranking humans above animals). To varying extents, evolution has produced in us this need to be close to some and to compare with and confront ourselves against others.

But how did this fragile, talkative, and egocentric Monkey become the master of the world?

We know that different *Homo* species coexisted in the past. According to the dominant thesis, around two million years ago in eastern Africa we could have come across *Homo rudolfensis*, *Homo habilis*, *Homo ergaster*, or its twin *Homo erectus*. The latter set off to conquer Afro-Eurasia between 2.5 and two million years ago. Fossils have been found in Georgia from around 1.8 million years ago (*Homo erectus georgicus*), in China from 1.6 million years ago (Lantian Man, *Homo erectus lantianensis*), and in Indonesia (Java Man, *Homo erectus erectus*) from more than one million years ago. This led to what we can consider new branches in the family tree, such as Flores Man, *Homo floresiensis* (nicknamed the Hobbit because of its small size), *Homo antecessor* in Europe, *Homo heidelbergensis* in Europe and in Africa, and Neanderthals and Denisovans in Europe and in western Asia. Geneticists suspect there are also other ancient branches of the *Homo* family in Africa waiting to be found.

Are all these *Homo* species or subspecies in our genus? The debate is heated. After settling across Eurasia, *Homo erectus* seems to have spread out from there some fifty thousand years ago. For some researchers, *Homo ergaster*, who remained in Africa, is simply an African *Homo erectus*, and humanity is descended from these two ancestors. Others disagree, arguing that there are too many anatomical differences between the two. They argue that *Homo erectus* died out without leaving any descendants, and we are all the African descendants of a small group of evolutionarily adapted *Homo ergaster*.

There are three competing theories explaining the emergence of modern humans around four hundred thousand years ago.

1. The Noah's Ark hypothesis (monocentrism): isolated evolution in Africa followed by migration between 300,000 BCE and 180,000 BCE, which led *sapiens* to conquer the world and wipe out other species of *Homo*
2. The candelabra hypothesis (pluricentrism): several regional evolutions of *erectus* toward *sapiens* in Africa and in Asia
3. The intermediary hypothesis (reticulate or network evolution): evolution stemming from isolated emergence of *sapiens* in Africa followed by migration out of Africa and a vast genetic exchange with other *Homo* species

Homo has been an omnivore since the beginning, which gives us rare opportunism. Much like the invasive species (animal or vegetable) that we are now fighting all over the world, we can rapidly conquer new ecological niches. Although we eat meat, it seems we have always had only weak molars and small canines. As we have seen, early *Homo* probably ground up their meat with stone tools, which helped digestion and meant our jaw could get smaller. Feedback loops between humans and nature were already in place here. Humans ate the plants and animals around them, and their teeth changed because the development of tools meant they no longer had to grind food with their teeth. Our organism absorbs animal proteins more easily than vegetable fibers, so this carnivorous diet contributed to the development of our energy-hungry brain. This organ doubled in size between *Australopithecus* (around 300 cubic centimeters [18.3 cubic inches]) and the first *Homo* (around 600 cubic centimeters [36.6 cubic inches] for *rudolfensis* and *habilis* between 2.8 and 1.8 million years ago). It doubled again between *habilis* and *erectus* (900 cubic centimeters [55 cubic inches] around 1.5 million years ago and 1,200 cubic centimeters [73.2 cubic inches] around 0.5 million years ago). This progress peaked with *Neanderthal*, whose brain of 1,500 cubic centimeters (91.5 cubic inches) wins hands down over our average brain size of 1,400 cubic centimeters (85.5 cubic inches).

But a big brain must be well fed. Humans' brains consume 25% of the calories we absorb, whereas our stomachs require only 2%. These proportions are the exact inverse for Koalas, which eat only the tough poisonous leaves of three species of eucalyptus. Their lazy brains require only 2% of the calories they consume, but their superhardworking stomachs need 25%. Think or digest—it's one or the other!

After Steak Tartare, Barbecue

At some point, hand axes gave way to fire. Traces of hearths have been found in Africa dating back 1.5 million years and in Indonesia from perhaps half a million years later. They suggest that *Homo* was able to rapidly exploit natural fires to collect and preserve the flame. Some authors even see a correlation between the use of fire and the reduction in the number of our teeth around two or three million years ago. The ability to perpetuate fire seems to be demonstrated by a site in a cave in Petralona, Greece, dating back around one million years, while the creation of fire is suspected to have developed around eight hundred thousand years ago in China and in Europe. There are two main methods of lighting a fire: either friction is used, rubbing two pieces of wood together at great speed until the softer of the two becomes hot, or specially selected rocks are struck together to produce sparks. The sparks are then fed with a highly flammable substance, such as dried moss. Both methods are still used by hunter-gatherers today.

The naked ape, master of fire, took a new step forward. Monkey became a cook. Meat is so much more digestible cooked. Many vegetables, too, although tough and fibrous raw become tender and tasty when cooked. Using open coals, hot stones, ovens, boiling water, or steam, all these methods probably also involved increased conviviality in food preparation. Fire also meant keeping warm at night, surviving colder temperatures, and thus later colonizing more inhospitable lands. It also meant being able to scare away predators. Wooden spears were hardened in the fire, which considerably improved their range and efficiency as projectiles. Fire modified the chemical and physical properties of certain minerals so that they could be crafted. For example, silcrete is a very sharp stone that can only be shaped by heating it to 400°C (752°F). Traces of this process have been found in South Africa from seventy-five thousand years ago. Fire also helped humans pursue and bring down the most powerful prey. It meant we could burn vegetation so soft young plants would grow in their place, prevent animals from hiding, and encourage the development of small game. Fire transformed humanity and ensured our promotion from animal to titan.

Monkey became an incendiary.

Opinions differ as to the exact moment when humans began to wield these flaming torches to bend nature to our will. Stephen J. Pyne,[4] a fireman turned historian of fire, suggests that humans are animals made by and for fire. Our cuisine modified our genes, our face, our teeth. Once cooked, all food becomes essentially predigested—less effort is needed to chew and absorb it. In some cases, fire can even make something inedible into a delicacy. As a result, the size of our faces shrank by 5% to 10% over the last two millennia, and this increased still further with the appearance of intensive agriculture and pre-prepared food half a century ago. Before that, we had to cook our food to properly assimilate it. But above all we "cooked" the landscape to extract what we needed to eat. Pyne argues that the species of flora that dominate in South Africa and Australia have to a large extent been naturally selected because they are adapted to human practices. Anthropic burning, repeated over tens of thousands of years, probably encouraged the development of hegemonic vegetation able to resist fire. Some species also evolved to need fire for reproduction, relying on the heat of the flames to release seeds, for example.

Once Giants Reigned

Let us take a closer look at what existed before the Pyrocene,* the era when Monkey became the master of fire. Imagine we can go back in time to the landscapes of the Pleistocene* one hundred thousand years ago. What would we have seen? Most likely megafauna—the name given to animals that weigh, according to the definitions of different authors, more than 100, 40, or 22 kilograms (220, 88, or 48 pounds)—like those in the children's film *Ice Age*. In the North America of the Pleistocene, we would have seen three different species of mammoth (including a miniature one). There were also two closely related elephant-like genera, *Cuvieronius* and *Stegomastodon*, and a mastodon, a kind of elephant with long straight tusks. These pachyderms lived alongside deer and elk weighing 2 tons, beavers 2 meters (6.5 feet) long, capybaras (giant rodents) of 100 kilograms (220.5 pounds), wild horses, several kinds of antelope, camels 3 meters (9.9 feet) tall, lamas, and saber-toothed tigers of 400 kilograms (882 pounds). There were also American cheetahs, bears that weighed over a ton, lions weighing 300 kilograms (661 pounds), glyptodonts (giant heavily armored armadillos), wild dogs, and wolves as big as a Saint Bernard. There were more than a dozen kinds of giant sloth, the biggest of which weighed over two tons. Some of these sloths lived on the land; others were able to dig tunnels 4 meters (13.1 feet) wide in the rock; there were even marine sloths the size of bears. A similar description could apply to any other region in the world according to the rule that convergent evolution leads to the development of similar animal species in equivalent biological contexts. This megafauna, prolific one hundred thousand years ago, has all died out. Only 15% of the species survive today, less than 1% in terms of the occupation of their original ecological niches. Who knew that two thousand years ago, lions could still be found in Greece, elephants in Syria, and aurochs in France?

Three theories have been developed to explain this wave of extinctions. The catastrophist vision of Cuvier, who believed in a geological event comparable to the biblical flood, was rapidly abandoned. The remaining two both still have their defendants today: the climate theory, initially developed by the British geologist Charles Lyell (1797–1875), and the human theory, sketched out by his Scottish colleague John Fleming (1785–1857). The first school of thought argues that climate change led these great species to extinction. The second argues that overhunting is really to blame, that human action was like a blitzkrieg. Many arguments allow us to favor the second theory.

Biologists believe that the normal rhythm of mammalian extinction is one species every eight hundred years, aside from specific events and circumstances. Yet between 70,000 and 10,000 BCE, several hundred species, dozens of genera, were wiped off the face of the globe. We must concede that these extinctions coincide remarkably well with the arrival of modern humans—they are particularly visible in America and appear only later on certain islands. For example, humans arrived in Madagascar one thousand years ago and in New Zealand over seven hundred fifty years ago, and both of these dates coincide with the near total destruction of megafauna in those areas.

Yet the debate still rages among paleontologists: humanity or the climate—which is guilty of this massacre? Those who support the climate theory argue that these large animals were affected by a warming that began fifteen thousand years ago. Deciduous forests slowly ate away at the steppes where the mammoths roamed, providing new opportunities for smaller mammals and forcing the older, larger grazing animals to move farther to the north. Those who support the theory of human responsibility object that these events had already occurred dozens of times over the previous million years. Their arguments are more convincing. Of course, the climate evolved during this period, but these extinctions occurred as much in times of warming as in cooling. Smaller animal species are ordinarily more vulnerable than larger ones (when their environment changes, larger species are better equipped to move toward more suitable climes). Yet this wave of extinction at the end of the Pleistocene had an excessive impact on large animals and spared small ones.

As archaeologists have investigated the past more intensively, it has become clear that species extinction is a complex phenomenon subject to multiple parameters. The rate at which the mammoths died out, for example, varies depending on the region. We previously believed that they had become extinct around ten thousand years ago. But we discovered that on Wrangel Island, in Siberia, they survived an extra six thousand years. Interestingly enough, their eventual extinction probably once again coincided with the arrival of the first humans in this desolate land around four thousand years ago. Here and there, small, isolated groups of animals survived this wave of annihilation.

Of course, the climate weakened the ecosystems that provided food for this megafauna, inducing stress that led to a decline in animal populations. But the arrival of humans in this fragile milieu was devastating. Since then, human expansion has been accompanied by sudden drops in biodiversity and the extinction of large animals who have had their habitat inexorably reduced.

So what is the legacy of Monkey the killer and his expansion against a backdrop of climate change?

Having survived at least ten cycles of alternating ice ages and temperate periods, the six species of North American proboscideans died out between 11,000 and 9000 BCE. They were rapidly followed to the grave by six species of wild pig, three genera of antelope, three genera of giant rodent (beavers and capybaras), musk oxen and other giant bovines, mountain goats, giant camels and llamas, at least three species of horse, the great elks, the slow-moving giant prehistoric armadillos, two or three species of bear, two species of wolf, two species of lion, one species of jaguar, three species of saber-toothed tiger, and a dozen species of giant sloth (whether they lived on the land, underground, or in the sea). In North America, out of the forty-nine known genera of vertebrates weighing more than 44 kilograms (97 pounds), thirty-three died out.[5] In South America, 100% of vertebrates over one ton (thirty-six species and eighteen genera) and around 80% of vertebrates over 44 kilograms (forty-six species from thirty genera) were wiped out. Most of these species had been around for more than a million years and had seen at least eight ice ages and temperate periods.

The animals that survived were those that were able to take refuge in areas where humans could not easily follow. They generally became smaller as a result, losing between a quarter and a third of their body weight. This was the case, for example, for both forest pumas and migratory bison (of the latter only one species survived out of four and at the cost of a severe reduction in size). The grizzly bear made the most of the disappearance of native bears and migrated from Asia, along with the elk and caribou that they occasionally fed on.

The Caribbean islands were spared this spate of extinctions that ravaged North and South America—until humans appeared there some five thousand years ago. After that, all the vertebrate species over 44 kilograms (97 pounds) were wiped out in less than one thousand years. Climate may have played a role in this by weakening these animals' ability to subsist and reproduce. But the main culprit in the extinction of the megafauna seems to be humankind possibly assisted by invasive species who accompanied our migration.

Four questions remain:

1. How can we explain the destructiveness of certain prehistoric human cultures such as the Llano culture (also called the Clovis culture, after the town in New Mexico near the archaeological

site), active fourteen thousand years ago in America and that was probably preceded by other waves of human migration?[6] Were the people of Clovis much fonder of meat than those who preceded them? Or were they simply better armed? Or were there just more of them? These hypotheses remain open.

2. Are we sure that humans could challenge ten-ton mammoths when their weapons were nothing more than wood and sharpened stones? Yes, we are. A single human armed with a stone-bladed assegai could kill an elephant. There are early twentieth-century reports of solitary pygmy hunters who could bring down elephants with one stab of a lance to a vital spot on the stomach. But collective hunting of elephants or mastodons using rudimentary weapons was more common, and prehistoric examples of this have been documented in China and in North America. Mammoths' thick fur may have given them extra protection, but a group of humans relying on traps or persistence could always bring down the biggest of them.

3. Why did megafauna survive better in Africa (and to a lesser extent in southern Asia) than they did elsewhere? South Asia and Africa are the last refuges of the surviving species of elephants, lions, and rhinoceroses. But they are also places where humans have been present since their very beginnings around two or three million years ago. In these places we coevolved with these animals. They have observed the evolution of our hunting techniques and learned to adapt to them over time. However, when humans arrived in a new place for the very first time, most animals were easy to approach (and kill) because they did not have the instinct to run. How could they know that this weakling was a hyperpredator?

4. Finally, where does modern humanity, *Homo sapiens*, the knowledgeable *Homo*, come from?

Monkey Who Knows and Kills Anyway

We almost did not exist. Geneticists make that perfectly clear. It is difficult to believe, given how many of us there are on Earth today. In fact, there is much more genetic diversity among the hundred thousand chimpanzees on Earth today (in passing, there were two million of them in the early twentieth century) than among the 7.4 billion humans. Moreover, the genetic diversity of the one hundred thousand San people of southern Africa (the antelope hunters in chap. 1) is equivalent to that of humanity as a whole.[7] Everything suggests that we narrowly escaped extinction, that our population experienced an "evolutionary bottleneck." For some geneticists this occurred somewhere between two hundred thousand and one hundred thousand years ago. The population of our ancestors would then have been little more than a few thousand individuals. For others, however, this apocalypse might be more recent, around seventy thousand years ago. Each hypothesis is supported by a scenario leading to the appearance of a creature with new abilities: a Monkey with a great capacity for cognition, which facilitated the expansion of the species and led to the eviction of all potentially competitive animals. Let us call these two hypotheses Eve (for the first) and Toba (for the second).

The period between two hundred thousand and seventy thousand years ago corresponds to what is commonly called the Cognitive Revolution. This second fundamental revolution saw humans master a range of skills that would enable them to increase their impact on the world around them: fire, language, and symbols. The theory of the mitochondrial Eve, also called African Eve, stems from the work of geneticist Alan Wilson and suggests that we all have genes inherited from a shared ancestor (it uses mitochondrial DNA, which are passed from mothers to children and thus traced back along the matrilineal line). According to this theory, our common ancestor lived in Africa between two hundred thousand and one hundred thousand years ago.

The other hypothesis is named Toba after the colossal volcano that erupted on the island of Sumatra in Indonesia about seventy-five thousand years ago. According to the geneticists who support this hypothesis, this eruption is the only potential geological cause for our brush with extinction around seventy thousand years ago. This theory is less well documented than the mitochondrial Eve hypothesis.

In June 2017 Jean-Jacques Hublin and Abdelouahed Ben-Ncer

announced the discovery of five *Homo sapiens* skeletons that were 315,000 years old (give or take 34,000 years). This discovery, at the Jebel Irhoud site in Morocco, has forced us to reassess considerably the date at which our species appeared. Up until then, the oldest skeletal remains of *sapiens* were of *Homo kibish*, around 195,000 years old, and *Homo idaltu*, around 160,000 years old, both found in Ethiopia. The tools found at the Jebel Irhoud site are so similar to objects from the same period found in other parts of Africa that it is now believed that *sapiens* had already spread over the continent 315,000 years ago. This discovery, along with others, suggests that the Cognitive Revolution took place over tens—if not hundreds—of thousands of years.

Unfortunately, the Cognitive Revolution occurred primarily in the mind, and thinking is not an activity that leaves many visible traces. The clearest archaeological evidence lies in burial rituals and cave paintings, two practices that demonstrate a symbolic response to the world. The oldest known sepultures have been found in Israel. They are around one hundred thousand years old and have offered up the remains of *Homo sapiens*. Indeed, the oldest paintings (in caves or on cliffs) are no longer considered exclusive to Europe. Cliff paintings from approximately forty thousand years ago were discovered around the same time in Spain, Indonesia, and Australia. There is also evidence from Blombos cave in South Africa that ochre was used for decoration at least 120,000 years ago, and possibly dating back as much as 160,000 years. A study from 2012 also raises the strong probability that Neanderthals used ochre between 250,000 and 200,000 years ago at the Maastricht Belvedere site in the Netherlands.[8] There are many traces of the use of ochre, carvings, and even jewelry made from shells from about seventy-thousand years ago. All this seems to suggest that even then, a preoccupation with aesthetics—decorating the body and the environment—was emerging everywhere—South Africa, Morocco, Israel, Australia, Europe—among *sapiens* as well as Neanderthals. From this time onward, Monkey clearly developed a sense of style, which supposes a highly developed awareness of the self!

What can we conclude from this? For most prehistorians, *Homo sapiens*, modern humans, the "wise," were born in Africa. Then, following in the footsteps of *erectus*, who preceded us by at least two million years, we left Africa to explore the world. Around 180,000 BCE *sapiens* was present in Israel and came into contact with Neanderthals. This was an ice age. Canada, Scandinavia, the Alps, Siberia, and the Himalayas were all covered by layers of ice several kilometers thick. This ice was therefore not in the oceans, which explains why the sea levels were much lower than today (up to 130 meters [426.5 feet] below current levels). Europe was covered in ice from Scotland to the Alps, the Mediterranean was a mere lake, and from China it was possible to walk to the Philippines or Japan without getting wet feet.

The hypothesis known as Out of Africa 2 therefore suggests that all modern humans descended from a group that came from Africa and colonized the Near East at least 180,000 years ago before spreading farther east. In their path they most likely came across Denisovans somewhere in western or central Asia. They reached India and moved on to conquer the Indonesian archipelago at least eighty thousand years ago. It is also possible that they interacted with *Homo floresiensis* in Indonesia, a pygmy population thought to be descended from *erectus* who had been isolated for a million years and who died out perhaps eighteen thousand years ago. From around seventy thousand BCE, *sapiens* began the maritime crossing that would take them to Sahul, the continent that would later become Australia. Finally, modern humans found themselves in China around 80,000 BCE at the latest, where they possibly coexisted with the last members of *erectus*.

Once again, this expansion was linked to climate. When the seas were at their lowest, Siberia and Alaska were connected by a large ice bridge. The only significant stretch of sea between Asia and Oceania separated Sunda (the continental block including Asia together with the Philippines and Indonesia) from Sahul, the island continent combining current-day Australia, Tasmania, and Papua New Guinea. Sahul was the flattest and most arid of the continents after Antarctica, and it covered a surface more than a third larger than it is today. Humans possibly appeared in New Guinea around seventy thousand years ago and then migrated south, toward Australia, shortly afterward. We know that these people must have mastered language because a crossing like this—over 200 kilometers (124 miles) of sea, probably using rafts—requires careful planning and coordination.

We Are All Hybrids

But what became of Neanderthals, Denisovans, the species *erectus*, the Flores pygmies, and all the other species of *Homo* that geneticists now believe existed? The partisans of the Out of Africa 2 hypothesis have long argued that they formed distinct species that died out in the face of competition from a more evolved form of *Homo*. Opponents of this hypothesis, by contrast, defend the idea of a candelabra-like evolution with these archaic *Homo* species all over the planet evolving separately but convergently toward modern humanity. This model depicts a family tree with multiple branches reaching out from the central trunk of *Homo erectus*.

Very recently researchers led by the Swedish evolutionary geneticist Svante Pääbo have proposed a third and very robust alternative hypothesis: interbreeding between archaic and modern humans. They have proved that we had genetic exchanges with all these now-deceased cousins. European and western Asian populations have between 1% and 4% Neanderthal in their DNA. For certain groups in Asia and Oceania, up to 6% of their genes are inherited from Denisovans, who lived north of the Himalayas around fifty thousand years ago, and possibly three hundred thousand years ago in Spain. In sub-Saharan Africa, studies show a much greater genetic diversity compared with the rest of humanity, which suggests other ancestral exchanges between non-*sapiens* humans. Studies currently underway will most likely provide surprising results in the years to come.

The conclusion therefore seems to be that we did not exterminate our cousins or drive them to extinction. We genetically cannibalized them through higher rates of reproduction and possibly a society better tailored toward competition. By far the best documented case of this is Neanderthal, whose territory stretched from Spain to the Urals one hundred thousand years ago. There are four factors that contribute to the fact that we have so few Neanderthal genes. First, they were less numerous than us, which probably meant that their scattered population was in demographic distress even before *sapiens* arrived. Second, the genetic distance between us most likely meant that a large portion of *sapiens*-Neanderthal offspring were infertile (just as mules, horse × donkey offspring, are). Third, we invaded their hunting grounds, weakening their way of life, forcing them into more inhospitable areas, or making them cover ever-larger distances in search of food. Finally, our more cohesive societies made it difficult for them to stand against us.

We were significantly less physically strong than the Neanderthals, physiologically less well adapted to the cold that covered Europe at the time, and no more intelligent than they were. In recent years, evidence has been found that suggests that Neanderthal also went through the Cognitive Revolution at the same time as us and independent of our influence.[9] They buried their dead, wore clothes, transmitted complex knowledge through language, invented and developed tools, made glue, built elaborate shelters, possibly painted the first cave art in Europe, and even constructed stone rings in a cave in France (Bruniquel) some 175,000 years ago. But we probably overcame them because of our ability to interact socially. And it is likely that our whole family, possibly a dozen species of *Homo* who coexisted with us one hundred thousand years ago, were wiped out like this, swallowed up by our progressive conquests.

Monkey is the heir to mass manslaughter.

All the branches on the tree of humanity are dead except for ours. This is what led us to believe that it was just a trunk, that we were unique. Humanity as a single fragile reed, alone in the universe. Conjuring the ghost of Neanderthal put an end to this illusion. Like some oblivious Cain, the last-born of the *Homo* family is most likely responsible for his Neanderthal brother's death, and probably others too. But there is no eye of God to remind us of this, just a few fragments of DNA that whisper a forgotten truth in a research lab somewhere.

And, without looking back, we went further than *erectus*. We went the way of the elephants.

Australia, or How to Burn a Continent

Around one hundred thousand years ago, we would have seen proboscideans everywhere. Mammoths and mastodons roamed all over America. Where it was warm enough, they shared the space with *Stegomastodon* and *Cuvieronius*. These were descended from *gomphotherium*, a kind of primitive elephant with two sets of tusks that emerged in America twelve million years ago and who had managed to migrate to where Brazil is today via China before the Ice Age arrived. In colder areas, mammoths reigned supreme over the steppes that covered the northern half of Eurasia and North America. Bare-skinned pachyderms lived in the warmer areas, including South America, southern Europe, South Asia, and Africa (which had at least five species of elephants one hundred thousand years ago, only two of which survive today). All these species became smaller when they were restricted to confined territories,[10] and fossilized remains tell us that dwarf elephants prospered on all the islands of the Mediterranean. Cyprus, for example, had a miniature elephant weighing only 200 kilograms (441 pounds) that died out with the appearance of humans some thirteen thousand years ago. There was also a pygmy mammoth on the Californian islands, and on the island of Flores in Indonesia there was even a miniature version of the massive *stegodon*, a fourteen-ton giant among the primitive proboscideans that once ranged from North Africa to North America, including Japan. The heavyweight record holder was the straight-tusked *paleoloxodon*, found everywhere from Africa to Japan. It weighed up to fifteen tons, which made it the largest terrestrial mammal of all time. Yet a prehistoric butcher's workshop discovered in England suggests that it was hunted to extinction in Europe by the ancestors of Neanderthal some four hundred thousand years ago. Even this giant had a miniature form, found on the island of Crete, which died out much later, around five thousand years ago.

Of the six genera and twenty-five species of proboscideans who lived around the world fifteen thousand years ago, only three species of two genera survive today: the Asian elephant and two species of African elephant (one that lives in the forest, and one that lives in the savanna). We long believed that the two African elephants were a single species, and it now seems clear that they sometimes interbreed.

How can we explain the fact that these pachyderms (and many other animals, like lions, bears, and wolves) managed to conquer the whole world (except for Australia, separated by sea) whereas *erectus*—although formidably opportunistic—limited themselves to southern Eurasia? The elephant family simply evolved from their beginnings in Africa. As they moved north into the colder zones, they put on fur coats to keep them warm and became mammoths. In hotter areas they remained naked—with so much body surface to cool, fur would be detrimental to their survival. When the climate cooled, elephants and mastodons migrated south, and mammoths moved into new territories. When the temperatures increased again, they moved back the other way. Evolution allowed the elephant family and all their cousins to live all over the world.

Humans, on the other hand, remained *naked* apes. As a result, the colder regions long remained beyond their reach. But Neanderthal was an exception. Their morphology was better suited to the cold, they could control fire, and they made holes in fur to use as clothing. Then *sapiens* appeared. We overstepped the bounds of evolution and, like Neanderthal, stole the fur of other animals to keep warm. But by using needles[11] made of bone or wood, we were able to make clothes that fitted closer to the body and were therefore warmer. From this point onward, culture became more important than evolution for us. Siberia whetted our appetite with its great reserves of meat, reindeer, mammoth, and hairy rhinoceros. *Sapiens* crossed the Bering strait, the Aleutian land bridge, or both at a date that remains a subject of some controversy (probably around 35,000 BCE) and reached Patagonia around fourteen thousand years ago.

We were already making profound changes to the world around us in at least two separate places: Australia and the Amazon.

The first humans crossed the land bridge separating Sunda (the landmass covering today's Southeast Asia) from Sahul (which included Australia and the surrounding islands) around seventy thousand years ago. They conquered New Guinea and wiped out its megafauna. At the time, what is now Australia extended much farther south. It was all colonized at least fifty thousand years ago. We know this because Lake Mungo, which is in southern New South Wales and is now dry, has revealed traces of encampments as well as the bones of an Aboriginal ancestor nearly 2 meters (6.5 feet) tall who lived to the ripe old age of fifty. Mungo Man's development was nourished by his carnivorous diet and probably physical exercise too. These early humans lived on a continent that was largely arid but where the steppes gave way in places to immense forests that had grown up around lakes and rivers that fluctuated with rainfall. Marsupial mammals prospered here. There were many large herbivores and a proportionate number of specialized carnivores. There were giant wombats weighing several hundred kilograms, kangaroos 3 meters (9.8 feet) high, 120-kilogram (264.5-pound) carnivorous marsupial "lions," 50-kilogram (110-pound) marsupial wolves (*thylacines*), and omnivorous and opportunistic predators like the 25-kilogram (55-pound) Tasmanian devil. There were also giant birds, like massive emu, and land crocodiles 5 meters (16.5 feet) long.

Mungo Man left his footprints in what is now solidified mud, and they reveal his long athletic stride as he chased after kangaroos. The partially cremated remains of Mungo Lady were also found nearby. These remains are between twenty thousand and forty thousand years old and provide evidence of one of the oldest cremations known in world history.

In a short space of time, humans drove the herbivorous megafauna and all the egg-laying species on land and sea (big eggs are such easy prey) to extinction. The large carnivorous marsupials, deprived of their prey, progressively followed suit. The bigger they were, the more meat they needed and the faster they starved.

With the extinction of the megaherbivores combined with global cooling that aggravated aridification in Australia, the steppes were vulnerable to repeated forest fires. Without the megafauna to keep down the vegetation, the balance was lost. With no one to nibble the tops of the trees and keep the vast grasslands in check, the bush grew uncontrollably. And then, during the dry season, it burned.

So, the last forests disappeared, and the great rivers vanished with them.

The landscape was now dominated by vegetation selected by and for fire. Spinifex grass, which needs fire to spread its seeds, and eucalyptus, which of all the woody plants is the only one to resist repeated fires (if it does not reach excessive temperatures that may occur because of the uncontrollable burning of dense bush), proliferated. Biodiversity declined and then reached a balance. Out of eighteen genera of marsupials over 40 kilograms (88 pounds), only one (*Macropus*, kangaroos) survives today.

Another human variable came into play in this fragile environment: the repeated use of partial burning off roughly every two years. This practice encouraged roots and edible plants to grow as well as the proliferation of small game. Evidence of the widespread use of this technique has been found from at least 4000 BCE (and is now suspected to have been used as early as 36,000 BCE), but the practice probably coincided with the arrival of humans on Sahul.

The result was that that early Aboriginal people tended the garden of Australia with fire. The importance of this was phenomenal. Although there was probably never more than a million people at any one time and their techniques were rudimentary, they left a mark on the continent as a whole and on its biodiversity. The first *Homo sapiens* and their fires destroyed all the megafauna, deeply altered the flora, and transformed the climate by aridifying the whole continent. From this example and others, we can see that environmentalist myth of the "noble savage" as the guardian of nature is unfounded. Of course, they could not know what they were doing. They killed the animals they found for food, sometimes also including them in cave paintings and making them a part of the Dreamtime stories. At the end of the glacial period, around ten thousand years ago, rising sea levels reduced the surface of the continent by one third. Driven by demography and lack of land, some Aboriginal tribes successfully adapted to life in the desert.

The megafauna also disappeared from the Amazon around twelve thousand years ago. At the end of the last ice age, the Amazon was not the tropical forest we know today. It was much closer to the African savanna, populated by the same kind of megafauna mentioned above: several species of sloths, armadillos, and pachyderms. A 2013 study by biologist Christopher E. Doughty demonstrated that these giant animals fertilized the Amazonian basin with nitrogen, phosphorous, and other nutrients in their manure.[12] The forest grew from the soil fed with their waste and with the help of the warmer temperatures of the Holocene,* which began around 11,700 years ago. The extinction of these species also left its mark in the sediments by radically decreasing the spread of nutrients in the soil. For Doughty, the large animals of the Pleistocene acted like "arteries" within ecosystems by dispersing nutrients around the world, and their extinction cut those arteries.[13] It is because most of these animals died out that the world has so many unfertile regions.

Some Primates Are More Equal Than Others

This war against the living reached new heights in what is now Russia among the mammoth hunters. Up until this point, hunter-gatherer populations were not able to accumulate excess because they had to be constantly on the move. This did not prevent the exchange of precious objects (such as eagle feathers, pierced shells, or tools of exceptional craftsmanship) over several hundred kilometers. But around thirty thousand years ago in Sungir, near where Moscow is today, people built long wooden houses partially dug into the earth as a protection against the extreme winter cold. We know that they hunted mammoth extensively because of the bones found around their encampments. Even their buildings were built from tusks and bones. Clearly, humans have considered living things a primary resource for a long time, although it is possible that this instrumentalization was tainted with a sacred awe of animal power. In any event, the human population increased thanks to the abundant sources of meat. A mammoth provided food for a whole family for an entire winter if they could protect the frozen meat from predators.

Inequalities began to emerge. Leaders appropriated resources, concentrated them, and redistributed them as they saw fit. This is particularly evident in one exceptional burial site in Sungir where a man and two children were buried in tunics covered with a total of thirteen thousand intricately carved ivory pearls alongside spears carved from mammoth tusks. The carving of each one of these pearls represents at least an hour of work. To make the spears, the natural curve of the tusks had been reshaped and straightened through a process that is still unknown to us. They must have been priceless. That children should be buried with such treasures is testimony to the ability of this society to produce wealth and the specific status that these young people occupied within the group.[14]

This type of society, where the beginnings of a hierarchical structure involved the concentration of wealth in the hands of a few, began to emerge particularly where hunter-gatherers lived in environments that were rich in natural resources. The two most well-documented cases of this are found in Japan and in North America. The Jōmon, for example, who lived in Japan between 14,000 BCE and 1000 BCE, survived on what the sea provided them as well as on fruit, nuts, and other things. The American Indian and First Nation tribes on the northwest coast of the United States and Canada (Chinook, Kwakiutl, Haida, Tsimshian, and Salish) exploited the seasonal migration of millions of salmon that they were able to preserve by smoking. These tribes invented an ingenious system to limit the concentration of wealth, the potlatch, an ostentatious festival of consumption that was designed to avoid any one chieftain monopolizing too many goods and too much power. At regular intervals the rich were obliged to empty their stocks of provisions and distribute valuable goods both to their own dependents and to guests from neighboring tribes. The idea was that these gifts would eventually be reciprocated. Societies functioning in a similar way also existed in New Guinea, in Southeast Asia, and in the Amazon basin.

The first discernable example of human elites therefore seems to be at Sungir, among the mammoth hunters. In his insatiable appetite for these gigantic animals, Monkey began to get greedy. We are pleonexic,[15] perpetually unsatisfied, always seeking more. Soon the Agricultural Revolution (see chap. 3) would allow us to fulfill our new dream and begin to hoard.

Today, this primate-turned-demigod leads a plethora of invasive animals in his path. One example among thousands of declining biotopes is the Asian python that is ravaging the Florida swamps. Because there is no native constrictor in this habitat, none of its prey is accustomed to avoiding this kind of hunter. The moral here is that when a superpredator leaves its area of origin, nothing can stop it because nothing is ready to fight back. This is our ancestor's dream of abundance, a land of plenty at their disposal in each newly discovered milieu. The Sungir hunters thirty thousand years ago (avid consumers of mammoth stew) and the Indonesian seafarers one thousand years ago (who specialized in omelets made from giant Madagascan birds' eggs) were the precursors of today's poachers who are massacring the last rhinos with their Kalashnikovs to sell horns at €60,000 (US$65,000) per kilogram (2.2 pounds) on the Chinese market for traditional medicines.

Humans are indeed the ultimate invasive species.

Species become invasive when they prosper in areas where they have no natural predators. And humans have no predators at all, particularly now we have overcome major epidemics. We are the guilty heirs of all the animals that have died out in this sixth extinction that began at least fifty thousand years ago and that has become even more destructive since the nineteenth century. Today there are biologists who dream of cloning the mammoth, bringing wilderness back into the world, returning part of the planet to the ecosystems of the Pleistocene.[16] This is a dream of the Monkey-god.

What if we concentrate on making sure the last elephants survive?

3

THE WHEAT DEAL

Monkey embarked upon the Agricultural Revolution around twelve thousand years ago. This momentum created civilizations, wars, and inequalities. It reshaped both the human body and the biology of the planet.

Five or six stone huts under a blazing sun. They are dug into the ground to stave off a few degrees: a meter underground, a meter above. Nearby there is a water source, a now-intermittent river or a dried-up lake. We are in today's Jordan. But it could also be Syria or Israel, where similar sites have been found and attributed to Natufian culture. In any case we are in the Levant,* which has been the crossroads of humanity since the first *Homo* left Africa to conquer the world. And it is among these stone huts, which archaeologists have reconstructed to give some shape to this prehistoric village, that the first traces of a fundamental evolution can be seen, an evolution that would deeply alter the world, its effects rippling out to transform our species beyond the point of no return.

Shaping Genes

It was here that Monkey became a farmer, a biotechnician, a wizard of the living world. This is where, for the first time, we altered nature on a genetic level. In this region thirteen thousand years ago, we made a pact with a plant: wheat. In fact, Monkey had been a miller and a baker for a long time, collecting grains, grinding them, mixing them with water, and cooking dough. When archaeologists opened their eyes to this, when they realized that our ancestors fed on grains long before they cultivated them, they uncovered the existence of grindstones dating back over one hundred thousand years in Mozambique and thirty thousand years in Europe. What was new thirteen thousand years ago was the nature of the wheat itself.

The plant had mutated. Before this, heads of wheat were made up of fickle flighty grains that blew away as soon as they were ripe, spreading out on the wind to ensure the plant's reproduction. As the climate became warmer, the Levant was progressively covered in prairie grasslands where game and grains abounded. Taking advantage of such rich surroundings, Monkey began to set up semipermanent encampments. We collected the grains we used for food and sowed them around our seasonal settlements. We chose the biggest seeds for sowing, those that would ensure a healthy descendance and produce the most flour possible.

In its wild state, wheat is subject to various mutations. One of these is a genetic infirmity affecting one head in a million that prevents the grains from separating from the head and blowing off to implant elsewhere and to await the rains and germination. This mutation prevents any chance of survival.

And yet thirteen thousand years ago this anomaly began to be more common. Someone modified the wheat, selected its genetic characteristics, made a deal: you entrust your reproduction to me, and I will eat you, but in exchange I will ensure you a legacy. Eventually, it became almost impossible for wheat to reproduce alone. Monkey modified it for our needs, domesticated it, literally brought it into the home (domestication comes from the Latin *domus*, meaning "home"). But who really benefited from this deal?

Before this step—the domestication of the living world—the human population had never been able to grow beyond 2.5 to 5 million people scattered over all continents (except Antarctica). This was an ecological threshold, a Malthusian check.* As hunter-gatherers, modern humans needed to cover substantial territories. Competition for resources led to the eradication of Neanderthal and the other species of *Homo*. Our population had reached its peak. We had annexed most of the world, from southern Africa to northern Europe and from Siberia to Patagonia and Tasmania. There were only a handful of islands left to conquer thirteen thousand years ago: Madagascar, New Zealand, and the archipelagos of the Pacific. But it had become difficult to increase the size of the human population.

Yet, as we have seen, all living things must submit to an iron law: reproduce to perpetuate the species.

Did humanity win out in the wheat deal? From a biological perspective, in terms of the strict reproductive success of the species, the answer is yes. Ten thousand years after the beginning of domestication, at the beginning of the Common Era,* there were between 200 million and 250 million people on Earth, most of them farmers. The population had increased around fifty- to one-hundred-fold! An absolute triumph!

Did wheat get anything out of this deal? This plant has been with us at every stage of our growth and expansion. It has subjected itself to all our whims. We have hybridized it, adapted it to hostile climates, made it thicker so that it would bear heavier heads, made it shorter so it would be the right height for our harvesters, and so forth. Some have ironically suggested that it was wheat that made a slave of Monkey, because from a biological point of view, in terms of the success of the species, it gained even more than we did from this exchange.[1]

A Critical Heat Wave

Since we first left Africa at least two million years ago, humans have been playing a game of freeze tag with the climate. These two million years have been marked by cycles of alternating ice ages and temperate periods, first rapid—lasting around forty thousand years—and then more spaced out. For the last million years, the cycle has been eighty thousand years of cold—covering two thirds of the earth's surface with a crust of ice several hundred meters thick—followed by twenty thousand years of warmer temperatures—during which the ice shrank back to the poles and tropical animals ventured north. In England, prehistorians are still uncovering the bones of hippopotamuses around London, remnants of these temperate times that saw pachyderms wallowing in the Thames.

Our ancestors, *Homo erectus*, and their various grandchildren were subject to these fluctuations in glacial tides. In warm periods, they followed their prey toward the north and multiplied as their hunting lands grew. When the ice returned, it decimated them, forcing them into exile in the south and isolating them from each other. This is how the Neanderthal, trapped in a colder Europe, managed to survive. They changed. They became stronger and stockier with more body fat, paler skin, and a fatter nose that allowed them to warm the air. They became specialized in resisting the cold, systematically using fire and fur to warm themselves.

When Neanderthal died out between forty thousand and thirty thousand years ago, modern humans adapted to the cold of Asia and Europe. They were in Sungir, Russia, but they also colonized the Americas and hunted megafauna to extinction all over the world. Some developed a gene allowing them to digest a predominantly carnivorous diet, a gene found today among artic peoples, like the Inuit and Aleut, who have made the inhospitable polar circle their home. These hunters had a massive impact on their environment in three ways: they ate the large mammals and consequently reduced fertilization of local flora through lack of manure; they made the vegetation vulnerable to fire for want of grazing animals; and they forced the surviving animals to adapt to the landscape they left behind.

Studies have shown that an awareness of predators—be it visual, olfactory, or auditory—is enough to inhibit the reproductive abilities of prey animals. When alert to the presence of predators, these animals produce only half as many young—smaller litters, more stillbirths, and higher mortality among babies. As a result, subsequent generations are smaller. Over at least the last ten thousand years, and possibly much longer than that, humans have appropriated land through fire. Repeated burning was used to mark hunting grounds, controlling biotopes at the risk of draining them of animals. The smoke from these burnings was the visual and olfactory demonstration of the reign of Monkey the arsonist, forcing the surviving animals to be ever more discrete.

Then the climate played a nasty trick on Monkey. From about 15,000 BCE it got hotter, which helped weaken the megafauna by modifying their ecosystems. Large animals thus proved to be more vulnerable to an increase in predation. Humans finished them off. But we were going through one of the first increases in population just as our food source was decreasing. We had to adapt. In the Levant, therefore, the decline in antelope populations meant people relied more and more on snails and wild grains.

But this warming was too fast. The melting in Canada's frozen north led to a colossal breakup of ice that submerged the North Atlantic with a sudden influx of freshwater. Lighter than saltwater, this freshwater drowned the North Atlantic Drift, the current of warmer water from the Caribbean that warms Europe and evacuates the heat of the tropics. The speed of this warming meant the climate machine launched into a feedback loop.

Around 12,800 years ago, the North Atlantic Drift stopped circulating warmer water, and the world was again swept with cold. In just a few years, average temperatures dropped by 7°C (12.6°F) in the temperate Northern Hemisphere. This thermal shock was felt all the way to equatorial Africa, where temperatures dropped by 3°C (5.4°F).[2]

This mini–ice age, known as the Younger Dryas, lasted little more than a thousand years. It was too fast for humans to escape to warmer climes. They held on, found ways to survive. And when it began to warm up again, around 11,500 years ago, they prolonged the process and put nature to work for them. It was thus at the end of a climate crisis, just as this ice age was making its last stand, that Monkey began to harness plants and animals.

The warming, interrupted by the parenthesis of the Younger Dryas, would last three millennia, which was long enough for humanity to begin its third revolution: agriculture. The steppes receded massively over the whole of the Northern Hemisphere, giving way to lush forests. The desert zone of today's Sahara was then rich and green, fed by a multitude of lakes, such as the giant Lake Chad, which then covered a surface equivalent to today's Germany. The plains of the Levant were covered in wild grains. By extension, archaeologists refer to the Fertile Crescent, a zone that stretched in an arc from Mesopotamia, the Tigris-Euphrates river system, and the coastal zones of Syria and Palestine.

This sudden abundance of resources was followed by a demographic boom. Rising sea levels caused by the melting ice probably forced certain populations to migrate inland, away from the coast. The densification of populations led to a transformation of subsistence living. Monkey progressively controlled the environment in order to be able to feed large masses of people in a limited area, many more than hunting, gathering, and fishing would allow. Humans already had experience selecting certain plants by gathering grain. But this practice had not caught on because the climate had either been too harsh or too variable. The Neolithic was the age of the farmer. Thanks to continual planetary warming it became a lasting revolution. Over a few thousand years, Monkey was able to impose a threefold process of domestication onto the world, controlling plants, animals, and eventually other humans through the creation of hierarchical societies.

The Agricultural Revolution was made possible by the ability to store food. It took place in areas that global warming had left rich in grains, fish, and game. People were able to produce a surplus and establish permanent habitations. As their populations grew, resources were likely to become rarer. Faced with the perspective of deprivation, Monkey took control. Good production had to be maintained. We were hunter-gatherers, sometimes fishermen, but progressively we became farmers and shepherds to ensure that we always had what we needed to eat.

Hunting and gathering allow for a population density of four people per 100 square kilometers (about 38 square miles), whereas domesticating plants and animals allows this ratio to be multiplied by one hundred or even more. Archaeological traces show that this was a very long process. Several centuries, even millennia, were required for clear signs of domestication to appear in living things. In the Levant, which is by far the most extensively investigated and thus best-known zone, the first villages began to live on a combination of grains (wild and cultivated) and other plants as well as on the increased predation of local fauna some eleven thousand years ago. The rarefaction of antelope led humans to hunt other animals such as boar, goat, sheep, and aurochs (large ancestors of today's cows). Over around four thousand years, the consumption of domesticated grains progressively overtook that of wild grains.

Three Grains, a Single Revolution?

French historian Fernand Braudel defined three zones of civilization according to the grains they each consumed: civilizations of wheat, rice, or maize.[3] These three grains remain the most cultivated plants in the world even today just after sugarcane. Let us now look at the social, economic, historical, and ultimately biological consequences of the large-scale symbiosis between these plants and human societies.

The wheat civilization is the most well documented of the three. This is logical because it was the civilization of all-conquering nineteenth-century Europe just at the time when it was normalizing scientific frames of reference, particularly in history and biology. In this, European civilization was copying ancient Greece, which saw the consumption of grains as the most tangible sign of civilized society. Barbarians were those uncultured foreigners who did not eat bread. Several signs of very early wheat domestication have been found in the Levant, possibly dating back thirteen thousand years. This was associated with the development of the first sickles (curved sticks with small sharpened stones driven into them) and the by then routine use of grindstones (ancient grains were protected by very strong husks).

It was a long time before we came to eat bread with every meal. Global warming had allowed generations of human communities to remain in permanent villages, which was the case in Anatolia, the Levant, and apparently in Iraq eleven thousand years ago. It then became necessary to domesticate several grains and pulses (lentils and chickpeas in particular) while complementing this diet with hunting, gathering, and sometimes sustainable selection (finding gregarious animal species that could be controlled and domesticated).

Then there was an upheaval in the composition of our meals. It has been calculated that fifteen thousand years ago our hunter-gatherer ancestors consumed 40% meat to 60% plants on average. But by seven thousand years ago, farmers were present on all populated continents except Australia, and for these populations, this ratio shifted to 90% vegetable to 10% meat. Omnivorous Monkey had (almost) (re)turned to vegetarianism. Culture determines food. Many populations, such as those in Asia, came to adopt almost exclusively vegetarian diets, and their bodies adapted very well to this.

In the Levant, the domesticated grains were emmer wheat, einkorn wheat, and barley. Nine thousand years ago, emmer wheat—which is probably already the result of the prior genetic fusion of two different plants—was hybridized with einkorn wheat. This apparently unnatural union, which meant that three genetic strains coexisted in the genome of a single plant, led to the development of common wheat, also called bread wheat, whose starch-rich grains are used to make our morning toast, among other things. Hard wheat, which results from another selection process, is rich in gluten and proteins. It can be cultivated in arid climates and is today essentially found in the form of pasta and semolina.

Over eleven thousand years of domestication, wheat, whether hard or soft, evolved into more than thirty thousand subspecies. This genetic diversity is the result of adaptations to the numerous environments where it was sown, repeated hybridization with more or less wild grains, and the prowess of hundreds of generations of anonymous farmers conducting agricultural selection. It was important to develop heads that were more compact and fleshier with finer, looser grain husks and stronger stems able to support these increasingly heavy heads. Plants also had to be able to confront disease, parasites, and predators, which increased as these plants took up more and more space. Much of this genetic heritage has already disappeared because of the homogenization caused by industrial agriculture, something that is true for all domesticated plants.

The discovery of pottery, which emerged in the Levant around ten thousand years ago, would play an important role here. Up until this point it is likely that grains were ground into a sort of gruel to which we may have added other pureed plants, such as beans. But now it was possible to cook wheat grains separately, to prepare porridge or pancakes. These unleavened flat cakes were initially quite hard, but they would come to be softer. According to the instruments used by geneticists, the oldest strains of yeast, the microscopic artisans of leavened bread, are around eight thousand years old. Starch becomes easier to digest when it is cooked, and our organisms adapted to a diet that was increasingly reliant on cereals.

If I have focused on wheat at some length, it is because this grain is of great interest to us today. But this was not always the case. Up until the first millennium of our era, barley—which was developed at the same time as wheat—and, to a lesser extent, rye were more important in human consumption. For the people in the Levant around ten thousand years ago, as for the ancient Greeks around 2,500 years ago, a basic meal consisted of barley gruel. It was only around one thousand years ago that wheat began its march toward hegemony with the benefit of the warm weather of the Medieval Warm Period* and the introduction of new farming techniques (deep plowing, succession planting). Indeed, up until the Industrial Revolution of the eighteenth and nineteenth centuries, it was eaten along with other cereals, such as spelt, or other grain plants, such as buckwheat.

Agriculture, A Global Invention

Let us continue this exploration and head toward a second site for the domestication of living things: China. The country can be roughly divided into two parts: a temperate or even cold zone to the north around the Yellow River (Huang He), and a tropical zone to the south around the Yangtze river. Although we might associate this country with rice, that is a filter imposed by our present. The northern zone in fact domesticated millet, while only the south domesticated rice. There is evidence that wheat noodles have been eaten in northern China for six thousand years, and even today people in the north tend to eat that rather than rice, which is a staple in the south. This is by no means an ethnic divide, however, as people in both areas may be from the ultramajority Han, which originated in northern China.

Another long-accepted stereotype is that the Neolithic, the Agricultural Revolution, was a straightforward linear progression. Observing what happened in the Levant, European archaeologists deduced that everything then followed a logical order. Monkey domesticated plants and animals, which allowed us to settle in villages, build houses, and increase population size, which led to the need to establish hierarchies to control others and direct them in their work. The result was a clear distribution of labor: peasants worked, priests and kings governed, soldiers controlled the former for the latter, and artisans fulfilled tasks that required specialization. People had to find a way to cook and preserve grains, so they invented pottery. This preprogrammed schema fits the case of the Levant, where ceramics were invented in a sedentary society that had gone through its Agricultural Revolution and built its first temples. But it cannot be seen anywhere else.

In China, for example, pottery was invented twenty thousand years ago.[4] In Japan, it appeared slightly later, around seventeen thousand years ago, but agriculture was not introduced from the continent until around three thousand years ago. Before this, Jōmon culture prospered in Japan without agriculture, its sedentary populations living off what the sea provided. Fishing was intensive, and shellfish were consumed in such large quantities that their shells were piled into veritable hills. In addition, these people also used forestry planning and planted hazelnuts, chestnuts, and oaks around their settlements to complement their seafood diets with gruel made from chestnuts and acorns cooked in ceramic pots. They built permanent villages even though they remained hunter-gatherers (or rather fisher-gatherers), and it seems that they had already developed a form of social hierarchy. All of this was accomplished without agriculture, relying only on careful consumption of nuts and shellfish.

Archaeologists have long believed that the Agricultural Revolution began in the Levant and then progressively spread out from there. Of course, Europe was largely "neolithicized" from the Middle East. But in fact, this Agricultural Revolution occurred in various places around the world simultaneously (or in the space of about four thousand years), beginning around thirteen thousand years ago. The oldest known sites, apart from the Levant, are the Indus River basin (today's Pakistan), North China, South China and Southeast Asia, Central America, the Andes, the northern Amazon, and Papua New Guinea. And the process of domestication continued. Sometimes there was diffusion, and crops and herds spread and became acclimatized to other zones, moving, for example, from Anatolia to Greece and from there toward Europe. Sometimes there was innovation, and new zones of origin appeared, particularly in the Sahel region of Africa and in Ethiopia.

The peasants in northern China benefited from an extraordinary heritage: loess. This is a kind of silt, the dust of fertile soil eroded by glaciers, carried by winds from central Asia, and deposited in northern and central China in layers that can sometimes be several hundred meters thick. This silt preserves moisture in the soil, making it ideally suited to growing grain crops. The Cishan archaeological site in northern China provides evidence of an 8,500-year-old village that was so productive that its inhabitants managed to store fifty tons of semidomesticated millet. The cultivation of millet was also practiced twelve thousand years ago in Nanzhuangtou, where it was combined with wild millet (one third domestic to two thirds wild). In Cishan nine thousand years ago, this ratio was half and half. Two thousand years later, harvests were only domesticated millet. This plant was also affected by the mutation that meant the heads remained attached to the stem, the sign of a new pact between the plant and Monkey.

Then the climate warmed again. Common millet was abandoned in favor of a related species, foxtail millet, whose rapid growth and resistance to warm dry summers worked wonders. The Chinese diversified their diet with various varieties of cabbage and mustard as well as domestic pigs and ducks, and they also hunted other available animals such as bears, raccoon dogs, macaques, and deer. Then around five thousand years ago, barley and soft wheat arrived from Iran, imported along with sheep by people living on the steppes. These two new arrivals would progressively replace millet, enriching the diets of people all over China. There was also sticky rice, the prodigal son of Asian rice, lacking in gluten and preferred for feast-day cooking, as well as sorghum and pearl millet, which had been introduced via India. A greater variety of grains available means a lower risk of famine, and the cereals domesticated in Africa (sorghum, finger millet from western Africa, and teff, native to Ethiopia) all have a good resistance to drought.

Northern China and southern China represent two distinct biotopes. The north was once covered by a temperate forest of pines and deciduous trees. The south was a tropical jungle. Since 2012,[5] we know that it was in the valley of the Pearl River, which irrigated the heart of southern China, that certain subspecies of wild rice were collected and possibly cultivated at least ten thousand years ago and perhaps even 13,500 years ago. This occurred at the same time people in the south were domesticating annual tubers such as yams while remaining seminomadic and relying heavily on fishing and hunting for their diet. Among other things, they ate rhinoceroses, elephants, crocodiles, domestic pork and chicken, and water buffalo. They were also able to ferment rice to produce wine.

The cultivation of rice differs from other grains in two important respects. First, it is more productive. For the same number of plants, you get more grains than with wheat, barley, or millet. Second, as rice is buried in mud rather than sown in dirt, you need fewer grains to reproduce it. But it is much more labor intensive. In the south of China it would be cultivated not in fields, but in gardens, first in small dry patches and then under water. This was the beginning of an endless cycle for peasants. Sow the seeds, collect the seedlings, plant them out, remove the weeds. . . . A continual supply of water had to be maintained by various means, from canals constructed to divert river water to rivulets and channels built to distribute the precious liquid. Five thousand years ago, this cumulative knowledge progressively covered hundreds of hills in southern China with terraces and submerged rice fields. The cultivation of rice in water was extremely successful. At the beginning of the Common Era, this technique was used in Korea, Japan, and the Philippines, and it had long been established in the Ganges Valley in northern India.

But the history of rice is more complicated than that. For the first six thousand years of its cultivation by humans, it was part of a diversified diet associated with various other crops. Societies that did not give way to the fashion for grand terraces (which required settled lifestyles) remained seminomadic. They collected tubers (yams and taro probably originating from northern India) and left parts of the roots in the soil to ensure a harvest the following year. There was no need for long-term storage; the forest fulfilled most of their needs with tubers, fruit, and game. The remainder was provided by semiseasonal crops such as rice and by trade with neighboring cities, and then states, in the plains. For the first six thousand years of its use by humans, rice was gardened rather than farmed.

The American Counterexample

Now let's cross the Atlantic to see what was happening in the Americas. Here, practices were necessarily indigenous, as there was no possibility for exchange or trade across the ocean. And nothing here happened as it did elsewhere. There is no sign of the tipping point observed in China or the Levant, the encounter between a grain and humans looking to increase their population. Yet three Neolithic sites have been found: one for maize in Mesoamerica,* one for potatoes in the Andes, and one for manioc in the Amazon jungle.

Although today maize is a domesticated cereal with very large grains, it was not predisposed for this. Its wild parent plant, teosinte, is a small grass with seeds that are too fine and husks that are too tough to be edible (its sweet stalk is chewed by Mexican children though!). In Mesoamerica, agriculture began at least twelve thousand years ago, with various cucurbits, pumpkins, squashes, and marrows (some of which were emptied out to make containers) as well as beans, chilies, avocado, and tobacco. The irony is that these cucurbits were destined for extinction because they relied on the mastodons we encountered in the previous chapter for their reproduction. These large elephant-like creatures, which died out with the arrival of humans, were the only ones to be able to eat these enormous thick-skinned fruits and spread their seeds as they roamed. Domestication saved the squash family from extinction and guaranteed it a glorious future.

Initially, the largest human populations remained in coastal areas, their diet relying more on crustaceans and fish than on the plants they cultivated. Gardening simply provided a supplement. Four thousand years later, the population had grown and was consuming more and more vegetables. And then a miracle occurred: teosinte, a tufty weed that grew all over the Mexican plateaus, became a source of abundance with a single stem bearing an enormous seed head. Biologists are still unable to explain how Indigenous American peoples brought about this miracle—transforming a weed into maize. They very tentatively suppose that dense planting encouraged certain genes favoring a strong straight stem that could rise above competing plants to seek the light. This domestication of maize, combined with the arrival of the tomato and the potato from the Andes, allowed the civilizations of Mesoamerica to flourish.

There was a problem though. Advanced metalworking techniques needed to make plowshares had not been developed in this region, and moreover there were no large domesticable animals with the physical strength necessary to work the land. There were no aurochs in Mesoamerica, and no buffalo. In the Andes, lamas and alpacas were domesticated some five thousand years ago, but as the former can only carry 40 kilograms (88 pounds) maximum, they were not candidates for draft work or riding. However, they were both sources of tasty meat, and alpacas provide excellent wool. The extinction of the megafauna discussed in the previous chapter thus led to a time-lag effect. If the American horses and large camels had not been wiped out thousands of years earlier, could they have changed the course of history?

But the people found ways to do without. They developed inventive gardening techniques that enabled large populations to be concentrated in cities. In 1520, the Aztec capital of Tenochtitlan probably housed 250,000 inhabitants just before the Spanish invasion. Unable to plow the land, these populations became masters of fertilizing and combination planting. They dug holes, applied fertilizer in the form of a mix of ash and human excrement (a good way of getting rid of it), and combined plants such as the famous "three sisters," known as *milpa*: maize, beans, and squash. In this system, the squash covers the soil, prevents evaporation with its broad leaves, and discourages grazing animals with its spines (which have been bred out of modern varieties, which is why only remnants of them are found on your garden squash). The maize can thus grow without disturbance and in so doing provides a stake for the beans that will fix nitrogen in the soil as they grow, which both the maize and squash need to develop. Other plants may be added to complement this perfectly balanced trio. Although it is possible to periodically obtain extremely high harvests (which then require the land to lie fallow to rest), *milpa* is an agricultural technique structured around family production. This "three sisters" system would progressively spread north, into the center of what is today the United States.

Over at least the last thirty-five hundred years, maize has generally been nixtamalized before being eaten. This process involves soaking the grains in an alkaline water-based solution (traditionally obtained by adding wood ash, today by adding lime or bicarbonate of soda) to soften the skin. As a result, the grain is more easily digested and more nutritious because the process creates additional vitamins. It is also easier to grind it into flour.

Flour made out of tuberous roots was introduced farther south at least ten thousand years ago in the north of the Amazon and on the coast of Ecuador to supplement the traditional diet based on seafood. In Panama, manioc was dug up and prepared on-site so that next year's harvests might grow from the scraps. It was pounded into a flour using grindstones and made into flat cakes. In Ecuador, sweet potatoes were used in the same way. Further south, in Peru coastal populations fished sardines in large quantities thanks to the abundance of fish at the confluence of the cold river and the warm sea. As well as peanuts, potatoes, and beans grown on a small scale, they were already cultivating cotton, which they were able to weave and make into clothing. Above all, they also made fine mesh nets essential for fishing. These nets and textiles were the greenback of the day, and they were traded north into Ecuador and then down the other side of the Andes into the hills of Bolivia. Large-scale cotton farming required the capture of the torrents of ice melt that flowed down the flanks of the Andes after each winter. Major irrigation works began. Important cultural centers were built on the Peruvian coast five thousand years ago. Along with the culture that developed in the Indus Valley at the same time, this is the only sedentary society in which archaeologists have found no signs of conflict over several millennia.

In the Andes, agriculture is thought to have developed at least ten thousand years ago. Beans and chili plants were domesticated locally, and potatoes were bought from the Chilean coast, where they were probably domesticated around the same time. Potatoes have to be cooked to make them edible, and they require complicated preserving techniques so that they do not develop toxins. Quinoa (a close relative of spinach, of which the abundant seeds are collected to be used as cereals) has also been cultivated in this area for at least five thousand years. Near the high-altitude lakes on the Andean plains, a very productive agricultural technique known as *waru* or *sukacollos* was developed. Farmers terraced the plains in lines radiating out from the lake, alternating 5- to 10-meter-wide (16- to 33-feet wide) shallow channels and raised garden beds up to 100 meters (328 feet) long. Various tubers (potatoes, oca, ulluco) were grown in these beds along with quinoa and amaranth. Plant remains would fall into the water and feed the high-nitrogen algae growing there, which in turn would be periodically harvested to serve as fertilizer on the beds. At night the water would release the warmth it had accumulated during the day and thus create a microclimate that prevented frost damage. This method, which is now coming back into use, relies on intensive manual labor. But in the arid ecosystem of the Andean plains, it is also three times more productive than modern industrial methods based on chemical fertilizers.

The Roots of All Evil?

Growing grains like wheat, barley, millet, or rice enabled the emergence of hierarchical civilizations around six thousand years ago. These crops required collective labor and thus social organization to bring in the harvest at specific times and ensure year-round storage (sometimes longer, in case the harvests of the following year were poor). These were civilizations of power. Their kings were sacred. They spoke to the gods who confided in them the secrets of time. In death they were accompanied by slaves sacrificed for the purpose. These societies were shaped by very precise calendars, ostentatious buildings, solemn proclamations of sowing and harvest dates, and taxes. Indeed, the regular harvest of these grains, along with their high production and annual storage, allowed for control, taxation at specific dates, and the development of a hierarchical society. Henceforth, everyone had their role. Peasants produced. Civil servants taxed, inventoried, and managed food stocks. Soldiers maintained order both inside and outside the city. Priests ensured the smooth symbolic operation of the whole and made up the court of the kings with whom they hoarded a significant proportion of collective wealth.

In 1957 historian Karl August Wittfogel (1896–1988) described this as "oriental despotism" or "hydraulic civilization."[6] His idea was that the agricultural system conditioned the political regime. In summary, Wittfogel argued that in the West, freedom had been able to flourish around the culture of wheat based on the model of an individual farmer working his own lands without help from anyone else. In Mesopotamia, Egypt, and China, hierarchical societies developed because hydraulic systems required collective labor and thus a king who could tell his subjects what to do. In practical terms, however, the use of the environment was never this deterministic, and agriculture was only ever possible in societies where people worked together.

Humans have some things in common with ants. They can't do much on their own, but together they are formidably efficient. The impact of agricultural societies on nature has been spectacular. Biodiversity plummeted with every new field. Old-growth forests were felled using fire and axes[7]—the term *Neolithic*, from the Greek "new" and "stone," reflects the fact that archaeologists were struck by the number of axes made from polished stone (no longer chipped as during the previous Paleolithic "old stone" age) in the sites from this period. The Mediterranean landscape changed dramatically seven thousand years ago. Forests of deciduous trees and giant heather gave way to land that, once cultivated, was abandoned. Now only plants able to resist drought could grow here. The same process has been observed in many regions as the land has been progressively cleared. Deforestation and the creation of pasture for sheep and goats have accentuated erosion and the decline of biodiversity. Monkey facilitated the expansion of domesticated animals and vegetables to the detriment of wild species to the point where some died out and others were subject to genetic alterations. Wild plants and animals hybridized with their domestic alter egos. Even as humans took on (un)natural selection and thus increased the genetic diversity of the living world, the "natural" part of this biodiversity gradually declined.

The Agricultural Revolution was based on the storage of food stocks as the foundation for wealth. This fed the jealousy of neighbors. Evidence of violence is increasingly common in this period. Fortifications were built as early as eight thousand years ago. Weapons not designed for use in hunting also appeared. All those hammer-like tools, for example, could only kill one animal: the one with a large, fragile skull. Shields also appeared, their sole purpose being to protect the body against fellow humans. War became a matter for professionals. The use of large troop formations and maneuvers, which became systematic in Mesopotamia at least five thousand years ago, greatly contributed to the efficiency of sedentary soldiers. Trained from childhood, they acquired an unparalleled knowledge of the art of war. The oldest massacre known to us today took place twelve thousand years ago in a sedentary society that lived on the planned cultivation of semidomesticated grains, like in the Levant. Here, in Jebel Sahaba, near the northern border of what is now Sudan, sixty-one people, predominantly women and children, were violently killed, probably by a shower of spears and arrows.

Hierarchical religion also appeared around this time, a sign of Monkey's new power over nature.[8] Archaeological evidence is eloquent on this point: the oldest stone building ever found is Göbekli Tepe, in Turkey.[9] This was a sanctuary, an ensemble of megaliths masterfully sculpted by hunter-gatherers some twelve thousand years ago, two millennia before agriculture was developed here. Fifteen centuries later, the semiagricultural village of Jericho in

Palestine built a tower that seems to have had a ceremonial function.[10] Two millennia after that, the large agrarian agglomeration of Sha'ar Hagolan in Israel produced a series of effigies of the oldest known divinities, steatopygic goddesses (with large buttocks) probably associated with fertility cults. The Turkish contemporary cousin of this site, Çatalhöyük, had hundreds of houses decorated with murals of goddess-like figures who were pregnant or giving birth under the protection of leopards and bulls, who we suppose symbolize gods.

Before the transition to agriculture, animals—and the living world more generally—were probably partners with whom hunter-gatherers had to negotiate. This was the role of the shaman. Once Monkey had bent nature to his will, power no longer circulated horizontally but was exerted from the top down. The spiritual manifestation of this hierarchical power was religion. Humans were no longer surrounded by invisible spirits who had to be appeased—we dominated nature. But we were also submissive and under the auspices of forces to which we had to pay tribute, make offerings, and say prayers. These forces had their worldly representatives, sacred kings and priests, who had to be respected. As the human population grew, so did the complexity of the hierarchies, and religions evolved alongside power as its faithful reflection. Thus, in Mesopotamia, like in Egypt and many other places, temples were where inventories were conducted and, by extension, where taxes were payed. Mythologies—which would soon be written down—laid out models of behavior that governed life in these ever-expanding societies.

This grain-based social model spread throughout the world. Although it has significantly evolved, it still governs our existence. Wheat, rice, and maize still provide the basis for our food, and we still live with war and religions. Yet this was not the only possibility open to us.

The hills of Southeast Asia make up a zone known as Zomia, which stretches from southwest China to Nepal through the mountainous regions of Vietnam, Laos, Cambodia, Thailand, and Myanmar. In the broad acceptation of this geographical zone it covers some 2.5 million square kilometers (about 1 million square miles), the equivalent of over a quarter of the United States. For the anarchist anthropologist James C. Scott, the highlands of Zomia provided a refuge for populations who wanted to be free of the authority of states and their taxes, census, conscription, and land registers.[11] According to him, these tribal societies established systems that prevented individuals annexing and monopolizing power and transmitting it to their kin and descendants. These societies lived on rice, vegetables, pork, and game. The rice fields were seasonal, cleared by burning off, and associated with horticultural practices based on tubers.

This social model demonstrates the potential contrast between agriculture (crops grown in fields and requiring cooperation of a certain number of people) and horticulture (gardening) as it has been practiced in Zomia, the Amazon jungles, or Papua for some nine thousand years. Horticulture encourages the propagation of plants using root fragments or cuttings of aboveground plants. Societies that were more mobile and egalitarian (notwithstanding the fact that women were subordinate, as they were everywhere) emerged in tropical environments and combined hunting, fishing, and gathering with seasonal vegetable growing.

The Submission of Animals

As we have seen, the Levant was by no means the only site of the domestication of the living world. If the Neolithic began earlier there it was because the region lies at the junction of East and West with no natural boundaries. Its position at the intersection of Africa, Asia, and Europe provided it with once-rich flora and fauna. Indeed, half the animals and a third of the plant species that we eat today come from this region. The people of Asia and Europe were thus able to trade, from hand to hand, while staying in more or less temperate regions, bypassing the Himalayas to the north (via central Asia) or the south (through the Indian peninsula). Chinese millet thus reached Iran around seven thousand years ago, two thousand years before Iraqi wheat was adopted in northern China. However, unlike Eurasia, which is organized on an east-west axis, the north-south axis of Africa and the Americas meant that the transfer of animals and plants (voluntary or not) was complicated by having to negotiate different climates. This geography of trade over long distances goes some way to explaining why most of the animals and plants that we eat today come from the Old World, and later, thanks to the Columbian exchange (see chap. 9), from another intercontinental intersection: Mesoamerica.

To illustrate the variety of relationships that Monkey would come to develop with his vassals, we will look more closely at the paths of domestication for four animals: dogs, cats, pigs, and aurochs.

We begin with the dog because the wolf was domesticated well before any of the others. This can be seen in different skeletal remains buried in Russia, the Czech Republic, and Belgium up to thirty thousand years ago. In any case, dogs were there eighteen thousand years ago, well before the others. How can we tell that an animal has been domesticated, that the wolf has become a dog? Their skeletons show clear signs of cohabitation with humans, and they are slighter than their wild brothers. Why did this domestication happen so early? Most likely it occurred through a progressive familiarization and friendship between the two species that led to a symbiosis associating animals who hunt in the same way, in coordinated groups relentlessly tracking previously selected prey. It is also possible, however, that wolves, having seen their hunting territories overturned by this new arrival, began to follow humans simply to feed on their leftovers. They may have occasionally left puppies behind who were then adopted by humans. In any event, the association must have benefited both parties, humans being able to plan ahead and follow tracks and wolves being able to hear and smell their prey from afar.

Dogs have been used for many purposes. They are pets and playfellows, as we can see in Israel, where a woman was buried with a puppy in her arms thirteen thousand years ago. Or they may be delicacies for special occasions, as was the practice of some Australian Aboriginals (dingoes came from Asia at least five thousand years ago and were used neither to hunt not to guard possessions, although they were considered sacred among certain tribes, and live ones were sometimes worn for warmth in some places). Of course elsewhere they were also guard dogs and companions in hunting, fishing, and later even in war.

From an individual perspective, what are dogs if not degenerate wolves? Man's best friend was primarily selected for docility. Only the least aggressive were allowed to reproduce. We managed to convince them that the naked ape was their master. Later, when we needed to create animals capable of fighting humans or defending herds against large predators, a few rare breeds appeared that could challenge wolves. The truth is that a wolf has infinitely more endurance, more independence, and more bite than a dog. In my book, *Homo canis*, I explain that our ancestors domesticated the wolf by selecting juvenile characteristics in it. A dog is just a wolf trapped in an eternal youth, and just like a child, it is far easier to push it into submission. Maybe humans were autodomesticated the same way from the Neolithic.[12] Civilization makes all of us soft, even youthful, and malleable.[13]

In fact, dogs were domesticated several times. The main strain, identified by its Y chromosome (carried by males) comes from central Asia, Mongolia, and Nepal. It has been domesticated for at least fifteen thousand years. At one stage it was bred with an older strain, because all our pooches have mitochondrial DNA (transmitted by females) directly stemming from the European gray wolf. Evidence suggests that this wolf was domesticated around eighteen thousand years ago in the Levant and in Europe. In other words, just as we are hybrids of *sapiens* and other ancient species of *Homo*, Fido is a cross between Asian wolves and European canids.

Around eleven thousand years ago in the Levant, Monkey made a very particular bargain with a new kind of animal: the cat. We suppose that cats moved closer to human habitations to hunt the many mice and rats that were attracted by the grain humans had begun to store. These commensal (as the species who prosper in Monkey's shadow are called) rodents would henceforth have to face the teeth and claws of our new partner. Cat lovers will tell you that cats have never really been tamed and remain independent even though their long cohabitation with humans has considerably altered their genes. In any case, Monkey found them useful in protecting harvests and perhaps also as an agreeable companion to the point where they journeyed with him to Cyprus eleven thousand years ago. Further proof of the cat's usefulness is that it, too, was domesticated several times. It was tamed in China 5,500 years ago, for example, although these native cats were ultimately supplanted by cats from the West. Today, the world's five hundred million cats all descend from the small population that helped Monkey protect grain stocks in the Levant.

The trajectory of the pig was quite different from that of the cat. It was initially domesticated perhaps fourteen thousand years ago in Mesopotamia and then in Anatolia and the Levant. Again, it was quite independently domesticated in China and Southeast Asia around eleven thousand years ago and then again in Papua New Guinea around nine thousand years ago. A final independent site of domestication occurred in Italy perhaps six thousand years ago. In the beginning, as is still the case in Papua New Guinea, swine were probably fenced in at night and, watched over by young humans, left to roam in search of food in the undergrowth during the day. Pigs were introduced to Cyprus around 11,500 years ago, where they immediately returned to the wild. This same scenario played out on all the Mediterranean islands. In practice, hogs (domesticated pigs) and boar (wild pigs) have always had close genetic ties to the point where up until recently, pigs resembled their wild cousins in everything except that they were slightly smaller. The big pink pig that we make sausages from today was imported from China in the eighteenth century. Breeders in the Middle Kingdom had managed to create a species in which females were very fertile (up to fifteen piglets per litter) and good at caring for their young, were resistant to disease, and had a propensity to fatten quickly. Their meat was also rich and exceptionally tasty. Contemporary industrial pig farming, which produces ridiculously low-cost meat, has been made possible by a thousand-year-old process of genetic selection initiated by Chinese peasants. The irony is that the countries who first domesticated pigs in the Levant no longer benefit from this. For the last five thousand years, from ancient Mesopotamia onward, various religious taboos have led to the prohibition of the consumption of pork in Jewish and Islamic cultures.

Our final example is aurochs, which were to domestic cattle what wolves are to dogs. They were a force of nature, at least 25% larger than today's bulls. Around one hundred thousand years ago, this lord of the megafauna could be found throughout Afro-Eurasia, from China to Mali. It was domesticated in a least three separate sites: in Iran around ten thousand years ago (the ancestor of today's cows), in Pakistan and India around eight thousand years ago (the local species, which evolved into the zebu, had a hump), and in West Africa around five thousand years ago. Populations of wild aurochs declined steadily after that. Toward 50 BCE, Julius Caesar noted that none could be found in Italy but that in Gaul they were considered a much sought-after quarry. The Merovingian kings hunted the last aurochs in Gaul and, century by century, the territory of these giant bovines shrank. Its poise and its rarity made it the sport of kings, and a hunting reserve was created in Poland in the sixteenth century to try and preserve the remaining survivors. But it was not enough. Despite the measures to protect them, a bovine epidemic contracted through interaction with cattle killed off the last aurochs. They died out in 1627. Since then, several experiments have attempted to bring the legend back to life. There are today several breeds of "reconstructed" aurochs genetically selected to resemble their wild ancestors. But just as the German shepherd or the wolfdog is not a wolf, contemporary aurochs are closer to cattle than to the bovine monsters that Caesar—somewhat exaggeratedly—described as being "a little below the elephant in size."[14]

This is a result of the domestication process itself. Aurochs were very dangerous. Naturally, the individuals selected for breeding were the least aggressive, the most docile, and the most gregarious. The most dominant were either eaten or castrated. Once domesticated, these bovines rapidly lost up to a third of their size. They also did less exercise because their movements were controlled by humans and they no longer had to escape from predators. Their bones became less thick and their meat more fatty. The steak tartare we eat today does not have the same taste as the ones our *Homo* ancestors had been eating for a million years. The same story is true of goats, sheep, and all the other animals under Monkey's heel. We made sure domestication meant smaller animals and bigger plants.

So many animals were domesticated. Chickens in Myanmar around ten thousand years ago. Horses in central Asia around 5,500 years ago. Dromedaries in Arabia five thousand years ago, rabbits in Europe around five hundred years ago (and before that in the city-state of Teotihuacan in Mexico 1,500 years ago). Ostrich and deer are still being domesticated today. Each story is unique and should be told. However, let us remember that for the last fifteen thousand years, humans have increasingly pursued new forms of interaction with the living world. Our interventions have had many consequences, such as epidemics (see chap. 8), which have become more frequent because of the concentration of animals in intensive farming and the ease of pathogenic microorganisms moving between species.

The most amazing thing is that one hundred thousand years ago, humanity made up an insignificant part of the biomass of all land animals. Today we know that the world's 7.7 billion humans[15] collectively weigh approximately five hundred million tons and their cattle around nine hundred million tons.[16] Do the math: anthropic biomass, which exists purely because of our presence and our actions in the living world, comes to a total of 1.4 billion tons. By comparison, mammals in the wild, whose existence is entirely independent of our own, weigh a mere 55 million tons. Thus, these animals make up less than 4% of the biomass of all terrestrial mammals, next to nothing. The rest is due to humans and our activities (livestock, farming, etc.).

The Price of Comfort

The analysis of human remains shows us that the people who lived through the Agricultural Revolution experienced a steep decline in size, muscle mass, and bone density compared with those who remained hunter-gatherers. The difference was not striking at first, probably because agriculture was adopted progressively. Over the first one thousand years of sedentary living it seems that the peasants in the Levant were healthier than their nomadic peers. But after three thousand years of harsh agricultural work the verdict was clear: people living off agriculture were now 12 centimeters (4.75 inches) shorter than those hunting, gathering, and fishing. They were also more likely to fall sick. New illnesses emerged, such as osteoporosis, the result of a loss of 20%–25% bone density!

It would be easy to imagine that food is to blame here and that a grain-based diet spells trouble for an omnivore. But this would be projecting contemporary concerns onto the past. We know that people's average height increased over the course of twentieth century. How many of your children are taller than you? We tend to think that our height has evolved constantly from Lucy's 1.05 meters (3 feet 5 inches), to basketballer LeBron James's 2.03 meters (6 feet 8 inches). Nothing could be further from the truth. Skeletal remains of the modern humans who arrived in Europe forty thousand years ago (whom we used to call Cro-Magnon Man, the ones who precipitated the downfall of the Neanderthals) are sizable indeed; the male specimens are 1.83 meters (6 feet) tall, on average! The skeleton found in the Cavillon Cave in Italy of a woman who died twenty-five thousand years ago was over 1.90 meters (6 feet 2.5 inches) tall. Her height led researchers to initially assume her skeleton was that of a man (until they examined the pelvis), hence the nickname Menton Man. Today the global average for the height of a male human is 1.75 meters (5 feet 9 inches), which, despite increasing steadily, is still well below that of our ancestors twenty thousand years ago. Moreover, a simple economic and social crisis would be enough to reverse this trend.[17]

This situation is partly due to the invention of agriculture. In the Levant around seven thousand years ago, a male hunter-gatherer measured 1.75 meters (5 feet 9 inches) on average, while a farmer from the same period measured only 1.63 meters (five feet 4 inches). Beyond the question of genetic predispositions, there are three significant factors that influence human growth and limit the size of adults. The first is child labor. If children work during key growth periods, before six years old and between nine and thirteen years old, their skeletons become compressed, and they will eventually be smaller. The second factor is nutrition. Lack of food (due to famine or mistreatment) during these same growth periods forces the organism to adapt to scarcity. As a result, it develops less, and this adaptation is transmitted epigenetically* to subsequent generations. The third factor—generally less decisive than the first two—is the presence of certain parasitic diseases that can be detrimental to growth (as well has having other disagreeable effects).

Historically, size has been a direct indication of environmental variations and social inequalities. In the late thirteenth century, a cold period known as the Little Ice Age (more on this in chap. 11) swept the globe and lasted until the nineteenth century. The average temperature dropped by just one degree Celsius, but average height dropped by 6 to 8 centimeters (2.35 to 3.15 inches). In nineteenth-century France, when conscription provided a justification for collecting complete statistics on the new recruits to the army, it became clear that officers (from wealthy families and the nobility) were a full 12 centimeters (almost 5 inches) taller than commoners. Finally, the continued growth observed in the twentieth century was of course correlated with an increased abundance of food, but it was also, and perhaps more importantly, to do with the spectacular progression of education that meant children were at school rather than in the fields or factories.

Of course these processes—child labor, poor nutrition during growth periods, parasitic diseases—are all cumulative. And farmers were the first victims. Obtaining good harvests meant times of the year that required all available manpower, so children were put to work very young. But sooner or later they would have a bad year, the harvest would be poor, and food would be rationed. Because they were dependent on specific supplies, farmers were much more vulnerable in times of crisis than hunter-gatherers were. Being sedentary and living in symbiosis with various animals (chickens, pigs, etc.) meant they were also more frequently ill. This was because any nonnomadic community has to resolve the problem of what do with its waste, particularly feces. Illness also flourished because of the concentration of populations and the close contact between species, which constituted the perfect environment for viruses, bacteria, and other pathogens to develop. Moreover, the starch in the carbohydrates that had become the basis of our diet unfortunately meant we were regularly filling our mouths with a nutritious mix that encouraged the development of the bacteria that cause cavities. Almost unknown to hunter-gatherers, these microscopic devourers of dentition began their mass destruction ten thousand years ago. The Agricultural Revolution thus provided a fantastic opportunity for humans and many other plants and animals (whether domesticated or symbiotic) to multiply, but it also paved the way for a substantial procession of pathogens.

If the Agricultural Revolution was so deadly, why did we pursue it? First, because its consequences evolved over the very long term, so they were not perceptible and were in fact appropriate adaptations to the constraints of the moment. The climate became warmer, and the changes that humankind implemented on the living world allowed us to better exploit our environment and went hand in hand with our demographic expansion. Once the process had begun, it was impossible to reverse without accepting high mortality rates that would have brought the human population back to its previous size. In this sense, the Agricultural Revolution was a trap.

But it was also a success. In hunter-gatherer societies women breastfed their babies until age three on average. Breastfeeding has two implications. It provides the child with increased immunity from diseases and provides the mother with a form of natural contraception (without being perfectly reliable and as long as it is practiced exclusively). This meant that women in these societies generally had a child every three or four years. In sedentary farming societies, the birth rate was closer to one child every year. This was partly because sedentary living encouraged women to give up breastfeeding to return to work. In both cases, high infant mortality rates served to curtail demographic growth (one child in two died before age five). But generational growth was extremely strong in agricultural societies, which explains how they spread so rapidly across the world.

From an individual perspective, a farmer was generally less strong and less capable than a hunter-gatherer. But for the former there was strength in numbers. Their population was growing constantly, and their societies were more cohesive, more used to collective action and technological innovation. Sound familiar? This same configuration had already allowed modern humans to overtake Neanderthals. History has been repeating itself for eleven thousand years. Agricultural societies assimilate hunter-gatherers, drawing them into the whirlwind of "progress." Today, the last surviving hunter-gatherers are found in the most inhospitable corners of the globe, from the Kalahari Desert to the tropical jungles of the Andaman Islands, the Amazon, or Papua New Guinea. They are forced back into the most challenging environments, reduced to eating the poor-quality food found there, and they are sometimes forcibly settled and exploited.

However, from an evolutionary perspective—the only one that matters in biological history—the Agricultural Revolution was a success. We saw our population expand exponentially and, eventually, even the individual benefited from this success. Today we are healthier and live longer than any humans ever have because thirteen thousand years ago, a far-off ancestor in what would become Jordan started picking wild wheat to feed her family.

The Agricultural Revolution was a crucial stage in Monkey's rise to power. It led to the domination of nature, the expansion of human populations, and the development of the first cities, which quickly became kingdoms, and then empires. This was associated with a multitude of inventions that also profoundly changed humankind: the division of labor, the invention of calendars, the development of written texts, the formation of hierarchical armies, the expansion of global and maritime trade, and the emergence of urban planning. These all pushed humanity into a new era 5,500 years ago. This period, that we will now look at more closely, is commonly known as the Bronze Age, because from an archaeological perspective it is characterized by the invention and use of metalwork. In addition to being a king, a priest, a merchant, a sailor, and a soldier, Monkey became a blacksmith.

4
COLLAPSE

Writing, cities, taxes, bronze, and horses allowed Monkey to reap the benefits of a stable, prosperous civilization that once extended from Ireland to China but then collapsed around 1200 BCE.

Some 1,300 years before the Common Era, a ship was wrecked on the southwest coast of what is now Turkey in a place called Uluburun, which means "Grand Cape" in Turkish. We don't know much about the ship. Its name and the origin of its crew are unknown. Whether it foundered because of navigational errors, was wrecked in a storm, or was intentionally sunk to avoid pirates remains a mystery. However, the fact that most of its cargo seems to have been loaded in Cyprus, further east, leads archaeologists to suppose that it was on a westward course.

A Treasure of Copper and Tin

And what a cargo it was! Lying on the seabed more than 50 meters (164 feet) down, this trading ship remains the oldest wreck ever explored. It was oval shaped, over 9 meters (30 feet) long, and it had a single mast and a potbellied hold filled with the most astounding cargo. Most of the treasure was made up of 354 copper ingots weighing roughly a ton and probably mined in Cyprus, as well as 100 kilograms (220 pounds) of tin ingots. There were also ingots of other precious materials such as cobalt, turquoise, and lavender-colored glass along with ivory from elephants, hippopotamus teeth, ceramics from Cyprus, and tableware made of porcelain, bronze, and copper. There were also various kinds of jewelry, mostly from the Levant, as well as a valuable golden scarab inscribed with the name Nefertiti, the royal bride of the Pharaoh Akhenaton who had reigned over Egypt some decades before the shipwreck.

But the heart of the treasure was the ton of copper and 100 kilograms (220 pounds) of tin, enough to make significant quantities of bronze, that much-coveted alloy made of precisely 90% copper and 10% tin. Enough to supply an army with sharp swords and solid breastplates. The destination, or purchaser, of this cargo was probably a Greek kingdom situated either in Greece itself or on the western coasts of what is now Turkey.

Uncovered in the 1980s, this shipwreck proves that during what is known as the Bronze Age, commerce involved the circulation of large quantities of merchandise. A system of international trade had been developed. Powerful states swapped embassies and sumptuous gifts. They sometimes went to war against each other or sent large amounts of grain to assist their allies in times of famine. From China to Ireland, from Algeria to Kazakhstan, elite groups operated networks of varying density. These ruling castes shared a particular lifestyle based on the same kinds of ostentatious behavior. They were able to go to war on horse-drawn chariots and fight with weapons made of bronze; their ceremonial objects were precious, finely crafted, and sometimes made continents away. At the heart of this system was the bronze alloy: its components, copper and tin, are found in mines vastly far apart from each other. Bronze metalwork therefore required long-distance exchanges and cultivated diplomatic relations.

Like agriculture, metalwork is an innovation that left traces in several archaeological sites around the world. Native copper was used sparingly in Mesopotamia from 9000 BCE.[1] Then it began to be extracted on a more industrial scale in the Balkans, for example, around 5000 BCE, where pits were dug 25 meters (82 feet) down, and ovens were used to purify the metal on-site. There is a famous mummy who provides an illustration of this era.

Ötzi: A Witness to an Intermediary Time

On September 19, 1991, a pair of German tourists were hiking up a glacier in Italy near the Austrian border. After taking a short-cut, they stumbled across a human head and torso sticking up out of the ice. The body was so well preserved that both the police and those who made the discovery believed it to be the body of a hiker who had died of cold. It was only when it was transferred to the morgue that it became clear that Ötzi (named after the valley where it was discovered) died nearly 4,550 years ago.

This mummified body confirms that violence was a feature of its time. Ötzi was killed by an arrow in the back, and the analysis of the blood on his weapons and clothing shows that he had killed or wounded four other people before he died. His clothing was quite sophisticated, reminiscent of what is worn by current-day Inuit or Sami people in polar regions today. He was dressed in three layers to optimize these materials' ability to preserve heat. His undergarments were made of sheepskin held in place by a calfskin belt, and he wore a large coat made of goatskin and sheepskin sewn together and a cloak made of woven plant fibers. His legs were protected by trousers made of goatskin, his feet by leather moccasins padded with straw for insulation. A bearskin hat kept his head dry. His equipment included various tools for making fire, a range of flint tools including a dagger, a large bow made of twisted yew, and fourteen flint-tipped arrows. He also had an ax with a polished and surprisingly pure copper blade around 10 centimeters (4 inches) long.

The autopsy revealed that when he died, Ötzi was in his forties, measured 1.6 meters (5 feet 3 inches) tall and weighed around 40 kilograms (88 pounds). He suffered from various parasitic diseases and seems to have carried mushrooms that might have helped with healing. His lungs were clogged with soot. His last meal was of domesticated grains with a stew of venison and wild goat. His hair contained large amounts of arsenic, copper, manganese and nickel. All these clues suggest that he lived in a complex society that had mastered a range of activities: agriculture, animal husbandry, hunting, and metalwork. Indeed, the presence of these metals in his hair can only be explained by copper metalwork.

Ötzi embodies an intermediary stage between two ages where the distant past of the first agricultural societies and the future of civilizations coexisted. The influence of the past can be seen in the food he carried—a mix of things gleaned from the environment (hunting and gathering) and produced by agriculture—as well as his flint arrows and tools. A glimpse of a near future can be seen in the fact that his society was clearly sedentary and with a hierarchy and a division of labor. In that society there were hunters, farmers, miners, and artisans who worked copper or flint, probably all overseen by religious and secular leaders. Ötzi was a perfect representative of this so-called Copper Age, which seems to have been a kind of transition period between the Neolithic, where we saw the emergence of agriculture and livestock farming, and the Bronze Age, in which societies in Afro-Eurasia would become more complex and develop a "civilizational" system. In this coming age, towns would extend their sphere of influence to become kingdoms and then empires, petty kings would become ever more powerful, and a tax system based on the control of land inheritance would lead to the invention of writing, a genuine cognitive revolution.

Native copper is a malleable metal. It can be worked without heat. In certain geological contexts it can even be found on the surface of the land. From about 9000 BCE it was used all over the world in all inhabited continents except Australia (apparently because of a lack of easily accessible minerals). From 5000 BCE, societies in Anatolia and then the Balkans intensified the extraction of this metal and planned the first mining operations. Copper ore—a combination of stone, various minerals, and copper—had to first be extracted from the ground and then melted several times in charcoal-fired kilns heated to over 1,000°C (about 1,800°F) to eliminate all impurities.[2] Substantial skill and knowledge were required to build kilns powerful enough to reach such high temperatures. These techniques were fueled by and interacted with the progress already made in the production of ceramics, which also involved important innovations in terms of firing and kilns.

Over the next two millennia, metalwork using copper appeared in the great city-states of Mesopotamia, then North India (in the Indus Valley, today's Pakistan), and finally in northern China. It was associated with intense deforestation because we had discovered that partially burned trees provide a high-performance fuel. Now we were no longer simply clearing new fields, we were cutting down forests to produce charcoal. For some authors,[3] this period between agriculture and industry was the beginning of the Anthropocene. They see the large quantities of carbon dioxide and methane that were then released into the atmosphere as a sign of humanity's new hold over the planet. However, the volume of the gases released is difficult to quantify, and it does not seem to have really affected the equilibrium of the climate at the time. This intermediary period is known as the Copper Age or the Chalcolithic. This term combines the Greek words for copper and stone, a reminder that copper remained a secondary material; people were still using stone tools.

Around 7000 BCE a new cold wave hit the Levant. This climate crisis led to population movements toward Europe, Mesopotamia, Egypt, and Turkey. Whole agricultural settlements were deserted; others swelled into towns with the influx of new inhabitants. Çatalhöyük, founded in 7000 BCE in today's Turkey, for example, had swelled to at least 13 hectares (about 32 acres) a thousand years later. By then, it was home to nearly one thousand households, each with a domestic altar, which suggests the importance of religion in the everyday life of its five thousand inhabitants. It was abandoned around 5000 BCE. Mehrgarh, in what is now North Pakistan, experienced a similar expansion and managed to survive for five thousand years.

This cold period also encouraged another type of adaptive response. A new economy emerged, that of the nomadic herdsmen who wandered the arid landscapes of western and central Asia, following their herds. The hunter-gatherer economy was over. There was no turning back. In environments where biodiversity had declined with every new interaction with human cultures, a return to that lifestyle was no longer an option for our species. Some humans continued to undergo mutations, which seemed to become increasingly frequent. For example, like most mammals, humans are normally unable to digest maternal milk once they have been weaned. But the animal herders in Caucasia, East Africa and central Asia each developed their own distinct genetic mutation that enabled them to digest lactose. Today, nearly half the human population carries these mutations. Given the globalized expansion of dairy products, the other half is likely to also develop this ability in coming generations.

Monkey became a rancher and a cheesemaker.

Further south, this disturbance in the climate led to a greening of the Sahara. Between 8000 and 3000 BCE this was a fertile zone, and the populations who lived there experienced their own Agricultural Revolution during this time. It was in this now-desertified region that sorghum, African rice, and the aurochs native to the region were domesticated. When the desert crept back, spurred on by new global warming, these populations took refuge in the Nile valley or on the Mediterranean and Atlantic coasts.

An Age of Bronze, Gold, or Silver?

The Greek poet Hesiod is said to have lived around 800 BCE, although there is as yet little historical evidence that he did (the same is true of his contemporary Homer). Hesiod is presented as the author of *Work and Days*, a text that has profoundly influenced our understanding of the past. After recounting the misadventures of Prometheus, the Titan who dared to defy the gods and grant humanity fire, Hesiod laid out the five Ages of Man. During the Golden Age, there was such abundance that humans were carefree—there was no winter or drought, no old age or war, and nature generously provided for all their needs. Then followed the Silver Age, in which humans ignored the gods and lived in hubris—the pride that is the source of all the great tragedies of our species—and were forced to work the land to survive. Next was the Bronze Age, which plunged humanity into the chaos of war and saw humans kill each other down to the last man. Then there was the Heroic Age, with the demigod protagonists of the Greek myths, and then finally the Iron Age.

Hesiod's semimythical classification is no longer used. The first three ages can be equated with what is now called the Bronze Age. In hindsight, the historically documented Bronze Age did indeed have a certain prosperity. In fact, it could be a candidate for the very questionable title of a Golden Age. It saw the expansion of long-distance trade because its economic momentum relied on the joint extraction of copper and tin, which had to be mined in sites far away from each other. The main copper mines were in Cyprus, in southwest Spain, in the heart of the Balkans, and in Afghanistan. The Mediterranean world brought tin in from mines in England, Brittany, and then Spain, while Mesopotamia and northern India imported their tin from mines in central Asia. Tin was so important that some authors have seen it as being the equivalent of oil for that period—the essential fuel for the geopolitical system. It now seems more relevant to compare it to rare earth elements, those seventeen strategic metals that all electronic industries rely on (the market for which is today controlled by China).[4] This was also an age of silver, of money and commerce, because trading in metals involved exchanging sumptuous gifts between leaders over very long distances according to the rare written archives that remain. It was also the age of bronze because this metal was intricately connected to the power of weapons, the power of farming tools over nature, and the power of the wealthy in the form of hoards of metal jewelry.

A combination of factors contributed to the rise of the Bronze Age, a period that left us many great legacies. If we were to choose one example (among so many) of the exaggerated pride of its leaders, it would be the construction of the Pyramid of Giza. Begun around 4,500 years ago, it probably involved more than one hundred thousand workers who had to pile more than two million tons of stone blocks, each weighing 2.5 tons on average, into a colossal monument 145 meters (476 feet) high. This was a record-breaking display of energy, suffering, and resources, which seems to have had no purpose other than to provide the Pharaoh passage to the afterlife! The Bronze Age enabled an unprecedented intensification of inequalities, which can be seen in the increase in the number of personal palaces throughout the Levant. These buildings sometimes covered large areas and must have required countless slaves and servants.

Egypt had been a unified empire since 3000 BCE. Mesopotamia, which had been fragmented between powerful city-states (Sumer, Ur, Lagash), began to follow its example under the reign of Sargon the Great, who established the Akkadian Empire in Mesopotamia around 2300 BCE by gaining control of the whole region. Although Egypt remained exceptionally stable (notwithstanding sporadic internal revolts) because the surrounding inhospitable landscape made it easily defendable, this was not the case for the empires that emerged between the Himalayas and the Mediterranean. They were more difficult to defend, and their existence proved precarious. The Akkadian Empire, for example, lasted only two hundred years.

Around 2200 BCE, bronze metalwork spread over the steppes and appeared in norther China. Here there was a succession of dynasties that are not well documented: the semimythical Xia (between 2000 and 1600 BCE) and the Shang (1600 to 1050 BCE), subsequently supplanted by their vassals the Zhou. Archaeology has also found evidence for the rise of city-states or kingdoms around this time, such as Erlitou (around 1900 to 1600 BCE) and Erligang (between 1600 and 1400 BCE). Over the course of the following millennia, southern China adopted state structures similar to those in the north and was covered in irrigated rice fields by 500 BCE. This agricultural boom was associated with significant deforestation due to continual population growth and the development of state structures that were larger and better organized and that consequently had greater impact on natural environments. Mines, deforestation, terracing, and demographic growth are all incentives for new populations to move into previously pristine areas. The stronger the state is, the more impact it has on the environment.

The Indus Valley saw the development of several large cities, such as Mohenjo-Daro and Harappa. Between 2800 and 1700 BCE, they boasted highly innovative urban planning, including monumental avenues built with uniformly sized bricks and extensive and efficient sewer systems. Around the same time, semi-industrial ceramic workshops appeared in Mesopotamia and in today's Bahrain. They were responsible for the large-scale production of pottery for export to the Arabian Peninsula and the Indus Valley. A system of international relations evolved—whether between city-states, kingdoms, or empires—alternating between bellicose conquests and commercial exchanges. The latter were most often carried by large donkey caravans but transferred to boats wherever possible. Boats have always been the most economical form of transport.

It is important to remember that the civilizational system of the Bronze Age began in Mesopotamia, before 3000 BCE, by bringing together both internal and external elements. Among the external elements was the horse, domesticated at the latest around 3500 BCE in the Botai culture in what is now Kazakhstan. Whether harnessed or ridden, this animal provided humans with exceptional mobility. The wheel, initially solid and made of wood, made its first appearance in eastern Europe around the same time. By around 2000 BCE wheels had become lighter and more robust because of the addition of a bronze rim, which increased the technical abilities of trade carts, pulled by oxen or donkeys. These wheels also improved chariots used for war, pulled by horses. Over flat terrain, the military superiority of chariots (and thus of those rich enough to have more of them than their rivals) was overwhelming. Between 1674 and 1548 BCE the nomadic Hyksos, who had come from the Levant, overthrew the Egyptian Empire using a combination of innovative weapons, including two wheeled chariots, which were extremely maneuverable and which carried both a charioteer and an archer using a newly powerful composite bow. The latter was made more powerful by the clever combination of more or less flexible materials. It took a little more than a century for the Egyptians to appropriate these new techniques and get rid of the invaders. Two centuries later, Ramses II used these same weapons to make Egypt the largest imperialist power of the time.

The arms race would now mark the beat of history.

A Perfect Storm

As the archaeologist Eric H. Cline notes,[5] most authors working on the late Bronze Age describe more or less the same process. Between 1600 and 1300 BCE, diplomatic and trade relations reached an unprecedented density, creating a cosmopolitan world-system.* Leaders in Assyria, Babylon, or Mycenae contracted matrimonial alliances and swore to come to each other's aid. Egypt exported Sudanese gold, imported Lebanese cedar, molded Cypriot copper, hosted artists from Crete who painted with Afghani lapis lazuli, and so forth. The importance of these economic circuits was such that certain archaeologists see them as a sign of protocapitalism in which merchants began to gain autonomy from royal elites.

All this went up in smoke in the twelfth century BCE. From Greece to Mesopotamia cities were destroyed, populations were decimated, and famines and epidemics ravaged previously prosperous lands. Egypt managed to survive but emerged from the crisis significantly weakened. Cline emphasizes the significance of this cataclysm. From the Andean city of Tiwanaku to the Angkor Empire in Southeast Asia, history is full of collapsed societies. But the collapse that occurred at the end of the Bronze Age was systemic, and in that respect it has only one equivalent in history: the fall of Eurasian societies—from the Han dynasty in China to the Western Roman Empire—after the third century CE (chap. 6). At the end of the Bronze Age, this catastrophe wiped out five of the six major powers in western Asia. The powerful Hittite Empire, which occupied what is now Turkey; the young kingdom of Assyria that had just conquered the Mitanni kingdom in eastern Syria; the Kassite kingdom, solidly anchored around its capital, Babylon; the dynamic Minoan civilization, between Greece and Crete, dominated by the city of Mycenae; and the kingdom of Elam in today's Iran. All fell, except the much-weakened Egyptian Empire.

So, what happened?

The authors who have studied this question all mention six possible explanations.

1. A change in climate. A significant cold wave rendered the arable land in the Levant, Mesopotamia, and North Africa barren. Cold winters and summer droughts wiped out harvests several years in a row. Even the flooding of the Nile, so important for Egypt, declined dramatically.
2. Famine. This was linked to disturbances in the climate and documented in archives and archaeological evidence; this weakened the populations.
3. Seismic upheavals. Between 1250 and 1150 BCE there were regular earthquakes that contributed to the destruction of cities.

4. Civil war. Both Greece and part of the Levant appear to have been beleaguered by internal upheavals.
5. Invasions. Throughout the Mediterranean region, so-called Sea Peoples (whose identity remains a mystery beyond this descriptive title) ravaged many towns and even led to the collapse of states. From Mycenae to Babylon, but also in Troy and Ashkelon, hundreds of cities were destroyed. Of those burned, some were rapidly rebuilt, but others were definitively abandoned. Twice Egypt managed to hold back these invasions on its beaches.
6. Broken economic exchanges. Trade routes fell apart, and leaders were no longer able to assist their allies either by sending soldiers or goods.

An individual society can be destroyed by any one of these events. But a stable international system should manage to be resilient. Eric H. Cline diagnoses what he calls "a perfect storm":[6] a conjunction of extreme events associated with a multiplying effect, each element amplifying the effects of the other phenomena. Ultimately, the complexity of the international system led to the downfall of the Greek kingdoms (Mycenaean, Cypriot, and Cretan) along with the Hittite, Babylonian, and Assyrian empires and many other states. Perhaps if these societies had been autonomous, they would have been more resilient. Their interdependence apparently led them to fall one after the other, like a house of cards. If they had remained united and helped each other, perhaps history might have been different.

Only Egypt, protected from famine by the fertility of the Nile and benefiting from a then dynamic and expanding society, survived the hecatomb. But the empire of the pharaohs was weakened by the cataclysm that brought down its neighbors. Having fought against two invasions, it shrank to a third of its initial size. Its power was little more than a memory.

There may be something to add to the list of causes contributing to the systemic collapse in the Bronze Age. From 1250 BCE, a new technology emerged in Mesopotamia: cast iron. This technique was also mastered around the same time in Africa, a few hundred years later in India, and then in China. This evolution is not mentioned by Eric H. Cline, yet it seems to have been decisive. Cast iron was made possible by the invention of more efficient furnaces able to reach temperatures of 1,500°C (about 2,700°F). Unlike bronze, iron requires no international trade as it is made of a single metal, one that is readily available nearly everywhere! The fact that iron was accessible to most would eventually encourage the spread of military capabilities—it was no longer necessary to be a member of the elite to have access to powerful weapons. And iron was far superior to bronze in terms of its cutting edge and its resistance to impact. War therefore became a potential resource for many small communities that did not have significant logistic means. In other words, it became "democratized."

What can we learn from the Bronze Age? A systemic collapse plunged a great number of interconnected countries into chaos, the result not of a single cause, but of a complex sequence of events. The unpredictability of the climate and seismic activity contributed to this chain of events by weakening the way of life of those who were the most vulnerable to famine. The self-interest of populations and leaders who chose to protect themselves or to prey on their neighbors rather than to cooperate and share resources clearly made the situation worse. Technological innovation can change power relations. The environment alone is not enough to destroy a society. The choices that a society makes in a situation of adversity determine its fate.

Three centuries of anarchy followed.

A Key Invention: Writing

The Bronze Age left us an invention that would change the world: writing. Reading and writing are activities that have changed us psychologically and physiologically.[7] No gene in our DNA codes these processes. The brain simply adapted. It attributed this new task to neural pathways previously used for other things.

Writing, like all other crucial inventions, was preceded by a long period of preparation. From zigzags etched into shells in Indonesia half a million years ago, probably by *Homo erectus*, humans have left symbolic marks on many objects and places. There are numerous cave decorations, constellations of dots and lines, that we still have trouble deciphering.[8] From 7000 BCE, marks on clay tokens were used to count cattle in cities in Mesopotamia. These marks gradually became more complex and led to the first writing toward 3600 BCE. A similar system emerged in Egypt around 3200 BCE. Was this actually an invention though, or a reinvention? Given that it existed among their trading partners, did the Egyptians create a system of writing spontaneously or did they copy it? The same question can be asked for other systems of writing. In China, one form emerged around 1300 BCE[9] (perhaps the concept traveled over the steppes?) and a second invention emerged in Mesoamerica around 1200 BCE.[10] Some researchers consider that the seals used in Harappa (in the Indus Valley) from 4000 BCE or the quipus (a sequence of colored cords tied together according to predetermined sequences) invented in the Andes in the third century BCE also constitute writing. These graphic systems indeed enabled the transmission and conservation of complex information.

In any event, writing emerged in hierarchical societies. A new cast of specialists was born with the emergence of scribes, or civil servants. Writing was both a symbolic activity and a pragmatic technique. The first task was to preserve information relating to tax and commerce. What volume of barley did a particular peasant harvest? How many clay pots did that merchant send to his client? The possible uses of this technology rapidly became apparent: account keeping, proclaiming laws, telling stories (myths, epics, histories). Writing meant privileging one vision among others; what is written down is increasingly difficult to question. Writing became an instrument of power. Those who mastered it had a polyvalent tool at their disposal, both an extension of their memory and a rule that could be imposed on others. Language itself was affected. Human society developed its first state administrations.

Monkey became a public servant.

The second millennium before the Common Era saw the spread of the invention of writing. This was when the first state archives appeared (in Mari, Syria, in 1800 BCE) and when the oldest myth was recorded (*Gilgamesh*, written in Sumer in 1800 BCE, telling the story of a king who wants to be immortal and has his eternal life stolen by a snake). It was also when the first legal code was written (the Code of Hammurabi in 1700 BCE) along with the first alphabet containing vowels (Ugarit, in Syria, around 1450 BCE). The latter led to a simplification and universalization of older graphic systems associating letters and sounds phonetically.

For many people today, writing is sacred. The appeal of the Bible, the writing of which began from the eighth or seventh centuries BCE, can be also explained by its wealth of myths. The first of these myths, Genesis, echoes the discussion of the Agricultural Revolution in the previous chapter. Adam and Eve, the most beautiful beings on Earth because made by the hand of God himself, tasted the fruit of knowledge and for an instant believed themselves equal to their maker. In punishment for their hubris they were expelled from Paradise. That put an end to their laid-back hunter-gatherer days; no more free food to be picked or scavenged in God's great garden. Their descendants would have to earn their pittance through sweat and toil. And they would become debased and ugly, commanded to go forth and multiply.

Hundreds of years later, we tell similar stories, although based on conceptually different beliefs; science and religion do not go well together. But our cultural references remain influenced by the Bible.

Adam and Eve left Paradise where they lived in harmony with nature. Their children would then have to learn to master nature and earn their food through hard labor. Cain the farmer would kill his brother Abel the shepherd in the prelude to an ongoing war between settlers and nomads.

This was a new kind of slavery for Monkey, who wanted to discover the secrets of the gods. The time of great religions had begun.

PART II
Monkey Dominates Nature

5

WHEN GODS GUIDE THE WAY

In the wake of the Bronze Age collapse, anarchy reigned. Then came the rise of the great universalist religions and their new ideal for humanity: do unto others as you would have them do to you. But this did not apply to nature. It became Monkey's plaything.

An account from the place where empires and religions meet passed down to us from the dawn of time. A universal tale that unfolds in our imagination and still feeds the culture of more than a billion humans. A narrative of fratricide that reveals the complexity of the human soul. A myth.

As the arrows begin to whistle past his head, the commander of the army of the Good begins to doubt. He looks at the opposing army and notices that his cousins and close relatives have joined to fight against him and his brothers. He drops his bow, gets down from his chariot, and begins to cry. His charioteer, who is none other than the god Krishna, reminds him of his duty.

A Warrior's Duty

The *Mahābhārata*[1] is an immense work, written possibly 2,500 years ago. It is the Indian equivalent of and is roughly contemporaneous with the Hebrew Bible or the Greek *Iliad*, although with its 328,000 to 380,000 verses (depending on the version) it is significantly longer then they are.

The main story focuses on a conflict that tears a family apart. On the one hand there are five brothers, the Pandavas, who are as united as fingers on a hand. On the other, there are their one hundred cousins, the Kauravas. A trick of fate has led the Pandava brothers to all marry the same woman, the wonderful Draupadi. But she has been humiliated by the Kaurava brothers and begs her husbands to defend her honor. After many trials, the brothers assemble an army with Arjuna, the oldest of the brothers and the best warrior in the world, at its head. But their cousins have also issued a call to arms, summoning a large number of soldiers and war elephants. The conflict draws on alliance after alliance until all humanity is involved. Nearly two million heroes are assembled to fight a cosmic battle that has consequences for the fate of the universe itself. This war has attracted the attention of the gods. Vishnu, supreme lord and master of the world, the god responsible for balance in the universe, has taken human form. As Krishna, he offers to be Arjuna's charioteer because—in true Bronze Age style— the fighting is done in horse-drawn chariots.

The *Bhagavad-Gita* (*Song of God*) is the heart of the *Mahābhārata*. It tells how Arjuna, the hero, feels himself falter, and in his doubt, his resolve wavers. What benefit could there possibly be in massacring his relatives when the opposing army includes—he counts them—a much-loved great uncle, a former teacher, even his half brother bound by a promise made to the other side, so many friends that honor demands he fight? The war would make him and his brothers into murderers. Better to turn tail and be seen as a coward but at least not be in the wrong.

Then Krishna speaks. A warrior's duty, he says, is to fight no matter what happens. Under Vedic law, existence is woven of successive reincarnations. The god reminds Arjuna that the soul can glimpse salvation far off, like a light on the horizon, but this salvation is only accessible if one accepts behaving in accordance with the demands of one's social position, inherited at birth. Indeed, this birth is itself associated with the merit accumulated in previous lives, and this capital of merit, *karma*, must by maintained by righteous action, *dharma*, which provides balance in the world. From this perspective, war appears to be completely impossible. *Dharma* advocates respect for all life, and yet warriors must kill. But Krishna reminds him that this war is necessary to preserve the cosmic order, *Brahman*, which will ultimately grant his soul fulfillment and allow him to be one with the divine universe.

Over the eighteen chapters of the *Bhagavad-Gita*, Krishna teaches Arjuna about the nature of war and duty. War can be seen as an ultimate moment of tension, a perfect metaphor for existence. We all fall prey to doubt and struggle, and there are several paths that lead us to free ourselves from them. Through Krishna's words, the anonymous authors of this ancient text present their vision of the world and list possible paths to wisdom: *jnana*, philosophic introspection that pushes us to seek wisdom; *karma*, the path of the righteous action of the warrior; the various forms of *yoga*, ascetic paths; and *bhakti*, devotion, that consists in completely entrusting oneself to divine beneficence. Krishna depicts a threefold hierarchical society in which warriors (*Kshatriyas*) fight, priests (*Brahmins*) have the monopoly of the sacred, particularly over sacrifices to the gods, and workers (*Vaishyas*), peasants, and artisans, as well as servants (*Shudras*), must all place themselves in the hands of the gods. "O Arjuna, the wise one, [he] to whom happiness and unhappiness are the same, is ready for immortality." He forgets to mention the untouchables and outcasts.

As a result, reinvigorated by his faith-fueled feeling of duty, Arjuna carries out the cosmic command. He slaughters everyone in good conscience—including relatives and friends—in order to preserve the order of the world. Krishna tells him,

> The wise do not mourn for the dead or for the living
> ....
> Just as in the body childhood, adulthood, and old age
> Happen to an embodied being
> So also he [the being] acquires another body.
> The wise one is not deluded about this
> ....
> For the born, death is certain
> For the dead there is certainly birth
> Therefore, for this, inevitable inconsequence
> You should not mourn.[2]

Has Monkey become a fanatic?[3] Yes indeed. Can we blame Arjuna's obedience on Vedic philosophy? Of course not. All religions have eventually constructed systems able to justify the most terrible actions. In the Bible, as the Hebrews are about to enter Canaan, the Promised Land, God orders them to "clean" it by killing all the male residents and taking the women and babies as slaves. And we know the tragic fate reserved for all those who were brought before the Inquisition as Cathars, or alleged witches, and the apostate Jews and Muslims who were forcibly converted, all in good conscience. During the Second World War, most Japanese Buddhist "sages" put themselves at the service of imperial propaganda and pedantically pushed the message that to kill an enemy of the emperor (and thus of Japan) was to do the victim great service, because his opposition to the emperor was an error that darkened his *Karma* with every moment of life, and it meant that each breath drew him unavoidably away from the prospect of salvation. Killing him would thus allow his soul to return more quickly to the path of deliverance.[4]

This particular passage from the *Mahābhārata* is widely known because its teachings are universal. In just a few phrases it sweeps away all the heart wrenching the human soul has done over the last twenty-five centuries. Concern for the other—roughly formulated in the golden rule, "do unto others as you would have them do to you"—comes up against a duty of responsibility: the warrior must kill because it is the necessary condition on which the order and good functioning of society is based. Our world stems from the tension between these two opposing poles: on the one hand, the great revealed religions, and on the other, the infrastructures of power known as empires. How can we reconcile Caesar and the pope? How can we manage the tension between the spiritual and the temporal? How can we transform this inexhaustible energy into social cooperation? How can we ensure humans are conditioned to cooperate while still retaining enough control over them to have them kill each other on command?

Monkey thus became a soldier and a missionary. He had to reconcile these two callings while also adding a third: he became a merchant. This was the Axial Age that saw the beginning of our fourth revolution, the Moral Revolution. We can remember this as "Monkey's three Ms": missionary, military, and mercantile. They correspond to the three unifying global forces—politics, religion, and trade—with the vocation of encompassing ever more people.

The Golden Rule

The term *Axial Age* was coined by the German philosopher and psychiatrist Karl Jaspers (1883–1969). He was writing a history of philosophy, and he wanted it to be open to other cultures given that previous comparable works had above all restricted themselves to the contributions of Ancient Greek or Judeo-Christian European thinkers. By broadening his conceptual frame, Jaspers realized that all the doctrines that govern our existence today emerged at the same time—between 800 and 300 BCE. Pythagoras, Buddha, Confucius, and Isaiah all lived in the same period. Greece, India, China, and the Levant all simultaneously experienced an efflorescence of rival schools of thought, all desperate to understand our vast world and conceptualize humanity as a whole.

For Jaspers,[5] this conjunction of philosophies had to be the result of a historical process connected to the emergence of similar societies. From Ireland to Korea, urban civilizations were living through the Iron Age. This transitional period marked a pause in the dynamic of empires. Afro-Eurasia was broken up into a mosaic of local powers, kingdoms, and tiny city-states that were constantly at war and wrought with internal conflict. Each region seems to have developed a culture of renunciation. Some ascetics and hermits headed off into the forest or desert to escape the tumult. Others wandered from town to town begging a pittance and bestowing pearls of wisdom as they went. Everywhere, religious fervor pushed these thinkers to test all possible visions of the world. In China, as in Greece and in the Levant, there emerged a whole range of schools of thought that applied the principles of reasoned consideration to the grand questions of human existence. Skepticism, idealism, atheism, and blind devotion had their champions everywhere.

The full spectrum of major philosophical questions seemed to have been covered, including the origins of the universe, human nature, the functioning of the mind, and the finality of death.

Just as strong centralized territorial powers were finally emerging in each region, these marginal thinkers were integrated into the system and put their ideas at the service of the masses. Philosophers in Greece, prophets in the Levant, sages in China, and saints in India; all agreed to collaborate with earthly power. Politics and religion would henceforth walk hand in hand toward a new paradigm known as the golden rule: "Do unto others as you would have them do unto you." This motto expresses a concern for universal empathy and equity and is transcribed, in various forms, in many of the ideologies of the Axial Age that accompanied the increasing power and interconnectedness of empires. The goal was to provide all humanity with a shared framework for interaction, rather than just focus on one group among others.

More than just advice, the golden rule is the basis for a system of reciprocal moral judgment. It is constructed so that the roles of the advisor and the recipient are inverted. The underlying principle is that we should put ourselves in the other's place and feel the consequences of our actions as they would. Two forces come together in this: empathy (found in all primates[6]) and fairness. Empathy transforms the egotism of living things into a force for helping others. Fairness guarantees the other that this benevolence will be reciprocated. The golden rule weaves extremely powerful social connections and constitutes the first (and apparently the only) universal ideal shared by most of humanity today.

The major innovation of the Axial Age is the appearance of universalist ideologies, which are associated with the development of power structures that also aspire to universality. The dynamic that resulted from this had severe environmental consequences because it allowed human societies to achieve a degree of cohesion hitherto unseen. Only Egypt had momentarily come close to this capacity for mobilization, which it demonstrated through the construction of the pyramids. The Axial Age would anchor this characteristic in our species.

Let us take a look at the great systems of thought that have emerged from this foundational brew and that continue to influence our reflections in one way or another. We will begin at the beginning, with their great founding figures. Although these figures have often been adopted by causes that would have been foreign to them, their ideas have progressively evolved with the various contexts through which they have flowed. This history saw a new value emerge in human consciousness around the world: individualism. Africa, the Americas, and Oceania remained on the margins of this development. From our perspective today this is because there is a lack of written exchanges and historical sources from these regions and also because they did not benefit from the same intense framework of societal interactions as did other parts of the Ancient world.

Four Chinese Thinkers

In 500 BCE China was a patchwork of expansionist political powers moving toward a war for supremacy. This period of conflict, known as the Warring States, lasted three hundred years. It took place in a context of significant technological progress and intense demographic growth. It seems to have contained the seeds of the political and environmental possibilities of contemporary China.[7]

Confucius lived between 551 and 479 BCE. He was a public servant in China and had wide-ranging ambitions. However, falling short of the highest positions of political power, he became a sage. In the *Analects*, a book of sayings attributed to him and his followers, he defends an ideology promoting the strict respect of hierarchy tempered where necessary with reasonable critical thought. A gentleman (*junzi*) must act conscientiously, but wherever possible he will be expected to conform to social norms and rituals, from ancestor worship to traditional practices, which guarantee the smooth functioning of society. He must study and try to do what is good. It is expected that everyone respects the instructions of their superiors: a son must obey his father, a vassal must obey his lord, a wife must obey her husband. . . . And if a superior forgets the virtue he should incarnate, it must always be possible to remind him of his duty.

This philosophy of harmonious coexistence does not claim the existence of gods, it simply lays down the principle that morality is intrinsic to humanity and essential to the social order. Confucius may have been the first to formulate the golden rule, in its negative form: "do not do to others what you would not wish them to do to you." Mencius (or Mengzi), who enriched the philosophy of Confucius two hundred years later, introduced the idea that humans are fundamentally good and drawn to cooperation. What happens if someone falls in a well? Everyone's first reflex will be to try and help them out. From the second century BCE, Confucianism would become the official ideology of most Chinese leaders. And today this thinker has become the figurehead of Chinese soft power. The People's Republic spreads Chinese culture by funding "Confucius Institutes" all over the world.

Taoism, often presented as contemporary to Confucianism in its origins, is an ideology passed down from Lao Tzu. This mythical figure is said to have debated heatedly with Confucius. He also wrote—as he was fleeing the world on the back of a buffalo—a collection of aphorisms open to myriad interpretations, titled *Tao Te Ching, The Book of the Way and Its Virtue*. This book, playing on the polysemic nature of Chinese characters, can be read in several ways: as a reflection on impermanence, as a guidebook for wisdom, as a tool for personal development, or as a manual for divination. At the beginning, Taoism was a shamanic philosophy advising withdrawal from the world. It rapidly led to individual quests, combining mysticism with individualistic ethics and aimed at achieving long life or even immortality through physical exercises, alchemy, or both. From the third century BCE, this ideology became a religion that structured the dynamics of many rebel groups in various Chinese states. This school of thought defends nonaction to avoid doing harm, the return to nature, the study of vital breaths, the perfection of the body through physical and breathing exercises. Today it is gaining renewed popularity through hybridizations with contemporary environmental concerns.

Legalism was another Chinese philosophy. Summarized by Shang Yang (around 390–338 BCE). This ideology advocated wiping the slate clean of the past. In opposition to Confucius and his followers, it saw tradition as a deadly poison. Like Confucianism, it remained indifferent to gods. It promoted only one objective: the survival of the state, because without a strong and respected state, individual interest would dominate, and everyone would fight and steal from their neighbors. According to this philosophy, the prince does not have to be morally righteous. Rather, he must incarnate the efficacy of swift justice, productive agriculture, and aggressive military strength. This kind of pragmatism, according to which the end justifies the means, can be seen as an ideological predecessor of more recent totalitarianisms. The first emperor, Qin Shi Huang,[8] who unified China in 221 BCE, was a disciple of this school of thought. It probably provided him with an excuse for the numerous massacres that punctuated his conquests and those of his father. Mao Zedong also gloried in the fact that he caused more deaths than Qin Shi Huang in the name of the national interest.

Mohism, the fourth grand ideology of ancient China, defends a pacifist communitarian utopia formulated by the philosopher Mozi (479–392 BCE). Over the course of the fourth and third centuries BCE, Mohism emerged as a military and religious movement. It was attractive to many peasants and marginalized people with its claims that it is natural that the general interest should dominate over the excesses of the rich. Its disciples collectivized lands and intervened in conflicts, claiming to want to resolve them. Mohism enforced strict discipline in which all actions were subject to a judgment of utility: if something was not useful, it was therefore useless. The worship of deities and mourning rituals were considered a waste of time, for example. Any breach of the rules was sanctioned. This ideology disappeared around the third century BCE, about the time Qin Shi Huang unified China. But its collectivist ideals would be mobilized by most of the Taoist messianic movements that would periodically impassion the Middle Empire over the two thousand years of our era. Finally, this school of thought would be again resuscitated in the twentieth century and held up as a precursor to atheist Chinese Communism; as is often the case, analogies between the two enable a certain number of anachronisms to be overlooked.

Both Confucianism and Mohism have little concern for the

natural world. Society is the only thing that matters; when faced with torrential rains we must keep building more dams. Legalism pushes this approach to its extreme, considering nature as an opponent that must be brought to its knees. These ideologies see humanity as master of the elements. Taoism's position is more ambiguous: its gods and mythic ancestors submit the natural world to their will. Yet the ideal of wisdom they promote is to become one with nature and to avoid intervention to avoid harm. However, while Taoist philosophers can content themselves with contemplation, ordinary humans must take from nature to survive.

India: Three Roads to Salvation

The shared basis of Indian religions lies in the religious ideology of Vedism. This is expressed in a corpus of four sacred texts, the *Vedas*, which stipulate that at the foundation of all things there is *Brahman*, the universal, the absolute. Probably written around 500 BCE, the *Vedas* are attributed to the Aryan people who are thought to have migrated from Iran toward India from around 1500 BCE. From *Vedas* comes Vedism, the cultural foundation of Brahmanism, then Hinduism, Buddhism, and Jainism. (Although the latter two deny any connection with the *Vedas*, they nevertheless developed within that conceptual framework.) All these schools of thought postulate that the universe is cyclical and that the soul is bound to a cycle of infinite reincarnations that must be brought to an end.

Brahmanism has no founder as such. If Buddhism and Jainism can be seen as a reaction against the caste system that developed in India around 500 BCE, Brahmanism seems to be an attempt to establish this cosmic order and see it reflected in social organization around sacrificial rites. The four *varna* (main castes) have definitive roles as depicted in the *Mahābhārata*: Brahmins (priests) sacrifice and teach, Kshatriyas (warriors) fight, Vaishyas (peasants and artisans) produce, and Shudras (servants) serve. From this basis, a multitude of beliefs were born and were grouped together under the title of Hinduism. This religion often suffers from a persistent stereotype that represents it as eternal, immobile wisdom.

Yet all religions evolve in keeping with society. An ideology that does not adapt will crumble, and this is true for Christianity as much as for Hinduism. Eternal India is a fable. The diversity of beliefs within Hinduism enabled the development of various philosophies including atheism, which was promoted by several schools of thought, some of which have been documented in India for 2,500 years. Indian institutions, beliefs, and traditions have always been the product of constant invention. Contemporary Hinduism is the result of many different events. For example, the emergence of Islam, from the eighth century CE, initially bellicose and then later proselytizing, forced Hinduism to adapt. Around 800 CE, the philosopher Shankara succeeded in unifying Indian thought in reaction to Islam and other heterodox movements of the time. To simplify, we might say that this thinker replaced a hitherto dualist philosophy (that saw a distinction between the spiritual world of the Brahman and the material world of existence) with a monist philosophy (promoting the idea that the world is a unified reality that simply has various forms).

Buddhism was founded in the Ganges Valley by Shakyamuni, known as Buddha, the enlightened one, around 500 BCE. This legendary figure is presented as a former prince of a small kingdom in the foothills of the Himalayas. Having spent his youth in ignorance and luxury, he was struck by the pain that governed human existence in the outside world. He turned his back on his carefree days to seek an end to suffering. He explored the path of extreme asceticism, nearly died, and was saved by accepting food from the hands of a woman, a sin for which his disciples abandoned him. He achieved illumination through determined meditation and discovered the intuition of supreme truth. As a natural pedagogue, he summarized this into "four noble truths": (1) all life is suffering, (2) suffering is born of greed and ignorance, (3) knowing the reasons of suffering allows one to transcend it, (4) in order to achieve this, one must take the path of moderation. Salvation can be achieved without excess or asceticism by turning to the Three Refuges (or Three Jewels): the Buddha, the *Dharma* (the teachings of Buddha), and the *Sangha* (the monastic community). Initially this solution was only accessible to men of noble birth who alone were able to become monks. Exceptions were soon made for women, who could also join specific orders on the condition that they remained subject to the control of a monk. Christianity would follow the exact same procedure several centuries later, a parallel that demonstrates the profound misogyny of these belief systems. Like all the ideologies of this Moral Revolution, Buddhism establishes the conditions for harmonious coexistence. Believers are asked to respect the Noble Eightfold Path, which instructs disciples to understand, to think, to speak, to act, to live, to work, to observe, and to be mindful, in harmony with all things.

Jainism was founded by Mahavira, the "Great Soul," also known as Jina, the "Vanquisher," who lived in the fifth or sixth century BCE. He established nonviolence (*ahimsa*) as the foundation of this perspective. Jain laypeople are vegetarian because taking life of any kind is not permitted. For monks, the rules are more restrictive. The "wisest" among them should have the strength to let themselves die of starvation, like Jina, who was thus able to break free of the curse of reincarnation.

Some religions derived from Vedism even reject the *Vedas* themselves, as is the case for Buddhism, Jainism, and the "atheist" Hindu theological schools. Yet along with Hinduism, they all share the same concern for other people along with something that long remained unheard of elsewhere: a concern for animal and plant life. This extension of empathy can be seen in the widespread practice of vegetarianism in southern and eastern Asia, although this is evolving with the progressive Westernization of these societies. To differing extents, the systematic application of the principle of nonviolence leads to the resolution of conflicts without military confrontation, with Jains being the extreme example of this. Some two thousand years ago Jains even opened hospitals for animals. In chapter 12 we will visit a Hindu community known as Bishnoi, some of whom sacrifice their lives to save trees. In Thailand today stray dogs are systematically cared for in Buddhist temples, which does not prevent other monasteries from being involved in the illegal traffic of endangered species, such as the emblematic tiger.

Finally, the principle of nonviolence and vegetarianism must not obscure the fact that these religions can justify war when it is seen as necessary, or that they impose a cosmic order in which women—to varying degrees—are systematically reduced to subordinate positions.

For a One True God

Where Vedic religions push disciples toward a deliverance that involves becoming one with the cosmos, monotheism promises the soul's immortality and everlasting peace in paradise. The history of monotheism is traditionally structured in five stages around five emblematic figures.

Akhenaten was a pharaoh who reigned at the height of the Egyptian Empire, around 1350 BCE. He is often presented as being the founder of the first monotheistic religion in history, which is a contemporary interpretation. The young pharaoh began his reign under the name of Amenhotep IV. Like his predecessors, he was surrounded by the powerful priests of the temple of Amun, the all-powerful God. The distinction between temporal and spiritual powers so dear to the Catholic Church had yet to be conceptualized. But the young king was ambitious, and the yoke of the priests weighed heavily on him. His religious reforms were a political coup d'état. He declared that there was one God above all others and that he alone, the pharaoh, received His instructions. There was no need to go to a temple to glimpse this God; He reigned supreme in the middle of the sky. This was Aten, the Sun. His temples would be open to the sky, religion practiced in courtyards. Akhenaten was king of a state that was wealthy, expansionist, diplomatically dominant, and militarily unparalleled. As the incarnation of the sacred, he was also the guarantor of the cosmic order. This is what allowed him to impose his religious reforms and marginalize the economic and political power of the Amun priesthood. He built a new capital, Amarna, which yielded the archives demonstrating the extent of his politico-religious reforms. His early death and the decline of Egyptian military power put an end to this theological experiment, however. His heirs returned to the old gods and their priests, and his name was virtually expunged from history.

Akhenaton's reform was probably not the beginning of monotheism. But it did represent a first step toward it in the form of henotheism, the worship of a single dominant god within a polytheistic pantheon. Monotheism probably evolved according to this schema. One of the Hebrew divinities, Yahweh, would see himself presented as the single God. One of the Arab divinities, Allah, would benefit from the same consecration sometime later.

Zoroastrianism, also known as Mazdayasna, stems from a reform of Iranian Vedism that seems to have emerged between the tenth and the twelfth centuries BCE. This religious renewal is attributed to a prophet known as Zoroaster, which may translate as "the one with aging camels." The sacred texts of Zoroastrianism defend the dominance of the mental over the material world in a cosmogony that pits a God of Light, Ahura Mazda, against a God of Evil, Angra Mainyu. This dualistic religion, opposing good and evil, was only belatedly likened to an early monotheism, the results of modern-day efforts by the Parsi community. These Zoroastrians, descended from Persians, settled in India in the seventh century CE and have sought to present a vision of monotheism for political reasons.

It is not impossible that ancient Zoroastrianism, the state religion of the Persian Empire, may have influenced Judaism. Indeed, the latter borrowed a certain number of concepts from the neighboring Egyptian or Persian-Babylonian powers. The myth of the Great Flood, which figures in the epic poem *Gilgamesh*, is borrowed from Mesopotamian cosmogony.

Judaism is the inventor of monotheism as stipulated in the first commandment: Thou shall have no other God besides me. This ideology seems to have been developed in the Kingdom of Judah from about the seventh century BCE. Hebraism may initially have been one Levantine religion among others, polytheistic or henotheistic, which connected the peoples of the kingdoms of Israel and Judah. In 722 BCE, the Assyrian Empire destroyed the Kingdom of Israel (the northern half of the current state of Israel). King Josiah reigned over Judah between 640 and 609 BCE. It was during this time that a priest apparently "exhumed" the Book of Deuteronomy from the Temple of Jerusalem, the most sacred site of Judaism. This text is a list of the commandments the one true God left to His people. The rediscovery of "lost" religious documents is a classic theme in the history of religions, also seen in Tibetan Buddhism. It allows disciples to affirm the sacred nature and prior legitimacy of "rediscovered" texts. King Josiah's reforms added legitimacy to the territorial aspirations of the small Kingdom of Judah, which dreamed of becoming a conquering power. From a range of different beliefs, the elites would elaborate a monotheism to reinforce their power around the only place of worship considered legitimate, the Temple of Jerusalem. The Hebrew Bible, a compilation of texts affirming that there is only one true God, proclaims a universalist message. But the neo-Babylonian empire of Nebuchadnezzar II put a stop to this in 587 BCE. The first temple was destroyed, and the Judaean elites were deported to Babylon. Some of them would return in 539 BCE, liberated by the Persian emperor Cyrus II, but many would remain behind.

A prophetic philosophy emerged from this exile and is symbolized by the vengeful preaching of Isaiah: the nation of Judah had sinned through its pride and been vanquished and occupied by unbelievers. Only a strict application of the faith would allow the people of Israel to return to God's grace. This would become central to the conceptualization of monotheisms. If you are struggling, then that is a sign that your actions have caused you to fall out of God's favor. In order to return to grace, you must turn your back on vanity and the cosmopolitan towns of which Babylon was the archetype.[9] Earn your pittance with dignity, through work in the fields, measuring yourself against nature, day after day. And wait for the end. Because just as time is cyclical in Vedic religions, in monotheisms it is linear. It unfolds in two stages. In the first, God creates the world. In the second, He destroys it, as announced by a Messiah or messenger. As it developed, Judaism would become a religion waiting for the end of the world.

Christianity introduced an intermediary stage between the beginning and the end of the world, that of the Revelation. This is in fact the beginning of destruction, or even a failed destruction. One Jewish prophet among others, Jesus was indeed crucified by the Romans in Jerusalem around 30 CE. His disciples would then claim he was brought back to life. The context for this was somewhat apocalyptic, because the Romans had occupied the region, now the Province of Judea, and a number of Jewish people were waiting for the arrival of the Messiah who would free them from this yoke. At the time, the disciples of Jesus, known as Judeo-Christians, were simply one Jewish sect among others.

The apocalypse was indeed triggered in 67 CE. A rebellion shook the Jewish provinces of the Roman Empire, which retaliated by conquering Jerusalem in 70 CE. The temple burned, and Judaism lost its only legitimate site of sacrifice. The disciples of Hillel the Elder restored the situation by collaborating with the Roman occupiers and agreeing to set up a school that would teach only theology. This enabled the development of a certain orthodoxy that would progressively unify the different components of Judaism and that would later become Rabbinic Judaism. In 135 CE a new messianic rebellion ended with the destruction of Jerusalem and the exile of many Jews. In the meantime, Judeo-Christians became persuaded that Jesus was Christ the Messiah announced by the prophets, that he had proved this by rising from his tomb, and that his return was imminent. For Christianity, during this wait for the second coming, time was suspended, and the religion was progressively institutionalized. It was forced to delay the end of time because the expected Messiah did not return, and it established churches to manage this waiting period.

Two branches of Judaism, Rabbinic Jews and Judeo-Christians, were therefore in competition with each other and managed to win over a small number of Romans because they provided spiritual comfort. Roman religions (public worship intended to strengthen the community, private worship to perpetuate the memory of ancestors) were by no means concerned with what happens after death. Various versions of Judaism and other eastern religions, such as Mithraism, promised resurrection, paradise, or both. In other words, they answered the question of "what awaits me after death?" Judaism and Christianity thus forged their theologies in competition with each other, determining the canons, laying down the exegesis of their holy texts, and vilifying each other to better dismiss each other's teachings. In the fourth century, Christianity would be consecrated as the state religion of the Roman Empire.

Once institutionalized, it made its authority durable by declaring that the return of Christ, precursor to the end of time, would occur at an unknown date.

Islam was born from the teachings of a mystic Arab merchant, Muhammed, in the early seventh century. Inspired by the many Jewish and Christian communities then present in the Arab world and after many years of tribal conflicts, he imposed a strict monotheistic reform. He made Mecca, formerly a site of pagan pilgrimage, into a sanctuary for the one and only God, Allah. His successors would spread his message over a vast territory in the century that followed (see chap. 7).

Monotheism established humanity as lord over nature. Judaism, Christianity, and Islam all draw on the same source of biblical writings according to which God gave the world to humans so that they would go forth and multiply, as is explicitly stated in Genesis, the first book of the Bible: "And God blessed them; Be fruitful and multiply and fill the Earth and subdue it, and have dominion over the fish of the sea and over the birds of the heavens and over every living thing that moves on the Earth" (Gen. 1:28). From this perspective, humanity dominates the world and takes whatever it needs.

Philosophical Reflections

The term *philosopher* comes to us from the Ancient Greek, *philos* (loving), and *sophia* (wisdom). Greek and Roman philosophers were collectively credited with the transformation of thought that changed the course of history. Classical antiquity gave rise to the scientific disciplines and key concepts of the rational thought that would later enable the West to dominate the planet. Without attempting to provide an exhaustive panorama of the classical philosophers' conceptions of nature,[10] it is worth revisiting the basis of the relationship that Western thought has had with animals over the subsequent 2,500 years.

Contemporary philosopher Elisabeth de Fontenay reminds us that in European antiquity animals were denied a voice. Most philosophers considered that, unlike humans, animals did not have the ability to express their moods. Like all complex civilizations in Eurasia, however, the societies of the Mediterranean basin did share the fact that animals were an important part of their systems of meaning. Sacrifices, although bloody, established animals as the messengers between the spiritual and temporal worlds. Whether doves or bulls, they were ritually slaughtered and gutted; their entrails were scrutinized for signs of divine will; their flesh was eaten to consolidate civic communities through sharing. With the triumph of Christianity from the fourth and fifth centuries CE, these practices come to an end. The "sacrifice" of the "sacred lamb," Jesus Christ, was henceforth seen as having washed humanity of its sins, and animal sacrifice, considered a pagan tradition, was outlawed. The Roman Church demanded a monopoly on communication with God, now considered unique. However, pre-Christian thought did not disappear. Some of its elements would be assimilated into the matrix of Christianity.

The foundations of European thought were therefore laid somewhere between the birth of Western philosophy and the triumph of Christianity, and today they wield cultural hegemony around the world. Stoics, for example, established a division between animals and humans. Although we can clearly see that the former is capable of suffering and able to express that through sound, the latter alone is able to use reason. From this perspective, only humans have an internal language that allows us to formulate thought, to be aware of its representations, and to act reflexively on them.

This logic still influences our preconceptions today, although it was sporadically challenged by Neoplatonists, such as Plutarch and Porphyry of Tyre (who, like the semimythical Pythagoras, defended the idea that we should refrain from mistreating animals or consuming meat). Animals cannot have rights because they are considered to be without language. From being the messenger of the gods, they have progressively become little more than meat on legs.

But history is full of crossroads and turning points, and we cannot accuse ancient philosophy, medieval Christianity, or modern humanism of having paved the way for the inhumanity of today's industrial farming and slaughterhouses. Let us take the example of Epicureanism, a philosophy that, had it risen to dominance in its time, may have changed the course of history. Today there is almost nothing left of it. Except a kind of philosophical testament written in the first century BCE by Lucretius, a distant heir to the Greek philosopher Epicure (who lived around 342–270 BCE). Far from the image of depravity that later Christian authors foisted on them, Epicureans advocated enjoying the pleasures of existence in moderation. When the Christian church came to power in the Roman Empire, from the fourth century on, it assiduously destroyed all the writings it considered contrary to its ideas. Plato and Aristotle, whose philosophy had a practical aspect, were spared, discussed, and occasionally acclaimed. But Epicurean thought horrified the church. Around 1420 an Italian humanist and manuscript hunter by the name of Poggio Bracciolini, unearthed a forgotten book from the depths of a monastery library,[11] an unfinished work with an enigmatic title *De rerum natura (On the Nature of Things)*. This is almost all that now remains of Epicurean thought; its author, Lucretius, we know almost nothing about.

De rerum natura is a long prose poem. Written during the reign of Caesar and Pompey—a dark time between civil wars, slave rebellions, and the twilight of the Roman Republic—it is filled with phrases that are much more pessimistic than those Epicure probably wrote. The book sets forth a philosophy that must have seemed like a long list of blasphemies to its Renaissance discoverers. Judge for yourself, keeping in mind that these ideas are more than two thousand years old: the universe has no creator and is infinite, as is time; life is governed by chance; the main characteristic of living things is their free will, and their nature means they are the product of random evolution; humanity is transitory, it will one day disappear; like animals, humans used inarticulate cries and gestures to communicate before inventing language and music; the soul dies with the body, there is no life after death; all religions are illusions, aimed at enslaving people; life is only worth living if it is devoted to the pursuit of happiness; the world is made of atoms—invisible, eternal, and indivisible elementary particles; emptiness is inherent to the entire universe; matter is eternal, but its forms are transitory and its components are subject to a cycle of disaggregation and reconstitution.

These assertions were unspeakable in the extremely Catholic Europe of the fifteenth century when this philosophy was rediscovered. Today they are much more familiar. It is not a totally original philosophy because it parallels certain elements of Indian philosophy. Two schools of thought in particular seem to have developed similar ideas, possibly 2,500 years ago: the school of Vaisheshika, the first philosopher to have presented the theory of the atom, and the Charvaka school, which had already affirmed that the world was an assembly of matter without divinity and which saw life as dedicated to the pursuit of pleasure.

The philosophy of Epicure and Lucretius miraculously survived over the centuries. It reveals the maturity of a rationalist philosophy possibly more adapted to the search for happiness than the religious systems that fought against it. But where Epicure's ideas survived, passed down to posterity via a fragile poem by one of his far-off disciples, how many anonymous thinkers have seen their ideas swallowed up by the ocean of time? From the prolific Ancient Greek playwright Aeschylus, who penned more than 110 plays, the classical canons selected and saved only seven of his masterpieces.

An emphasis on the contributions of Greek thinkers can lead to a mistaken belief that the air around the Mediterranean helped people think better than, for example, the sea breeze of the Indian Ocean or the China Sea. In the West, we have sanctified this history. We have done our utmost to explain by what miracle philosophy was able to emerge and prosper in Greece and how it was saved in Rome, Constantinople, Baghdad, and Córdoba before returning to Western Europe with the Renaissance. No philosophy can grow in a void, without roots, but the great oak of Greek wisdom prevents us from seeing the fertile forest of philosophers from elsewhere. Ancient Indian and Chinese texts were just as diverse and theoretical. But they were often simply less well preserved. Millions of books were written in western Asia even before the invention of the printing press in the eighth century CE. However, thousands of book burnings accompanied the rise of new powers, and the vast majority of these corpuses will remain forever unknown to us.[12]

The Time of Exchanges

This brief panorama of the spiritual landscape between 1500 BCE and 600 CE shows that there was a decisive peak around 500 BCE.

Very different conceptions of the world were in competition with each other: polytheisms against monotheisms; ideologies promising resurrection at the end of time against cosmogonies defending the idea of an infinitely reincarnating soul. Each was looking for answers to existential questions with an intensity that means our own philosophical questioning has barely produced anything really new since. All aspects of spiritualties were pursued and refined in India, China, Greece, and elsewhere. This overall perspective on the ideological-religious dynamic that still drives us today leads to a single lesson: whatever the ideology, particularly religious, whatever its conceptual system, it will evolve. Specifically, it will join with political power. At some point its spokespersons and leaders will bow pragmatically to the requirements of the moment, play on the complexity of the founding texts, and occasionally abuse their authority over the sacred or over knowledge in order to instrumentalize the ideals that they stand for in the name of speculative ends.

The terms Axial Age and Moral Revolution allow us to identify conceptually this key stage in the evolution of societies, this moment when religion, understood as an evolving arrangement of myths, became the main mechanism for ensuring the lasting cohesion of very large groups. Religion created a superreality that only existed in minds but that transcended individuals, placed them within a collective, and radically increased the abilities of our species. This characteristic reached its peak with the Moral Revolution because it corresponded to a new revolutionary phase in human societies: commerce. This foundational moment was to determine the history of the next two thousand years.

These would be missionary exchanges in which ideologies left their initial birthplace, had universal aspirations, and spread wherever they could in a competitive market. Of course, these exchanges would also be commercial. We will see that money, as we know it, appeared at the same time as universalist religions. They would be military, too, because the temporal aspect of missionary religions would take the form of universal empires. Taken together, these exchanges would provide humanity with the fuel to create ever more profound upheavals in our world.

6

ALL EMPIRES WILL FALL

Monkey's dreams are missionary, military, and mercantile. Powerful empires sprang up alongside the spread of universalist religions, but between 200 and 400 CE, the system once again faltered and collapsed.

A nervous trumpeting breaks through the mist. The Greek foot soldiers freeze, dripping with sweat under the weight of their armor. To their right is a fast-flowing river that they had struggled to cross upstream before daybreak. To their left is a hostile land and its unfamiliar vegetation. From the sky comes an uninterrupted downpour. It is the first time they have been confronted with the monsoon. Before them lies the powerful army of the Indian king Porus, reinforced by two hundred colossal creatures with wrinkly skin and iron-tipped tusks, substantially larger than even the most monstrous chimera of their folklore. And despite being outnumbered and having no war elephants of their own, Alexander the Great's soldiers have no choice. They must win—or die.

Elephants, a Double-Edged Sword

Once again, Alexander proved himself an exceptional strategist. Reading the tale of his exploits, it almost seems as if he enjoyed scaring himself, as if he was used to carrying out certain battles on the role of a dice. Heir to the kingdom of Macedonia, which his father Philip II had united with Greece after the defeat of the coalition of Greek cities in Chaeronea in 338 BCE, he decided to conquer Asia. He went from victory to victory, even bringing down the king of kings, the Persian Achaemenid emperor Darius III. He conquered an empire that stretched from Greece to Afghanistan. He had himself made a god in Egypt. He encouraged his generals to take wives from among the conquered populations to make this superficially Hellenized Persia a cosmopolitan world. Commercial exchanges were already fueling a prosperous and peaceful economy. Alexander was not yet thirty years old and still dreamed only of continued conquest. North India, merchants told him, was overflowing with riches. And the first powerful kingdom that lay in his way was that of King Porus, in today's Pakistan.

Were elephants a threat? Darius had already used them against him. The first time was in the Battle of Gaugamela, in the north of today's Iraq. There were twenty or thirty beasts whose tremendous mass overshadowed the countless troops of the king of kings. But the danger came from the horses first. On the flat ground, Darius unleashed his secret weapon, giant scythed chariots that dismembered the formidable Greek infantrymen and wreaked havoc on their legendary phalanx formation. Then the elephants moved forward. However, few of them, if any, seemed to have been trained to fight. They were simply ceremonial animals used for parades and trained to kneel before the emperor. After their initial shock at seeing these beasts, the Greeks pulled themselves together and counterattacked, chanting their war cries. In all the clamor, the giants apparently took fright; in any event they did not fight. Once the chariots had been defeated, and the king of kings had fled, Alexander and his armies took control of the surviving pachyderms.

Knowing that these elephants came from the very same East he dreamed of conquering and aware that he might have to face some that were better trained for war, Alexander used his new menagerie to ensure his soldiers and their horses were accustomed to the presence of these giants. He maneuvered them so that everyone was able to evaluate their speed, and he made sure that every soldier knew where their weak points were—tendons, stomach, trunk, eyes, or ears.

It was thus at dawn, one July day in 326 BCE, that ten thousand of Alexander's elite soldiers, cavalry and infantry combined, stood firm against an army that far outnumbered them. They aimed at

King Porus's elephants, which stood like a rampart protecting the troops resisting the Greeks' advance. These ones were trained for war, for tearing apart criminals, and were not likely to take fright at the sound of war cries. Goaded by their drivers and the orders of the noble warriors on their backs, they stood their ground. The first projectiles wiped out the drivers, then they targeted the elephants' most vulnerable points. Driverless and blinded by pain, the elephants went mad.

Training an elephant is a delicate procedure. Making an elephant a killer means making it a perpetual danger to its masters as well as their enemies. Alexander's soldiers were careful to maintain an exit passage in their maneuvering, but Porus's troops were assembled to create maximum resistance against the Greeks, and so they had nowhere to hide. The uncontrolled and enraged elephants forced bloody paths through the ranks of the king's army and left it tattered and disorganized. The twenty thousand Greek infantry and thirty thousand Persian soldiers that Alexander had kept in reserve could now cross the river and join in the carnage.

Victory was paid for dearly, however. Once the elephants had been eliminated, Porus's troops continued to fight and almost won. The Macedonian king's expeditionary force was decimated. Alexander's favorite horse, Bucephalus, a veteran of numerous battles, was killed. Yet his master was determined to continue his conquest. His next objective was Nanda, a state that covered the immense Ganges Valley and was reputed to have an army of six thousand war elephants. His generals warned the young conqueror that mutiny threatened to snuff out his ambitions. The Greek army, which had fought countless battles since leaving Europe, yearned for peace. Alexander turned tail on the banks of the Indus. Three years later he died of a fever, although it might have been because of his alcoholic excesses or a poison one of his courtesans slipped into his drink. . . .

Although this battle in many ways resembles the monumental confrontations described in the *Mahābhārata*, it is drawn from history and not from myth. But how did Greeks come to face off against elephants in the middle of what would become Pakistan 326 years before the Common Era?

The Plight of Pachyderms

Five thousand years ago, various species of elephants could be found from the Levant to Korea, from Iran to South Africa, from the Himalayas to the Maghreb. The whole of China, except for the mountainous western regions, was home to flourishing populations of pachyderms. It was around this time in the Indus Valley that the first attempts were made to domesticate, or rather tame, these animals. Because elephants do not reproduce well in captivity, these "breeders" generally removed young animals from wild populations and tamed them using other already tamed elephants. The idea that elephants may be used to serve humans is therefore only possible through interaction with remaining wild populations.

It was therefore a very strange bargain that Monkey struck with the largest of the surviving terrestrial megafauna. This seems to have occurred independently in three separate places: in the Indus Valley in 3000 BCE, in Mesopotamia a few hundred years later, and in China around 1500 BCE. Often their use was reserved for kings, for who but the elite of the elite could be worthy of the services of the emperor of the animal kingdom? The elephant carried the monarch on his journeys and knelt before him in ceremonies. It seems that the males were also trained for war quite early on, a difficult and dangerous task.

And then the climate changed. Around 3,200 years ago it began to get cold. The human population expanded and cleared ever more forests. Elephants lived only where they could find dense vegetation on which to feed. When there were no more trees, they turned to crops. They began to die out in Mesopotamia and declined in northern China. There, the peasants hunted them—sometimes with the help of kings. They were also killed for their trunks, a much sought-after delicacy. According to the historian Mark Elvin, although the changes in climate weakened the elephants' ecosystems, it was the progression of agriculture, demographic growth, and the decline of forests that saw the decimation of the population of Chinese pachyderms along with tigers and rhinoceroses.[1]

By around 1000 BCE northern China had no elephants left. By

around 500 CE they had also disappeared from the center. Today the only remaining Chinese elephants are found in the extreme south in the dense forests of the mountainous border with Myanmar. China once domesticated elephants, even using them to pull carts. But they died out. And in China people forgot that these animals could be tamed.

In India and in Southeast Asia there were still forests. As a result, the tradition of using elephants for everything, even as bulldozers, remained. Some kings boasted that they possessed thousands. Prefiguring the creation of today's national parks, they established royal forests reserved for elephants, who were both a symbol of their authority and a powerful tool on the battlefield.

And then came Alexander. His successors established Indo-Greek kingdoms in today's Pakistan and Afghanistan. The Seleucid dynasty, descended from one of Alexander's generals, controlled Persia. The Ptolemaic dynasty, descended from another general, took Egypt. The two quickly clashed, and each had war elephants. But the Seleucids were able to replace their dead animals with new ones bought from their Indian neighbors. The Ptolemies had no access to this source and so undertook one of the most substantial hunting endeavors in the history of the world. For want of Asian elephants, they would have African elephants.

At the time, although elephants had all but disappeared in the Levant, they could still be found in Ethiopia, Somalia, and North Africa. The Ptolemies built a network of trading posts along the western coast of the Red Sea and sent whole armies across the Horn of Africa with the sole purpose of procuring these precious pachyderms. And they tamed them. There are ethologists who argue that zebra and elephants have not been domesticated in Africa because those local species are much too rebellious, but history has proved them wrong. It was just that these societies chose not to tame them. Perhaps they did not need to? But it was possible.

Are elephants actually useful in war? The Greeks used them frequently to reinforce their phalanxes and their cavalries, inventing the *howdah*, a carriage strapped to the giant's back to carry several archers. Yet the highly mobile legions of the new Roman power quickly managed to get around the threat of these enormous animal auxiliaries. They learned that if you kill the driver (whom the Greeks called *indos*, "the Indian") and wound the animal, there is a 50% chance that it will turn against your enemy in a terrifying charge. The Greeks had made the elephant a crucial weapon in their armies. But the Romans quickly abandoned it, judging it far too unpredictable.

The epic journey of the Carthaginian general Hannibal—who was almost able to take Rome after successfully crossing the Pyrenees, the Rhône, and the Alps with a few dozen elephants—did not help the Romans change their minds. And yet the massive mammals could bring victory simply by virtue of the terror they inspired the first time soldiers encountered them. The last elephant to fight in Europe was apparently used as a psychological weapon by Emperor Claudius in the first century of the Common Era. Preparing to invade Britain, the emperor crossed the channel with a few elephants hoping to provoke awe and terror among the enemy.

But the Romans quickly found another use for these animals: the circus. The last European lions and North African elephants were massacred in the arena for the amusement of the crowd. In 55 BCE, upon his victorious return from Syria, general Pompey built the first stone amphitheater in Rome. Five hundred lions were massacred there in five days. But there was more to come. In one event, twenty elephants were pitted against a group of African warriors ordered to kill them. Their plight was so dramatic that the naturalist Pliny the Elder wrote that "when Pompey's elephants had given up hope of escape, they played on the sympathy of the crowd, entreating them with indescribable gestures. They moaned, as if wailing, and caused the spectators such distress that, forgetting Pompey and his lavish display specially devised to honor them, they rose in a body, in tears, and heaped dire curses on Pompey, the effects of which he soon suffered."[2] Seven years later, the curse was fulfilled. Pompey was assassinated, and his head was sent to his rival Julius Caesar as a gift.

During the reign of Augustus alone (27 BCE–14 CE), 3,500 elephants were killed in circuses in the city of Rome. In 79 CE, to celebrate his inauguration as emperor, Titus staged lavish games for the people lasting one hundred days. He inaugurated the gigantic

Colosseum, which could seat at least 50,000 spectators and which was destined to break all records in animal slaughter—5,000 on its first day! Bears, lions, leopards, wild boar, or more often stags and bulls. The animals were kept in darkness, starved, and tormented to increase their aggressiveness. They were then released directly into the arena and the deafening roar of drums, trumpets, and screaming crowds through a system of communicating cages. They were sometimes forced to kill each other, sometimes pursued and brought down by gladiators using pikes or tridents who performed as though they were hunting. In another event, depravity reached new heights with the carnage of three hundred terrified ostriches killed by means of custom-made projectiles specifically designed to decapitate the giant birds.

Around this time, both the Mediterranean and the Chinese worlds experienced a rapid drop in biodiversity marked by the disappearance of their forests and large animals. These processes evolved alongside the simultaneous increase in the power of both these empires on opposite sides of Eurasia. These were two worlds that would come into contact over 10,000 kilometers (about 6,213 miles) of hostile territory.

Three Aspects of Money

Before going into the tenuous relations woven between China and Rome 2,200 years ago, we must first explore the driving force that was behind this contact: metal coinage. Money, as we think of it today, appeared around the same time as universalist religions. Evidence of the first minted coins has been found from around 600 BCE in three distinct places: Asia Minor (that part of western Turkey that was then populated by Greeks), the Ganges Valley, and northern China.

For Aristotle, money, in a broad sense, is a tool that has three distinct functions. It serves as a standard, a universal measure for setting prices and thus estimating the value of debts or credits. It is also a means of exchange and a way of accumulating wealth. These three functions can be quite separate from each other.

The shekel, for example, before being the modern currency of the state of Israel, was that of the city-states and then empires of Mesopotamia between the third and first millennia before the Common Era. The word *shekel* is etymologically linked to the Sumerian unit of weight for grain. It was not a minted coin but rather a standard that allowed the vast royal palaces and temple complexes to determine the volume of grain they received in taxes or that was stored or available. It also meant they could calculate the amount of bronze, or the number of days work a specialized artisan might be expected to provide in exchange. Our inheritance from the ancient Sumerians and their accounting continues to shape our existence today. They were the ones responsible for dividing the year to twelve months, the day into twenty-four hours, the hour into sixty minutes, the sky into twelve astrological signs, and the circle into 360 degrees—all 5,500 years ago. According to them, once converted into silver, the weight of the shekel was equivalent to a bushel of barley. Just a bushel could be divided into sixty rations, a mina could be divided into sixty shekels, each corresponding to one meal. A bushel was equal to one month's work, sixty meals, based on two rations of barley per day. In ancient Sumer, money was not used for facilitating trade. Rather, it was a convention developed by public servants to systematize taxes, keep records of which resources were stored, and convert these resources into other resources (e.g., labor into food). Most of the cuneiform tablets from this period are in fact accounting documents tallying up shekels.

Many other forms of money have existed over the centuries. Well before it occurred to the Greek principalities and warring Chinese kingdoms to forge currencies that would allow them to tax trade by obliging merchants to resort to an intermediary that was produced and controlled by the state, there were myriad distinct currencies in circulation. The anthropologist and economist David Graeber demonstrates that barter was probably never an economic reality except in periods of crisis, when monetized societies saw their currencies collapse.[3] All around the world, articles were bought and valued in shells (throughout Africa and a large part of Asia), in cocoa beans (among the Aztecs), in salt (Roman legions received their pay in salt, hence the origin of the word *salary*), in ingots of various metals (often in times of war), in sacks of grains, or in rolls of fabric (for a long time silk was a base currency in Chinese provinces, as was cotton in Andean economies).

Yet about 2,500 years ago, three monetary zones emerged simultaneously around three state currencies. They correspond to the three global centers of economic and demographic development. This emergence does not have a technological foundation because the minting procedures used are very different in each area. Instead, this development was most likely a response to a politico-social objective; monarchs decided to replace the previous system of exchanges and credits by a new one, rightly calculating that by giving themselves a monopoly on issuing money, they could better finance their military and architectural endeavors. Prosperous empires go hand in hand with money.

Imagine that you are a ruler before the invention of minted metal coinage. Bad luck! You're going to find it difficult to raise a substantial army without it becoming a threat to society. It is crucial that an army is well fed; allowing it to starve would mean hungry but well-armed men among wealthy unarmed compatriots. You must draw on the grain stores to feed them and then fill up those stocks again, by threats if necessary, creating social tensions in the process.

Now imagine yourself in the same situation except that you now have money. You can mint as much you need and distribute it to your soldiers as payment. If you are a credible source of authority, you will transform your kingdom into a market, suddenly generating a mass of unimaginable wealth. Let your subjects know that taxes should be paid in coin. Soldiers will pay coin for their meals to farmers and innkeepers, who will then pay it to you. Every civilian must now find a way to sell something to the soldiers—food, clothing, or jewelry for courtesans—in order to pay their security debts to you. The state has an army, soldiers are well fed, and families are working toward earning the money that they will pay the state. An economist would say this is a virtuous cycle. For the last 2,500 years, this process has led societies to create more wealth than was previously conceivable, and as a result, to tax the natural environment ever more persistently to satisfy needs that have become commodifiable.

Minute fragments of precious metals, stamped with nominative symbols guaranteeing their value, began to circulate everywhere. This invention spread rapidly over all Eurasia, thanks in particular to Alexander. The Greeks' taste for trade was boosted by the stocks of gold accumulated among the Persian treasures. From western Asia, cargoes of gold were shipped to the Ganges Valley, China, and the Mediterranean. The historical emergence of state currencies is closely linked with the conception of universalist ideologies (understood as being valid for all humanity). When people trade over long distances and when empires encourage them to do so, gods must incarnate universal ideals in order to accompany these exchanges.

Monkey's character became missionary, military, and mercantile. Although this threefold role began 2,500 years ago, it is still ours today. We remain animals of power. And power is held by states (the apex of politics and the army), the wealthy, and those with religious knowledge.

The emergence of iron, after the collapse at the end of the Bronze Age, helped break up the great powers of the era, weakened as they were by the "perfect storm" of environmental and social circumstances (see chap. 4). When the empires pulled themselves together and reorganized, they were more powerful than ever and apparently more aggressive too—the weapons market was thriving. New powers emerged first in western Asia with the neo-Assyrian Empire (911–609 BCE) and then Achaemenid Persia (550–330 BCE). Their armies were far more efficient than those of their predecessors. They benefited from a major technological innovation. With iron, peasants were better equipped and more productive. Miners, artisans, and merchants all had more opportunities. Henceforth, the priority of states was to build infrastructure, particularly roads and bridges, canals, and fortifications to facilitate, protect, and tax trade.

These new states would try to preserve their longevity and experiment with different social models. The first millennium before the Common Era saw the emergence of three concepts of imperial governance intended to preserve these political entities as much as possible. The challenge was substantial. Harmony had to be maintained over immensely large territories. An empire generally covers more than a million square kilometers (about 0.4 million square miles), and the largest of them, the Mongol Empire in the thirteenth century and the British Empire in the early twentieth century, covered thirty times that much—controlling a quarter of the land on Earth! These lands were inhabited by a multitude of peoples, ethnic groups, religions, and languages typical of the melting pot that existed before the great homogenization of the modern era, which saw more than half the tribal and linguistic identities documented five hundred years ago disappear. For 2,200 years, empires were the dominant political form of humanity's history largely because they could establish social contracts that were able to manage differences between subjects harmoniously.

Simplifying somewhat, we may say that three types of social contracts were established. We can identify these with Roman, Chinese, and Asian ideal types. In reality, of course, empires used these models as references to draw ideas from. Conditioned as they were by their conceptual environments, they took from them the various administrative principles they assumed to be the most appropriate to their particular situations.

Rome, An Empire of Citizens

The Roman model of imperial governance was pragmatic. It was based on the Greek concept of "civilization," defined as a large internal space held together by civic belonging and opposed to "barbarians" who do not belong to the community.

The Roman model evolved in four stages. Rome was first a typical Mediterranean city-state. Like its Greek rivals, it relied on the involvement of citizens, free men able to pay taxes. They ensured the defense of the city by taking up arms. After a century-long pause during which the Etruscan kings reigned, the Roman Republic was founded in 509 BCE. It then took two hundred years for Rome to control Italy; one by one the Etruscan, Italian, and Greek city-states fell, unable to unite their defenses. Then Rome turned its attention to the Mediterranean. In the first century before the Common Era, this, too, came under Roman control. During the last phase of conquest, the appetite of Rome's military leaders was whetted still further, and a series of civil wars broke out ending in the decisive victory of Augustus in 27 BCE. He was appointed the first (Roman) emperor and made a god. This was important because it ensured imperial harmony. Civil wars, Augustus proclaimed, were impious. Under his reign, prosperity returned, and the emperor presented himself as the sacred guarantor of the concord between gods and men. He used religion to set himself apart from mere mortals and to shore up his power and that of his successors.

His right-hand man, Agrippa, remained in the shadows building the infrastructure that would consolidate the empire. Initially, Augustus reigned peacefully over the lands that had been conquered under the now-dead expansionist Republic. To make the most of these vast conquests, Agrippa established a network of roads. Henceforth, "all roads would lead to Rome." He was also responsible for making euergetism (public philanthropy) a systematic part of urban architecture so that the rich would ostentatiously spend part of their wealth on ensuring social harmony. Every Roman town was organized around two central perpendicular avenues that converged toward the center where the public buildings incarnating imperial authority were on display. Temples (for the gods protecting the city of Rome), amphitheaters, and public baths were key places of sociability.

But Agrippa had a problem. Citizens had paid with their blood for Rome's prosperity, which was linked to its military domination over an immense territory.[4] But Roman society, which included some sixty million people at the beginning of the Common Era, was highly inegalitarian. One third of Italy's population was made up of slaves (from conquered populations), roughly another third was made up of freemen from around the empire (who had certain rights under Roman law), and the final third were citizens of the empire of Latin origin, or of Rome itself. (The former, known as *latini*, were inferior in status to full Roman citizens, *cives*; the latter made up only a sixth of the population until the second century, when this group increased significantly.) At the very top were a handful of "ultrarich," the nobility, who controlled political power and property. The administrative success of the whole relied on a few dozen highly trained elite public servants. As for Rome's economy, it depended on large-scale farms worked by squadrons of slaves on behalf of a few landholders who monopolized public offices. Peasants who were removed from lands they had had to borrow against in order to buy food in difficult times often left the countryside to swell the population on the towns' edges.

Social inequalities, crowds of slaves and paupers, veteran citizens often lacking in resources—all this added up to a potentially explosive situation. Agrippa had to try and defuse it. The expression "bread and circuses" sums up this social pact. The generosity of the elites guaranteed the pillars of the regime—the citizens—all the food and distractions they needed. Their philanthropy consisted, for example, in funding the circus games that saw the massacre of elephants and lions in between gladiatorial fights along with the very first attempts at special effects. Or it was used to finance the thermal bathhouses; in the first century CE, the city of Rome, with its 750,000 inhabitants, had up to eight hundred bathhouses. These were immense spaces of cutting-edge technology, with saunas, tepid baths, cold baths, perfumed baths, complex plumbing with taps and mixing faucets, mechanisms for controlling ventilation, heated flooring (invented, along with sewers, in the early cities on the Indus around seven thousand years ago). In Rome for the first time in history, these public buildings could have high ceilings and natural light and be well heated. Up until then it had been either one or the other, but the Romans had learned to melt glass panes, which allowed them to construct veritable walls of transparent glass. They also borrowed recipes for waterproof cement from Eastern civilizations, providing new building opportunities by adding clay, ashes, and powdered volcanic stone (pozzolan) to lime. This allowed them to carry out great hydraulic projects. They altered watercourses to feed mills, which seem to have been relatively frequent machines in the Mediterranean landscape at the beginning of the Common Era.

The Romans received silk from far-off China. This was an extremely valuable commodity, and China guarded the secret of its fabrication jealously. The silk trade involved dozens of transactions, and the precious textiles had to cross all Asia, whether by the continental silk road that ran north of the Himalayas or by the sea route that crossed the Indian Ocean. In the last stages of its expansion, Rome dreamed of conquering the source of this wealth. But the Parthian Empire quashed these ambitions in the Syrian desert with their victory in the battle of Carrhae in 53 BCE. After that, certain merchants unwove and rewove the material more loosely so that they would be able to sell apparently greater amounts and increase their profits. Conservatives, such as the philosopher Seneca, criticized men who spent fortunes in gold to pay for their wives to have cloth so thin it was shamelessly transparent.

The center of the empire was fed by maritime trade routes organized around the Mediterranean. A boat moves ten times more cargo than a land caravan for the same cost, and it does not need the paved roads that Rome had laid around the sea that was the heart of its empire. Rome's power ensured its expansion and influence, but it destroyed its environment. It required ever more wood to heat the bathhouses and grand villas, to extract minerals from the earth, to melt glass, to build boats that would carry wheat back to Rome. Indeed, the whole Mediterranean region, including densely populated Italy, was fed on the grain produced in Tunisia, Libya, Egypt, and Syria, countries that were then far more fertile than they are today.

China, an Empire of Public Servants

The Chinese empires and kingdoms adopted a meritocratic model. It was founded on an administration of public servants who were recruited through competitive exams from 605 CE on and a shared awareness of the opposition between what was cooked (civilized, altered by culture) as opposed to uncooked (barbarians, savages).

In this model, the emperor is the guarantor of stability and the recipient of the mandate from Heaven. As a result, he is a sacred figure for as long as prosperity reigns, providing the administration with formidable efficiency. But in times of trouble, his opponents can argue that chaos means that Heaven has withdrawn the emperor's right to govern, which allows them to put an end to his regime.

A certain number of fundamental inventions were developed in China. That is not surprising given that China has made up a significant portion of the global population, fluctuating between one third and one sixth, over the last three thousand years. One of these fundamental inventions, along with gunpowder and the compass, was paper. Made from a paste of ground mulberry bark mixed with linen and hemp fibers, paper was perfected around the sixth century BCE. It was initially used primarily for clothes. Thick coats and even amour was made from hardened cardboard. Between the first and second centuries of our era, paper was refined enough to be used for writing. Bound books were invented simultaneously in China and in Rome (with the expensive Egyptian production of papyrus). In China, the abundance of cheap paper progressively enabled the development of an extensive administration. For part of the second millennium BCE, that administration included up to 10% of the active male population. Public offices were reserved for literate elites during the Qin (221–206 BCE) and then Han (202 BCE–9 CE and 25–220 CE) dynasties. From the seventh century CE, with competitive exams that rewarded a knowledge of classical texts and a mastery of Chinese characters, the recruitment of imperial public servants increased.

Politically unified in 221 BCE by the first (Chinese) emperor, after the nearly three-hundred-year-long Warring States period, north China had already experienced significant problems with erosion. Winds from central Asia whirled across the plateaus, and high-altitude farming was abandoned for the more intensive cultivation of the valleys. This was made possible by a large population, attentive recycling of organic materials (compost and human waste) to constantly improve the soil, intensive farming in rows associated with systematic weeding and hoeing, and other innovations. The most spectacular of these was a plow with a metal blade, which was invented in China 2,500 years ago but which would remain unknown in Europe until the fourteenth century and only become widespread in the seventeenth. The idea of slanting the axis of the plowshare meant the Chinese facilitated the evacuation of clumps of earth and reduced the effort required of the animal pulling the plow. Other Chinese innovations that increased labor productivity and trade included the blast furnace, which made it possible to forge series of plowshares and other objects made of high-quality iron. They also invented the wheelbarrow, which meant workers could carry more; the water mill, also widely used among the Romans, which captured hydraulic power to grind grain or make pulp for paper; and the rudder, much more effective than oar-based steering systems used elsewhere, which would not find its way to Europe until the thirteenth century.

The intensification of farming meant completely terracing the very fertile alluvial basin of Huang He, the Yellow River, which was now channeled between colossal embankments that were shored up generation after generation. But the river continued to carry along ever more silt, increasing its bed year after year. This situation was particularly dangerous because floods could be devastating, sometimes able to completely submerge these hydraulic works in extremely densely populated areas (see chap. 7). This impressive terracing project would transform Chinese society by encouraging centralization.

In the second century BCE, the Han dynasty conquered a vast territory in the northwest (now Xinjiang) out of fear of incursions by the people of the steppes. Chinese officers were involved in collecting information on these nomadic states. Their explorations took them to the edges of the Caspian Sea. They were concerned about the Xiongnu, a confederation of nomadic peoples who had formed an imperial military power to counter their rising hegemony. The Han then sought an alliance with the Yuezhi, another nomadic confederation, who had been recently defeated and pushed to the west by the Xiongnu. Although this offer of military cooperation remained unfulfilled, these diplomatic missions did raise the Han's interest in what lay on the other side of the Himalayas. This was around the time that the silk roads began to be laid out. Central Asia was overflowing with precious goods. Among these were the finest-quality horses and molten-glass trinkets, whose green color reminded the Chinese of jade, which they associated with immortality. This precious stone was traditionally obtained in the Himalayas, but so much of it had been buried in the tombs of the wealthy that it had become rare. Glass, originally invented in Mesopotamia, came from the Roman Mediterranean. There it could be produced for as long as there was sand on the beach, a resource that then seemed inexhaustible. Glass, jade, silk, horses, spices, sugar, ceramics, precious metals; the silk roads began their history with the accumulation of long-distance trade. Little by little, these roads linked Rome to the Chinese capitals Chang'an (today Xi'an) and then Luoyang.

The Chinese, trusting in tales told by the merchants ferrying these precious goods across Asia, thought the Mediterranean world was a paradise, full of peaceful cities governed by wise philosophers whose authority was unchallenged. They tried to reach these lands and poured significant resources into conquering the steppes. At the beginning of the Common Era, the Han dynasty faltered, weakened by a civil war and a devastating flood of the Yellow River. The empire ultimately revived, but it was forced to abandon the lands to the west that it had conquered from the nomads.

The imperial currency minted at the height of the Han dynasty, from 119 BCE, became a strong currency standard that would survive the dynasty and remain in place for eight hundred years. To fund its military campaign, expansion toward the west, and colonization (the state helped peasants who wished to move into these new regions as the armies progressed), the government nationalized the production of iron and salt in 117 BCE. These state monopolies were sold to private entrepreneurs 150 years later. In the first century CE, a government census recorded sixty million subjects in the empire, which made it the exact demographic equivalent to Rome. Its successive capitals, Chang'an, then Luoyang, each boasted half a million inhabitants.

India and Central Asia, Delegated Empires

India is the third pole in this imperial dynamic. Its capital, Pataliputra (now Patna) rivaled Rome or Chang'an in size. The Asian imperial model (let's call it that because it would be adopted by central Asia too) advocated a laissez-faire approach. Power was delegated to local intermediaries. The cohesion of the whole was guaranteed by equal treatment for different religions and communities in the public space.

The Maurya dynasty conquered the short-lived Nanda Empire, which had struck such fear into Alexander the Great's soldiers. Under their reign, India was unified roughly following today's borders with the addition of Pakistan. Around 267 BCE, the third Maurya Emperor, Ashoka, who was fascinated with the Alexandrine model, claimed the throne and converted to Buddhism. Legend has it that this conversion was provoked by a sudden bout of compassion after the bloody and merciless conquest of the kingdom of Kalinga. Henceforth, according to hagiography, Ashoka is said to have devoted himself to easing suffering in this world. He set up charitable institutions, defended vegetarianism and abandoned sacrifices, forbade hunting, and established a significant infrastructure of roads and reliquaries to encourage pilgrimages. He also sent ambassadors to all the Greco-Asian kings all the way to Egypt and Macedonia. He initiated Buddhist missions in all directions, encouraging the various currents of Brahmanism to do likewise. In the following centuries, most of Southeast Asia converted to one of these two competing Indian religions, Buddhism and Hinduism. The eastern part of Iran, and then central Asia along the silk road trading routes, also saw large numbers of conversions to Buddhism.

Indian prosperity was by no means disconnected from this, because it was the merchants rather than the state-sponsored religious missionaries who seem to have successfully spread these beliefs. India produced rice, sugar, textiles, spices, and precious stones for export. Yet the Maurya Empire was not able to impose a currency standard despite the quantities of gold it managed to drain through trade. Its failure to do so can be seen in the Roman money that exchanged hands in the wealthy ports of the northwest, such as Barygaza (today Bharuch); large amounts of Roman coins have been unearthed by archaeologists in these areas.

The size of the empire, disorganized every year during the

monsoon rains, led to a decentralization. When Ashoka died in 232 BCE, the regional kings quickly seized power, kept hold of their tax revenue, and began legislating independently. The port cities and private shipowners prospered.

In central Asia and in Iran, a new kind of military power emerged based on cavalry. Chariots had seen their period of glory and then declined in the face of increasingly heavily armed and armored infantry operating in more and more cohesive formations. Greek hoplites, Macedonian phalanxes, and Roman legions were stages in this process, which can also be seen in the evolution of armies in India and China. The world of the steppes was an exception. In this region, political organization was linked to the control of cavalry. In the sixth century BCE, the first nomadic confederation emerged, the Scythians, which was followed by similar kinds of "empires." Let us call them empires bearing in mind that they differ from sedentary or settled empires. Scythians, Xiongnu, Yuezhi, Xianbei, Turks, and Mongols—all were nomadic empires. In the early days, they saw themselves more as a shifting power dynamic linked to the charisma of a particular chief who managed to bring several groups together around a shared objective. Their political organization was based on a system of government by delegation. The political formations of this kind that we know of today are those that lasted several generations and interacted with sedentary powers, thus leaving records in the archives of the latter. They were therefore nomadic empires that managed to preserve their territorial homogeneity when one or several of the founding chiefs died.

These people of the steppes shared the same fighting techniques centered around an archer on horseback shooting arrow after arrow and thus avoiding direct combat with the foot soldiers of the enemy. Because of their mobility, they were also able to choose the times and modes of battles. Soon elite corps of riders appeared. Equipped with armor, they were able to force their way through infantry formations. These armored riders were called cataphracts.

Empires, however, whether following the Roman, Chinese, or Asian models, had two main priorities. First, engage in warfare only as a last resort (because the deployment of armies is expensive, and a major defeat may be irreparable) and second, encourage merchants to prosper in their businesses (so as to be able to tax them). Roads were therefore essential, and they would outlive the empires themselves. The fate of the world would be played out on the margins of this commercial system on the steppes of central Asia and Arabia (chap. 7), where there were people who were in contact with this prosperity without participating directly in it.

But before the margins asserted themselves, a second systemic collapse (after the first that occurred at the end of the Bronze Age; see chap. 4) led to the destruction of the Roman Empire and the Han dynasty in China.

A Succession of Crises before the Apocalypse

Having now arrived at the very beginning of the Common Era, it is time to take stock of our environmental impact. The demographic growth of humanity was substantial. The global population was now around 240 million humans. It had doubled in less than a millennium and would now increase more slowly. At the dawn of the Common Era, China, Rome, and India each covered one quarter of humanity. By 1000 CE there would be three hundred million people on Earth, one third of them Chinese.

Monkey has a significant impact on the environment. Civilization, which had fueled our species' ability for coordination, was responsible for this. Deforestation due to the expansion of farmlands and grazing lands and the use of charcoal to fuel kilns for ceramics and metalwork all left increasingly obvious archaeological traces. Everywhere that agricultural expansion occurred by cutting and burning, the fertilization was only temporary, and the soil was soon left barren. In the Mediterranean basin and a large part of East Asia, demographic growth produced soil erosion. This led to a rural exodus among farmers, with those remaining obliged to find ways to cultivate their land more intensely to feed an increasing number of city dwellers. This phenomenon can be seen during the reign of Alexander. The Greeks migrated en masse toward Asia in the wake of their army's conquests simply because they could no longer produce sufficient food in areas where erosion had left only bare rock.

Aware of what was at stake, people progressively implemented methods to fight against erosion. They added organic matter to the soil, used crop rotation, maintained hedgerows and ditches, built terraces and irrigation systems. This meant that North Africa, the eastern Mediterranean, and the Ganges and Huang He valleys could produce very large amounts of grain, but at the cost of an intense amount of labor.

Human health remained fragile. In Eurasia, city life facilitated the development of pathogens that wreaked havoc in the intestinal flora of the urban populations; typhus, for example, decimated Athens in the fourth century BCE. Killer epidemics of smallpox, Ebola-like viruses, mumps, and many others struck periodically. Populations progressively developed an immunity to many of these diseases, which, although costly in demographic terms, little by little relegated these mass killers to the rank of childhood diseases, but only for those humans whose ancestors were repeatedly exposed to them.

The Han Empire and the Roman Empire were long lasting, although they faced significant upheavals. Both empires benefited from a climatic optimum, a period of mild temperatures favorable to agriculture. Although they were able to maintain their hold on power, their trajectories were quite different. Their downfalls were separated in both time and space, yet they both sent shockwaves throughout Eurasia.

The Han Empire collapsed first, at the end of the second century CE. It had already come close to collapse between 9 and 25 CE. A nine-year-old child was proclaimed emperor, and one of the concubines of the previous monarch, who had become regent, entrusted the power to her nephew Wang Mang. He had the young heir poisoned and took control of the throne. Scathingly critical of the decadence of the Han, Wang Mang advocated the legacy of Confucianism while implementing policies inspired by the philosopher Mozi (see chap. 5). He began by breaking up the large agricultural properties and distributing them to the peasants before going back on this three years later when the nobles threatened civil war. He blocked the mechanisms of free trade, nationalized several basic products, devalued the currency, and forced public servants to provide interest-free loans to the poor. Wang Mang openly favored the people in order to force the rich aristocrats, who continued to support the surviving members of the Han dynasty, into submission. Nature was not on his side. In the years 2, 5, 9, and especially 11 CE, the Huang He experienced exceptional flooding. It broke its dikes several times, wiping out thousands of villages and their fields. Famine raged, stoked by the disorganization that resulted from chaotic reforms. Dying of starvation and convinced that this chaos was a sign that Heaven had retracted its mandate from the "usurper," peasants protested throughout the countryside. United around the Red Eyebrows rebellion (so named because the rebels painted their eyebrows red) allied with the nobles who supported the defeated dynasty, they made short work of the armies of the new emperor. Wang Mang was decapitated in 22 CE. After three more years of fighting, the Han dynasty returned to power.

From 165 CE, epidemics and famines further weakened the Chinese state. The Xiongnu nomads—themselves chased off their lands by another tribal confederation, the Xianbei—increased pressure at the borders. But the worst danger always comes from within. Peasant revolts broke out, often structured around Taoist movements that were becoming more institutionalized (the Yellow Turbans rebellion, the Five Pecks of Rice rebellion), seeking to establish a "Great Peace" that would benefit the peasant masses. A fifty-year-long civil war ended in the disintegration of the Han Empire.

Against a backdrop of a slight global cooling marked by a period of drought-ridden summers, a series of fatal dominoes would fall for Eurasia. As the Xianbei nomads pushed back the Xiongnu, the latter divided into two groups that the West would call the Huns. about 250 CE, having moved south, the Hephthalite Huns overthrew the Kushan Empire founded in today's Pakistan. Three decades earlier, their Persian neighbor, the Parthian Empire, had collapsed under the weight of the Roman legions. The Western Huns moved up to the borders of the Roman Empire, causing the various peoples that the Romans called "barbarians" to flee west.

Although Rome was struggling at the same time as the Han dynasty, it did not collapse until later. Between 165 and 180, CE, the Roman Empire saw the outbreak of a killer epidemic, probably smallpox. The Antonine Plague wiped out between one tenth and one quarter of the population. The dense network of silk roads seems to have been the source of this disaster: along with merchandise and religion, they also involuntarily carried germs. The third century CE was a period of civil war and the incursion of Germanic peoples, whose leaders sometimes dreamed of joining the power and comfort of the empire and sometimes of stripping it of its wealth to satisfy their troops. Roman leaders reacted by increasing the size of their armies, mobilizing ever more resources.

Although the fall of the Roman Empire has been widely discussed since Montesquieu, it is difficult to pinpoint a decisive cause, although many conservative historians today blithely make the analogy between ancient barbarians and modern refugees. The barbarian elites invaded the western part of the empire from Gaul to Tunisia and replaced the former masters. This process was accompanied by a marked Christianization of Western Europe, which made the church a major power. The Eastern Roman Empire, on the other hand, would survive for another ten centuries around its capital Constantinople.

The imperial model, although it survived here and there, was eclipsed for several centuries. But the trade that these empires established persisted. Road and sea transport forged stronger and stronger links throughout Eurasia. And universal religions would triumph once again in the creation of their own empires.

7

AFTER SUMMER COMES WINTER

After the fall of the Han Empire in the third century and the Western Roman Empire in the fifth century, the Islamic world went through a planned agricultural revolution and China a period of preindustrialization. But from the thirteenth century, the Little Ice Age, the Mongol invasions, and the Black Death redealt the cards of power. Monkey suffered but still grew stronger.

"The past resembles the future more than one drop of water resembles another."[1] The idea that history is an endless cycle of flowering and decay was developed at length by the scholar Ibn Khaldun (1332–1406), and it is a good reflection of the general tendencies of the millennium after the fall of the Roman and Han empires. Although they may have believed themselves eternal, these colossal state formations were in fact merely temporary. But the ideologies that prospered in their protective shadows, Roman Christianity and Chinese Confucianism, outlived them.

The Fate of Empires

Ibn Khaldun was a Muslim, a universal thinker, and a man of action. He lived in a troubled time. He could see that in seven centuries of existence, Islam had already outlived three great empires and a multitude of minor ones, all of which ended up as dust. From this perspicacious observation he derived a "universal" law according to which all empires go through three phases: emergence, zenith, and decline.

Khaldun described empires as entities controlled by social

forces. At the center there was an elite overseeing a prosperous state that was only able to ensure lasting peace by disarming and pacifying its subjects. Around this center are mobile social groups that are accustomed to deprivation, such as nomads or herdsman, for example. Tribal life is governed by a single rule: solidarity. The Arabic word for this is *asabiyya*. Survival is so difficult that everyone must grit their teeth and help their neighbors so that the group survives. This solidarity, *asabiyya*, is also valid in wartime. These groups were in permanent conflict with each other. Every male member of society had to know how to ride a horse (or a camel) and how to fight. *Asabiyya* can also be understood as military solidarity, an ethos of combat, as much as a community co-existing under the governance of the tribal elders who incarnate moral authority.

In the first of Khaldun's phases, the *asabiyyas*, established but peripheral communities, look hungrily toward the heart of the empire that becomes weaker day by day, relentlessly depriving its subjects of their skill as warriors. Then one of the *asabiyyas* becomes bolder. It uses a combination of coercion and negotiation to dominate the others. It then overthrows the imperial power, replaces it, and founds a new empire.

In the second phase, the conquerors develop a taste for luxury, reserve military power for themselves, and placate their subjects with food. A generation passes. The tough conquerors of yesterday grow fat and lazy. New *asabiyya*s loom on the horizon. The empire holds them back with threats. And then when the power balance shifts, they try to buy time. They try to provoke divisions between the *asabiyyas* so that they weaken each other. The central power pays some to be mercenaries while others try to take advantage of the center when they see it weakened.

In the third phase, inevitably one *asabiyya* again becomes bold enough to challenge the center. It defeats the old empire and creates a new one. Back to square one.

Ibn Khaldun is perhaps the first modern thinker. He saw that we cannot rely exclusively on sources for writing history. Instead we must focus on analyzing the social contexts in which the events reported by previous chroniclers took place. Only then can we see which facts are distorted, what is exaggerated, and what is left unsaid. Khaldun's logic and powers of observation even influenced natural sciences. He even implied—in the barest hint, so blasphemous was the idea—that humans may be a close relative of monkeys!

His thoughts on imperial cycles and the violence of communities are a good summary of the dominant trends that took place in the millennium that followed the fall of the Roman and the Han empires. Despite periods of dissipation, this was a time marked by the perpetuation of religions, the frailty of empires, and the maintenance of long-distance trade.

And Europe Became Christendom

To understand the origins of Islam, we must go back to the fall of Rome. This coincided with the global rise of Christianity. The story begins in Jerusalem in 70 CE. The Judeans had risen up, and the future Roman emperor Titus laid siege to their capital, Jerusalem. During the pillaging, the temple was burned to the ground. Judaism was left in crisis. The temple, which was the only place where legitimate sacrifices to Yahweh could be made, was destroyed. As a result, Judeo-Christianity began to rise. Under the influence of Paul, a Jew who had studied Greek, the Church of Christ had already begun active proselytism directed at non-Jews. Some Roman citizens, perhaps one in ten, converted to Christianity or Judaism between the first and third centuries. Although the elites of the two religions cursed each other, there seems to have been little distinction between the two for ordinary people.

But at the beginning of the fourth century, everything changed. After eighty years of military strife and internal divisions, the Roman Empire fortified itself around Emperor Constantine (r. 306–337). This emperor established a strong currency and a capital worthy of his name. He favored Christianity, perhaps hoping to be able to rely on the bishops' networks to consolidate his empire's administration. He imposed an orthodoxy on the new religion. Those who resisted were persecuted as heretics. The power of the church was reinforced by the Emperor Theodosius (r. 379–395), who proclaimed Christianity the state religion. The last pagans were hunted down, among them one of the few female philosophers and scholars, Hypatia, who was lynched in Alexandria in 415 CE.

The fall of Rome meant the end of comfort.[2] In the Italian town of Pompeii, life was cut short in an instant under the ash of Vesuvius in 79 CE. Archaeologists digging through the petrified remains of the food market (something like a food court we might see in any mall today, with stalls selling food to be eaten on-site) have made a list of all the things the populace, the *vulgum pecus*, could buy there: smoked fish from Spain, fresh sea urchins, olive oil, Indonesian spices, nuts, even giraffe meat. Citizens in this affluent town could eat what they chose and eat well. This was a luxury not permitted in many periods of human history. The archaeologist Bryan Ward-Perkins wrote, "a north-Italian peasant of the Roman period might eat off tableware from the area near Naples, store liquids in an amphora from North Africa, and sleep under a tiled roof."[3]

Why did this civilization collapse? It would be tempting to say that it altered its environment beyond a point of no return, and perhaps that is true. An exhaustive reading of the books dedicated to the reasons for the fall of Rome provides a list of more than two hundred possible contributing factors. And yet the environmental causes (global cooling,[4] overexploitation of resources, lead poisoning, deforestation, etc.) seem decisive. But complex societies are resilient, and their collapse is always linked to a number of social, economic, and environmental reasons (chap. 14).

Did Rome really "collapse"? At the end of the third century the empire was divided in two to ensure the survival of the whole. Its most politically stable wing, the Western Roman Empire, came unstuck in the fifth century after the barbarians repeatedly conquered Rome. Its most economically dynamic wing, the Eastern Roman Empire, survived for a thousand years more until the fall of its capital Constantinople to the Ottoman Turks in 1453. For some historians, Rome did not actually fall apart.[5] The barbarians who flooded across its borders dreamed of having access to Roman comfort. They replaced, or mingled with, the elites. The barbarian kingdoms were obsessed with rebuilding the Roman Empire, as Charlemagne partially and temporarily did in the ninth century.

From the end of the seventh century, the bishop in Rome established himself as the pope, the father of Christendom, a moral authority able to show political powers a thing or two. The papal network was based on monasteries as an institution. The idea had come from Egypt in the third century: persuaded that the end of time was nigh, fanatics went off to fast in the desert, and a generation later they grouped together to form communities. After another two hundred years, the monastic institution set up rules, hierarchies, and objectives of evangelization. Monastic orders were the spearhead of the Catholic Church. Catholic means "universal" in Greek and fully expresses the ambition of the institution.

The monastic orders accompanied the social evolutions of the Middle Ages in Europe. Founded in the early tenth century, the Order of Cluny provides a perfect illustration of these processes. It was based on the teachings of St. Benedict, which stipulate that a monk must divide his time equally between work and prayer. The church convinced its flock (nearly all Western Europeans with the exception of Jews, who benefited from a temporary period of tolerance) that they should donate all their wealth to the church when they died to pay for prayer that would ensure the everlasting peace of their souls. In so doing, it confiscated capital and prevented individuals from inheriting substantial estates. The Abbott of Cluny became immensely rich, paying thousands of monks to devote themselves entirely to praying for the souls of the dead. At its height, and in a gesture of upmost excess, this order built the Cluny III Abbey in France, a magnificent monument 187 meters (613.5 feet) long, then the largest building in Christendom.

This ostentatious wealth was not to everyone's liking. Some tried to live according to the evangelical ideals of poverty and often did not refrain from criticizing the church. These dissidents met with one of two fates. Some, like the Cathars and the Bogomils, were persecuted as heretics. This was also the fate of the Waldensians, followers of Pierre Waldo, a twelfth-century exemplary Christian from Lyon who gave all his assets to the poor. Other dissenters were reclaimed by the church, which evolved along with them. The Cistercians, born out of a conservative split within the Order of Cluny, lived in poverty and adopted the ethic of hard work. Their abbeys were built in wild places, and Cistercian monks worked to drain marshes, clear forests, and triumph over nature. This order evolved along with the demographic growth that occurred between the ninth and thirteenth centuries, the result of the Medieval Warm Period.

Founded in the thirteenth century, the mendicant orders (the Franciscans and Dominicans in particular) moved into towns where populations were growing steadily as a result of renewed prosperity. These new orders gave themselves the task of helping the ever-increasing poor. In so doing they relieved social tensions and, incidentally, were responsible for the Inquisition, holding special court hearings to rein in heretics.

In the West, Monkey donned a papal tiara. Sources of spiritual power worked hard to make their influence felt by sources of temporal power, although these struggles would have high points and low points. More than once, the emperors of the Holy Roman Empire or the kings of France managed to impose their will on St. Peter's successor.

In the East, the situation was very different. The Eastern Roman Empire had righted itself and began expanding again under Justinian (r. 527–565 CE). This emperor proclaimed himself to be the "lieutenant of God on Earth." He compiled a civil code that was to influence all the later legal corpora of Romance-language-speaking Europe. He retained firm control over the Levant and was about to reassert his authority over Italy, the Maghreb, and Spain—in other words, reestablish the Roman Empire—but . . .

The Year of the Elephant

The first historically documented plague struck in 541 CE, under the reign of Justinian. Brought from either sub-Saharan Africa or India, the pandemic devastated Egypt before spreading along trade routes toward Palestine, Syria, and Constantinople. The Byzantine capital, with its half a million inhabitants, apparently lost more than one third of its population during the summer of 542. The town was losing citizens at a rate of ten thousand people a day, and the scholar Procopius of Caesarea lamented that "slaves remained destitute of masters, and men who in former times were very prosperous were deprived of the service of their domestics who were either sick or dead, and many houses became completely destitute of human inhabitants. . . . And work of every description ceased, and all the trades were abandoned by the artisans, and all other work as well, such as each had in hand. Indeed, in a city which was simply abounding in all good things starvation almost absolute was running riot."[6]

The plague had already contaminated all the countries surrounding the Mediterranean. It followed roads and navigable waterways in all directions. The Sasanian dynasty in Persia, sworn enemy of the Byzantines, was infected in turn. The two superpowers were now sick. The plague was cyclical. It struck in bursts and then returned when people thought they were safe. When a third of the population had died, some trades had been decimated to the point that corporations fell apart and knowledge was lost and could not be passed on. Once a critical mass of members of the elites were dead or had fled to safety elsewhere, the whole administration collapsed at a time when organization was more critical than ever. Society had to wage war on the epidemic and its funeral procession—the dead had to be buried and the living had to be fed. Famine and social strife spread. Those in power, who claimed to govern in the name of celestial delegation (the Christian god in Byzantium, the Zoroastrian god Ahura Mazda in Persia), were considered responsible for the chaos. People like to have someone to blame. A solution was found to calm them: war. Focusing popular resentment on an external enemy is always an attractive option to try and distract people from a simmering internal revolt.

The trouble spread to the Levant and Arabia. Legend has it that in 570 CE, Abraha, an Ethiopian Christian conqueror from the Jewish kingdom of Himyar (in today's Yemen), began raiding in the direction of the pagan site of pilgrimage at the oasis of Mecca. The pillar of his army was an enormous elephant named Mahmud, which apparently stopped on the border around Mecca and refused to go forward. Abraha's soldiers saw this as a bad omen and apparently forced their commander to turn tail. They then clashed with Persian soldiers who decisively quashed Abraha's ambitions by briefly annexing Yemen. This legend of the white elephant is above all a way for the Islamic tradition to proclaim one of the miracles attesting to the exceptional time around the presumed birth year of the prophet Muhammed.

Another sign of the times, 570 CE was also the year that the Great Dam of Ma'rib, a feat of ancient engineering that had made agriculture possible in part of Yemen for centuries, was breached. All western Asia was struck by a conjunction of catastrophes, wars, and epidemics. It found itself the subject of disputes between warring tribes reduced to pillaging their neighbors to ensure subsistence that commerce and agriculture could no longer provide. The Byzantine and Sasanian empires became concerned and sent troops or made alliances in Arabia, the Levant, and Egypt. The situation worsened. Chaos spread through the region. In the three first decades of the seventh century, the Byzantines launched two major offensives against Persia, destroying many towns and threatening their enemy's capital Ctesiphon. The Sasanian Empire took revenge by laying siege to Constantinople and completing the devastation of the region. Most of the rural communities that the plague had spared would never recover.

Which of these two was the most destructive? The incessant war or the recurring plague? Inhabitable zones that had been carved out of the desert by monastic communities or villages thanks to irrigation techniques were now totally emptied of their inhabitants. The main sources of grain for the Italian peninsula—central Syria, the Negev desert in Israel, and Libya, all previously fertile regions— were reduced to sand. The combination of these two scourges, war and plague, led to the massive advancement of the desert, the collapse of commercial trade routes, and the depopulation of urban centers. It was not until the end of the nineteenth century that the population of Maghreb and Mashriq (the Arab world to the east of Egypt) were restored to their levels before 540 CE!

Certain authors, most notably David Keys, see these events as being partly produced by a series of substantial volcanic explosions.[7] Identified by traces in polar ice and dendrochronological measures (tree-ring dating), these eruptions are said to have occurred in 536, 540, and 547 CE.[8] Similar in scale to the volcanic eruption of Tambora (chap. 14), they apparently led to the "Late

Antique Little Ice Age." During the second half of the sixth century, average global temperatures dropped by 1°C (1.8°F) or even 2°C (3.6°F). This had wide-ranging environmental consequences, including the destruction of crops and the decline in population due to the ongoing ravages of the Justinian plague. But this cooling is also thought to have caused an increase of rain in the Arabian Peninsula, making the steppes more fertile, encouraging demographic growth, and increasing camel populations used for military activities. The weakening of the Byzantine and Persian empires and the subsequent expansion of Islam may therefore have been facilitated by our planet blowing off a little steam.

With or without the intervention of a *deus ex volcano*, the Byzantine and Persian empires were exhausted in this fight to the death. Each of them had armed large Arab confederations to fight across the vast deserts of the Levant while alienating the local populations with their taxing, pillaging, raping, and murdering. In 632, Muhammed died after conquering Mecca. He had spread a prophetic reform of the biblical religions, Judaism and Christianity, over the western half of the Arabian Peninsula. This new religion was called Islam. In the eyes of the Arab tribes, Byzantium and Persia must have seemed like fruit ripe for the picking.

After two years of consolidating alliances, the new religion federated the Arab mercenaries of the Levant. In 636 the emerging power defeated the Byzantine Empire in Yarmouk and the Sasanian Empire in al-Qadisiyyah. This latter battle involved a large Persian division of armored elephants, which almost wiped out the armies of Islam. The pachyderms were probably bought from Indian kings and clearly benefited from improved technology in terms of armor. They seemed now to be invulnerable to arrows. But three days of fighting and the sacrifice of the elite troops who threw themselves against the animals to cut the girths holding the towers for the archers on their backs eventually made the animals run away. The elephants took out most of the Persian army in their panic. Eight years later, all Iran fell to the Muslims. Arab troops laid siege to Constantinople in 674 and then again in 717 but did not manage to take it.

This second siege was interrupted by the plague, which made any conquest impossible. Indeed, the Qur'an forbids Muslims from entering contaminated zones. It also orders believers not to run from the disease if they do encounter it, unlike the prescriptions in the Christian world, where advice attributed to Hippocrates (460–370 BCE) advised people to "leave, quickly, far away, and for a long time." This advice probably contributed to limiting the spread of the disease in the new Muslim Empire.

This empire, which founded the Umayyad Caliphate in 661 CE, aimed to conquer the whole of the known world. It made Damascus its capital, halfway between Persia (already conquered) and Byzantium (which it hoped to eventually control). Initially, the Arabs did not seek to convert the new populations they controlled. These people, Christians in the Levant and in the Maghreb, were tired of Byzantine and Persian excesses and welcomed the conquerors without much resistance and probably with relief. Their new masters took the place of the old elites, even involving them in power. They needed help administrating such vast spaces and preserving diplomatic ties. Often conquests were administered from new cities situated on the edges of older urban centers. Not living among conquered populations meant greater security in case of rebellion. Moreover, fighting in urban areas was often the worst scenario that an army could envisage. The small spaces were good for ambushes, the enemy knew the town much better than they did, and weapons that were powerful in open battle (cavalry and later artillery) were much less effective here.

The Umayyad Empire rose dazzlingly fast. In 711, at the height of military operations, its armies simultaneously captured northern India (today's Pakistan) and Spain. As a result, it controlled an uninterrupted expanse of territory between these two regions. In 750, an eschatological rebellion took control to the benefit of a new dynasty, the Abbasids. The Umayyads had reigned from Damascus and were open to Byzantine influences. But the Abbasids had their capital, Baghdad, built in the east, near the old Babylon and Persia.

The Arab Agricultural Revolution

In 751 an unlikely battle took place on the edge of the world. The armies of the Chinese Empire of the Tang dynasty confronted large expeditionary forces sent by the new Muslim Abbasid Empire. This clash of the Titans was initially a stalemate. Legend has it that the battle, in the Talas River basin in today's Kazakhstan, lasted five days.[9] However, the defection of Turkish mercenaries from the Chinese army to the Muslim army, rallied by the Arab Turkish reserve, was decisive. For the Tang, this would be the farthest they ever pushed west. For the Muslims it would mean a change of paradigm. Henceforth, expansion would no longer mean military conquest but conversion through trade.

The Abbasid Empire encouraged trade and religious freedom. Theological and intellectual schools preserved the knowledge of ancient philosophers and enriched it with their own thought. Astronomy, optics, and hydraulics all made spectacular progress fueled by the spread of the invention of paper. At the Battle of Talas, the Arabs took a large number of artisans prisoner, including some who specialized in the making of paper. The Chinese army was an army of colonization and had moved westward with all the personnel required to found a Chinese city. The first paper production workshop to be built outside China or India was thus set up in Samarkand shortly after the battle. Another was built in Baghdad in 794, then in Cairo and Damascus around 850, and in Sicily and Morocco (Fez) in the early eleventh century. By the twelfth century, these Chinese techniques had reached Muslim Spain. This opened up a profitable market to the rest of Europe until Italy discovered the secrets of paper production in the thirteenth century. A fundamental requirement for bureaucratic administration, paper was now part of the arsenal of the Afro-Eurasian bureaucracies (India had been producing it since the seventh century).

Technology circulated, but so did biological organisms. The expansion of Islam went together with the spread of agronomic knowledge reaching Asia, Africa, and Europe. The historian Andrew M. Watson has baptized this phenomenon the Arab Agricultural Revolution.[10] For the first time, the space between Morocco and Pakistan was unified under a single power structure. Even better, it was unified by a single language: Arabic. The emergence of a shared Islamic market facilitated human interactions, including the spread of a considerably large number of plants. Watson, writing in 1974, listed rice, sorghum, watermelon, lemon, orange, artichoke, spinach, sugarcane, eggplant, mango, coconut, banana, and cotton, among others. Although specialists have since concluded that rice, cotton, and sugarcane were already developed in Persia and the Levant before the Arab invasion, Watson's thesis still holds: the Abbasid Empire intensively and intentionally spread the use of these plants over its whole territory.

There were two propitious forces involved here. First, the environmental context was favorable because the adaptation of new crops took place in equivalent latitudes. Although the environment was quite different in the Indus and Tagus valleys, the number of sunny days and temperatures were similar, which meant that it was relatively easy to adapt plants to these new lands. The second decisive element was that a powerful empire planned this adaptation. Its policy involved sending public servants in all directions. These travelers had instructions to compile botanical catalogs, take note of the ways in which the local people cultivated the plants, and collect numerous seeds and cuttings. Private landowners, often responsible for vast domains, were also involved in this policy. Technology was an important aspect too. Irrigation techniques used all around the empire were analyzed, adapted, and spread. Dams, canals, reservoirs, aqueducts, and qanats (underground passages found in central Asia and Persia carrying water with access shafts used as wells) became more frequent. Hydraulic techniques were improved with various machines, which had often been long known but only used in Mesopotamia, Egypt, or Persia. Now *norias* (waterwheels), *shaduf* (a counterpoise lever used to lift buckets of water), Archimedes's screws (a tube bent spirally around an axis used to raise water from wells), cisterns, and windmills all became common from India to the Maghreb. Irrigation meant being able to extend the harvest period on a large scale and produce a surplus. The spread of new plants and hydraulic technologies led to a demographic boom. The zones of the desert that had been abandoned during the Justinian plague period were not reconquered, but the oases did expand somewhat, which meant trade routes could be put in place with sub-Saharan Africa, some of them even crossing arid expanses of the Sahara. Population density increased considerably around the towns experiencing rapid growth. Although it was on the margins of the empire and rebelled against Baghdad's authority, Spain also benefited from these technologies. Córdoba became the largest city in eleventh century Europe. Its libraries housed some four hundred thousand works.

This horticultural melting pot was the source of what is now thought of as Mediterranean cuisine. It was these exchanges that meant olive oil (from the Levant) could be used to flavor eggplant (from India). The great civilization of knowledge and comfort was now an Arab one. Cotton clothes were more comfortable to wear than those made from rougher wool or linen. They were also better at resisting epidemics, because cotton has fewer irregularities and therefore is not so prone to sheltering disease-carrying parasites like ticks and fleas. Sugar was a sought-after spice and a driving force for the economy (see chap. 9). During the Crusades, Europeans were in raptures over perfumed roses, sugar sorbets, fine delicate cloth, and the quality of so-called Damascus steel made from iron ore refined in the blast furnaces of India. Knowledge and power now both lay in the East. The Islamic world, probably inspired by precedents in India, began to build the first universities and hospitals. The West would adopt and develop these models after the twelfth century.

In the early ninth century there were only two towns in the world with more than a million inhabitants: Baghdad and Chang'an, the respective capitals of the Abbasid and Tang empires. Constantinople, the seat of the Byzantine Empire and Córdoba, in Muslim Spain, came second, with three hundred thousand and four hundred thousand souls, respectively. In Western Europe, however, the largest towns struggled to reach fifty thousand inhabitants.

Chang'an and Bagdad also shared the fact that they were new and entirely preplanned cities designed according to a functional and orderly plan dictated by astrology and ancient knowledge. They were structured around wide avenues, large gardens, and all the kinds of infrastructure required for distributing water and evacuating the waste of their immense populations. These two megalopolises were planned around their architects' vision of the universe: a circular structure for Baghdad; a square for Chang'an divided into 108 segments (108 being the sacred number of Buddhist harmony). These cities saw themselves as urban paradises built for all eternity. In Chinese, Chang'an means "immortal city." Although through usage the name Baghdad, which means "God's gift" in Persian, dominated, the original name for this city was Madinat al-salaam, which means "City of Peace" in Arabic. Chinese accomplishments in urbanism were copied by Korean and Japanese kingdoms, who built their new capitals, on a lesser scale, in the quadrilateral style of Chang'an. Heian-Kyō, which is now Kyoto, means "Peaceful Capital" in Japanese.

The Mysterious Golden Rhinoceros[11]

It is impossible to precisely date the agricultural use of the tropical forests in equatorial West Africa, but humans have been clearing them for three thousand years at least. They grow yams and other vegetables there, hunt elephants and game, and raise cattle. They also collect fruit, leaves, honey, medicinal plants, and wood for building. In so doing, they act on the composition of the forest itself, cutting clearings, spreading seeds, and encouraging certain types of plants. At the beginning of the Common Era, the demographic growth of these populations led to a great migration through Africa now known as the Bantu expansion. These movements were accompanied by the spread of certain types of agriculture, animal husbandry, and technology (particularly ironwork) from the forests of the equator to the savanna of South Africa. These populations came into contact with the San, who had lived in southern Africa for 120,000 years. Part of the San adapted to this new lifestyle with a bit of farming and lots of livestock; they became the Khoisan. The San who remained hunter-gatherers were progressively pushed into inhospitable lands, withdrawing, at the very end of the process, to the Kalahari Desert.

The South African landscape was transformed. The Bantu peoples burned the savanna so that herds could eat the new grass.[12] However, the environmental impact of these herds would be tempered by the devastating effects of certain parasites like the tsetse fly.

The Bantu migration stopped somewhere around 1000 CE. The introduction of iron all over Africa meant that farmlands could be expanded, and that led to significant demographic growth similar to what Europe and Asia were experiencing at the same time. Many Indian plants, such as banana trees, contributed to this process. As the Bantu nomadic herders adapted their economy to the wooded hillsides of central Africa, the farming peoples who lived in the Great Lakes area—who seem to have developed ironworking techniques independently more than two thousand years ago—joined with them to create a new culture. This cultural ensemble then migrated toward the eastern part of Africa, and demographic expansion pushed back the forests and swamps. Great cities and kingdoms were built. The most well known of these is the Great Zimbabwe, which maintained trades links with Arabian and Indian ports between 1250 and 1500.

An eight-hundred-year-old golden statue of a rhinoceros, extracted from a royal tomb at Mapungubwe (in South Africa) in 1934, is testimony to the little-known trade history between India and Africa. This gold-plated animal has only one horn. Does it represent the symbol of the region, the powerful but two-horned African rhinoceros, or is it an image of its Indian cousin, which has only one? In any event, the figurine symbolizes the dynamism of the Mapungubwe kingdom in the thirteenth century, an interface between the inland peoples and the Arab sailors. Large quantities of gold as well as ivory, ostrich eggs, and animal skins flowed out from the center of the continent and were exchanged for glass beads and Chinese porcelain. The trade networks of the Indian Ocean therefore stretched from South Africa to the coasts of China.

The developmental boom of the Islamic world was fueled by the example of the ex nihilo construction of Baghdad. It swallowed up an astronomical amount of wood. The precious mangroves, the coastal forest that guaranteed marine and land biodiversity and protected the shoreline from erosion, had already suffered from centuries of overexploitation in Persia. Those in East Africa were now also taxed. The Baghdadi colossus also swallowed up slaves to feed its sugar plantations (chap. 9) in Mesopotamia. They were bought in eastern Europe, Africa, or elsewhere and sold along the same trade routes as spices, precious metals, ointments, medicines, silk, and porcelain. As proof of the expansion of these trade networks, black slaves, apparently from Africa, were found in the ninth-century Chinese imperial court.

Monkey periodically enjoyed the luxury of civilization. But this would have increasingly significant impacts on the environment.

Song of China: From Riches to Rags

After four hundred years of fragmented rival kingdoms punctuated by a devastating plague epidemic in 610, China was again unified under the Tang dynasty (618–907). Initially, the empire experienced a period of prosperity. The imperial competitive exams, developed in the early seventh century, established an administrative cast that shared both a writing system and a set of values, consolidating an empire that was linguistically highly diverse. The Buddhist monasteries were then in full growth and sought to spread the teachings of Buddha far and wide. They improved their techniques for wood engraving and developed the first printing presses, producing millions of pages of text using woodblocks from as early as the seventh century. Chang'an became the cosmopolitan nexus of world trade. Merchants from all over Asia could be found there. Muslims, Nestorian Christians (considered heretics in the West), Manichaeans, or Buddhists; they all arrived via the very active silk roads or the bustling ports of South China.

From 755 a series of civil wars shook the empire. Its borders were threatened by the appearance of powerful Turkish confederations—a direct reaction to its dominion over the people of the steppes—and by the expansionist kingdom of Tibet. The weakening of imperial authority went hand in hand with the reinforcement of local powers, which encouraged urbanization, trade, and industry. The wreck of an Arab *dhow* that sank in the first half of the ninth century was found off the coast of Belitung (Indonesia). It contained large volumes of gold and silver coins and tens of thousands of Chinese porcelain objects, a sign that China mass produced these ceramics for export to the Islamic world. China had the monopoly on porcelain, ceramic so fine it was transparent.

As the prophet had forbidden Muslims to eat from gold or silver plates, Chinese porcelain became a status symbol for Islamic elites. They were shipped by the maritime silk road via the Indian Ocean. These elites favored blue porcelain, obtained from cobalt. The best cobalt was mined in northern Persia. Brought to China by the traders of central Asia, it returned home baked into porcelain, having gone all the way around the Himalayas.

At the end of the first millennium, both sides of the Pacific were struck by a series of significant droughts, perhaps linked to the whims of El Niño.* The Mayan civilization abandoned its religious cities, which required a large production of corn, a plant that is particularly vulnerable to repeated droughts. They rapidly adapted to the new climate situation. Their warrior society evolved toward a less hierarchical tribal system with more autonomy and less conflict in which groups produced what they needed in their gardens. The Tang Empire, more centralized and more complex, was not able to preserve its unity in the face of fluctuations between sweltering summers and freezing extended winters. It came apart at the beginning of the tenth century under the combined influences of the climate, peasant rebellions, and invasions from the steppes in the north and the mountains in the west.

The Song dynasty restored order from 960 CE. Benefiting from the economic growth of urban centers achieved under the Tang regime, Chinese civilization reached an impressive level. The amount of land devoted to agriculture was doubled, and its productivity was doubled, too, thanks to the spread of technological innovations and the state-planned introduction of new rice species from Indochina that provided two annual harvests.

Both North China and (for the first time) South China experienced very significant demographic growth, which meant that the Song dynasty became a superpower. In the eleventh century, a third of the world's population was under the direct authority of the Song dynasty! The Chinese economy was boosted by the printing of the first banknotes, guaranteed by the state authority and printed in factories that employed hundreds of workers. This was the first economy in the world to receive more than half of its urban taxes in paper money, and its copper coins were used throughout East Asia.

More and more blast furnaces were built drawing on technology that had been used for 1,500 years. Black coal was mined now, replacing charcoal to achieve the high temperatures needed to make steel. An industrial revolution had begun (see chap. 13).

Archaeologists have uncovered thousands of standardized iron objects from this period in China including farm tools, weapons, religious artifacts, and money. Statues weighing several tons were cast in one piece. The Chinese were the first to use metal cooking pots, because shaping iron into sides that are both thin and flat requires sophisticated metalworking skills. The foundry records available demonstrate how dynamic this market was. Near the iron and coal mines there were seven hundred foundries in northern China producing a flow of cast-iron objects that were then traded by merchants. Trade benefited from state infrastructures such as imperial roads and, above all, the Grand Canal, a feat of engineering punctuated by one hundred canal locks that crossed China from north to south. Chinese merchants founded dynasties, accumulated capital, and transmitted their wealth to their descendants. The urban bourgeoisie, made up of wealthy landholders and rich merchants, grew substantially in the wake of increased demand for manufactured goods. The large number of artisans created a market supplying luxury goods. Porcelain and silk were exported all the way to Africa. China was (already) the workshop of the world.

According to Robert Hartwell, the first historian to document the extent of this technological revolution, the Chinese produced 114,000 metric tons of cast iron per year in the eleventh century![13] By comparison, England produced only 68,000 tons in 1788. Although this often-quoted figure may be exaggerated, China nevertheless produced a considerable volume of cast iron and steel.

Yet by 1150 CE, this prosperity was no more than a distant memory. The blast furnaces had been abandoned. The prestigious capital of the Song dynasty, Kaifeng, was in ruins. What happened? There are three possible explanations.

The first suggests that the greed of public servants nipped the seed of capitalism in the bud; officers confiscated goods from merchants and then found themselves unable to keep businesses afloat.

This explanation may be a projection of current-day concerns about the People's Republic, in which entrepreneurs are subject to the goodwill of an all-powerful and nitpicking bureaucracy. In reality, the Song dynasty delegated many of its prerogatives in public authority to the private sector. For example, the grain stocks, essential to preventing famine, were frequently traded between states and merchants.

The second explanation is environmental. The demographic expansion that took place between the seventh and the tenth centuries began in a context that was often described as idyllic, with clear rivers and luxuriant forests, by the authors of the period. Toward the west, upstream of the Yellow River (Huang He) basin, massive deforestation led to erosion. Felled trees could not hold back the soil, and their roots no longer drew up the water from the torrential rains. The steppes were home to increasing numbers of cattle and horses that destroyed vegetation. The dust moved toward the east, both on the wind and in the water. The Huang He saw more frequent and more devastating floods. From one per decade in the ninth century, these floods increased to three per decade at the beginning of the eleventh century. The final blow was dealt in 1048. The Huang He broke its banks and dikes and flooded the Hebei plain, altering the watercourse and drowning one million lives in its fury.[14] In the century that followed, similar catastrophes struck eight more times. A vicious cycle had begun: arable land became rarer while the farming population continued to grow. Peasants had to go farther to find land and were forced to terrace hillsides and cut down more trees. In short, they had to aggravate the conditions that had already brought them disaster.

The third explanation involves the invasions from the north. The Jurchen flowed across the borders from Manchuria and took control of Kaifeng in 1127. The Song dynasty moved their capital to the south and dedicated all their resources to resisting the continual pressure from the people of the steppes. In 1128 as the Jurchen prepared to cross the Huang He to pursue their conquests, the Song intentionally released the dams so that the flood would drown the opposing army. (The Chinese National Army of the Kuomintang would use the same tactic in 1938 to slow the advance of the invading Japanese army, drowning half a million of their compatriots in the process.)

Fleeing both the Jurchen and the floods, the Song withdrew farther south. Between this exodus and the rising population, records suggest that between 980 and 1290 CE the population density of villages at the mouth of the Yangtze River was multiplied by twenty. New land had to be found, and it was carved out of the sea using polders and dikes. The blast furnaces and coal mines were in the north, now in the hands of the northerners. The latter divided up the conquered territories, became more settled, and founded new dynasties, imperial courts, and bureaucracies modeled on those of the Song. But their economy never attained the complexity of that of their predecessors. In this context there was no point investing in the steel industry. The mines and furnaces progressively fell into disuse.

The second and third explanations were probably combined. Floods led to famine, peasant uprisings, and a weakening of the state. The invasions were the final blow to a process that, had it not been interrupted, might have led the history of humanity down a different path.

Genghis Khan or the Wrath of Allah

"I am the punishment of God. If you had not committed great sins, God would not have sent a punishment like me upon you."[15] In 1220, this speech was delivered to the terrified inhabitants of the formerly prosperous Muslim city of Bukhara (today in Uzbekistan). Their city had fallen in just a few hours, and they had become the prisoners of these seemingly invincible riders from the east. At the head of these riders was Genghis Khan, who delivered this apocalyptic sermon (via interpreters) from the *minbar* (pulpit) of the Great Mosque. The following day, his soldiers would pillage the city and organize the deportation of its inhabitants, in the same way as they herded their animals, to be used on the front line of the next attack. Forced, upon pain of death, to attack neighboring cities to weaken the opposing soldiers before the final assault, the inhabitants of Bukhara must have indeed thought the wrath of Allah was upon them.

Born under the name Temujin in 1162, it took forty years of fighting for him to unify the Mongolian tribes and have himself proclaimed "universal ruler" or "Chinggis Khan" (a title that is generally anglicized Genghis). These nomads were seasoned fighters trained by hundreds of years of war with China. Protected by armor and equipped with powerful composite bows, they fought on horseback. Each took with him—according to his means—a stock of dried meat and between four and twelve horses so as to always have a fresh charger for an attack. Highly disciplined and freed from numerous logistic concerns, this mounted hoard proved to be the most mobile army ever until the internal combustion engine was invented. It overcame the states in the north of China, the Western Xia, the powerful Jurchen Jin dynasty, and farther west, the Qara Khitai Empire. It also destroyed the Khwarazm Empire, previously in full expansion across central Asia. It shattered the Turkish-Arab armies and razed Baghdad to the ground in 1258. All the land between Poland and Korea was either devastated or under their control. They provoked a rout among the West's best horsemen in Hungary while fighting thousands of kilometers from their base. The Mongols conquered an empire of more than 30 million square kilometers (about 11.5 million square miles) in less than a century.

These conquerors remained convinced that wealth lay in grazing land. Their conquest of central Asia involved the destruction of the infrastructures for irrigation. Rumors of massacres, knowingly amplified to terrify the enemy and push peasants to leave, spread like a cold winter wind. Their goal was to return farmland to pasture so that their horses would have somewhere to graze, and they could to push their conquests still farther. But they also established the basis for the administration of the empire. They constructed many bridges and caravansaries where merchants (caravans) could rest as well as an extensive network of posthouses improving on the Chinese postal system and developing it at an intercontinental level. These indispensable auxiliaries of power would soon be imitated in both the Islamic world and in Europe.[16] Under the reign of

Genghis Khan, the Mongols imposed a system of writing, adapted from the Syriac alphabet, originally used by the Uyghur Turks. This was a hybrid writing system written vertically, like Chinese. During the thirteenth century it was used to codify the state apparatus required to administrate the immense lands that had been conquered.

When Genghis Khan died in 1227, his children divided up his empire into four parts: China, Russia, central Asia, and Persia. These were passed down to his grandchildren. The new elites administered their goods as their former masters had done, encouraging trade and agriculture to increase tax revenue. During this period of *pax mongolica*, it was said that a virgin carrying a golden platter on her head could have crossed Europe and Asia alone and in safety. This peace enabled a large number of travelers to move around Europe and China, including merchants such as Marco Polo, diplomats such as the Flemish Franciscan William of Rubruck, or Giovanni da Pian del Carpine as well as mercenaries, artisans, and others.

It was only in 1272 that the Mongols of Kublai Khan, supreme leader of China, broke through the defenses of the Song Empire and invaded the south of China. They ended the rival dynasty in 1279. But Genghis Khan's heirs did not know how to preserve what they had won. By the early fifteenth century there was not much left of the greatest land empire the world had ever known. All that remained were stories, somewhat embellished by travelers like Marco Polo.

The speed of the Mongol imperial cycle is a perfect illustration of Ibn Khaldun's thesis. Khaldun lived through the black death, the second plague epidemic (chap. 8), which spread through Asia along the roads made peaceful under the Mongol Empire. This philosopher could not help but be struck by the frailty of empires compared with the permanence of religions, clearly evident in the myriad prayers and pilgrimages of people praying that God would deliver them from the plague. Ibn Khaldun was one of the first modern thinkers shaped by the rationalism of Aristotle and Averroes. But he was also witness to the dusk of an empire. The Mongolian invasion of the Islamic world almost destroyed Sunnism.

With the condemnation of Averroes and his rationalist philosophy as heresy in 1197 and Ibn Taymiyya's push for conservative Islam—seen as a precursor to Salafism a century later—Islam chose to rely on a fixed orthodoxy. The interpretation of the Qur'an (*ijtihad*) was now set. The *ulema* (Islamic scholars) intended to impose their dogma on secular powers much like the pope in Christendom.

The Impact of Hot and Cold on History

As the techniques for reconstituting past climates improve, it is now possible to trace the substantial temperature fluctuations that occurred. In addition to the period of cooling that occurred around 1200 BCE that marked the end of the Bronze Age (chap. 4), traces of significant cooling are also visible in the Northern Hemisphere around 3150, 2200, 1950, and 1750 BCE, dates that are all marked by wars and the ends of dynasties in the Levant. It is important to bear in mind that collapse is always the result of several factors and that social, political, and economic components are fundamental. A dynamic society generally survives climate disruptions, although its foundations may be rattled, and this is particularly true for the agrarian societies of the past up until the mid-nineteenth century. Two other cold periods, around 850 BCE and 200 BCE, wiped out crops, and yet societies proved quite resilient.

These periods of cooling—which often took the form of consistently harsh winters that destroyed seeds when they should be sown and dry summers that reduced harvests—seemed to occur at regular intervals. They may be linked to the conjunction of different factors, the three most important being the rhythms of solar activity, the frequency of El Niño episodes, and major volcanic eruptions. The cycles of cooling were associated with a decline in monsoon rains. As a result, China, Southeast Asia, India, Persia, and East Africa experienced more or less intense aridification, even causing the Nile to dry up at least twice. Continental Asia cooled, but at different rates in different regions. There were droughts in some places and repeated torrential rain in others. The cold often forced the nomads off the steppes. In the third century CE a period of cooling saw the nomadic confederations move toward the rich lands in China, India, Persia, and the Mediterranean, which were already considerably affected by internal unrest. Similarly, the collapse of the Tang dynasty at the end of the ninth century was correlated with an intense period of cold apparently associated with an increase in volcanic activity. At the same time, at the other end of Eurasia, the Carolingian dynasty broke up. Between the two, the authority of the Abbasid Empire became meaningless as the victorious Turks, converted to Islam, dictated their instructions to the caliphs.

In the thirteenth century, a major eruption of the Samalas volcano in Indonesia initiated a long period of cold.[17] This brought the Medieval Warm Period to a close and tolled the end of the golden age of agriculture that had seen the spread of agricultural techniques through Eurasia and the growth of the world population to three hundred million people around 1000 CE. The new era, known as the Little Ice Age (chap. 11), ran roughly between 1275 and 1820 (specialists are still debating the exact beginning and end dates).

According to the historians Timothy Brock and Mark Elvin, this cooling period apparently began in China in the very early thirteenth century, roughly a hundred years before it struck Europe. This early cold wave may explain the agitation of the people on the steppes and may have encouraged the Mongol expansion: their grasslands were crippled by the cold just as their population and their herds had benefited from four centuries of consistent growth.

From the year 1315, the cooling was well established across the Northern Hemisphere, and it has been documented across the globe from the 1370s. Poor harvests seem to have spread from west to east from the 1320s. Europe was plunged into an unprecedented famine between 1315 and 1319, while Egypt and Persia went through one between 1318 and 1320. India also saw famine a few years later, as did China and central Asia. Many states had suffered from years of war, and Eurasia was considerably weakened. The final blow came in the form of the black death that unfurled from the steppes from the 1340s.

The French historian Pierre Chaunu jokingly said of world history that the Ottomans wanted to control the world but could not, the Chinese could have but did not want to, and the Europeans did because they wanted to and they could.[18] We have seen that China could have controlled the world. The idea probably would have pleased its leaders. But the northern invasions and the floods of the Huang He prevented it from pursuing this possibility. The Islamic world (as a geographic zone) could have controlled the world, but the *ulema* opposed it. Europe was able to control the world because geography protected it from the nomads and its secular powers were able to silence religious ones. But above all Europe benefited from the unexpected assistance of a formidable cohort of allies at a crucial time: plague, smallpox, measles, chickenpox, flu, malaria, and yellow fever. All these helped it to take the Americas.

8

BIOLOGICAL HAZARDS

Christopher Columbus stumbled upon the Americas while looking for Asia. The European explorers carried germs that decimated the Indigenous people and thus faced little resistance in taking control of the two continents. But why does Monkey live with germs?

1542. A group of Spanish conquistadors were lost in the jungle and starving. They had left newly conquered Peru and crossed the mountains following the paved roads laid long ago by the Andean civilizations. They went down steep slopes into increasingly luxuriant jungles and then stumbled on an immense body of water. They had lost their four thousand Andean porters, most of whom had deserted when their leaders were fed to the dogs for not having found the promised cinnamon, one of the many spices sought after by Europeans. The Indigenous people promised the Spanish anything and everything, having learned, at their expense, not to resist the whims of a conquistador. The dogs had since been made into stew, but this had not prevented 140 of the physically weakened Spaniards dying from starvation and infection. Upon reaching the river, the survivors separated into two groups. Fifty-seven of them used their remaining strength to build a small boat, a full-bellied vessel that then drifted off on the river. The swell of the river became unimaginably strong and the river wider still. They called it the *Rio Grande*, Big River. Later, it would be recognized as a tributary of the Amazon, the most powerful water course in the world. Its average flow is equivalent to the flow of the six next largest rivers in the world combined.

That night in June, the Spanish soldiers sailed for some islands assuming them to be uninhabited. To their surprise they were met with two hundred extremely maneuverable dugout canoes each carrying twenty to forty finely dressed Indigenous men. These islands concealed a small town with a central square, community buildings, family houses, and a high wall providing protection from outside. The town was surrounded by fields and woods, and nearby there was a swamp where turtles were bred for food.

The Jungle of Myths

Luckily for the Europeans, these Indigenous people were not hostile. There would be many occasions in their long expeditions through the forest when they were not greeted so warmly. But for the moment they wept tears of relief. Their hosts were joyful and enthusiastic, welcomed them with the sound of trumpets and drums, and fed them abundantly. At least this is what the Dominican priest Gaspar de Carvajal wrote in his records of the exploration led by Capitán Francisco de Orellana.[1] This meeting, along with a few others, provided the foundation for the myth of El Dorado, the land of gold. People as prosperous as this, whom the European explorers had barely glimpsed, had to be the subjects of a rich and powerful king. The problem was that subsequent travelers never found them. They had entered history, were documented in the blink of an eye, and then disappeared without a trace.

Today the archaeologists who explore the immense Amazon Basin, which is almost the size of all forty-eight contiguous US states, hope to put an end to another myth: that of the "virgin rainforest." Everywhere they dig, they uncover traces of villages; fragments of their history buried in the soil. Brazil's *terra preta* (black earth) is the most fertile humus on the planet and once covered a surface roughly equivalent to two and a half times Great Britain. We can tell that this *terra preta* was made by humans because of the large amounts of pottery found buried in it. These lost people of the Americas were the best composters the world has ever seen. This *terra* is extremely rich in carbon, phosphorous, and nitrogen resulting from a skillful combination of wood charcoal (from burning off and cooking fires), animal waste (particularly from fish), and vegetable scraps. The microorganic activity in this soil is so strong that the substrate, when there is enough of it, can renew itself—it is one of the only soils on Earth that can produce crops intensively and not become barren. The oldest of these soils has been around for three thousand years.

Putting together these clues, we can estimate that when Columbus arrived in the New World in 1492, the Amazon Basin was home to somewhere between five and ten million people—roughly the same as the population of Spain and Portugal at the time! There were tens of thousands of villages and hundreds of towns, all connected by a dense network of roads. Toward the west, where the alluvial plain becomes hills, great dikes and artificial lakes covering hundreds of hectares were used for cultivating edible aquatic plants. This was a civilization that was dense, innovative, and well adapted to its environment. Through their hunting and farming practices, these humans had had an impact on the biodiversity of the forest for at least seven thousand years. They were adept at large-scale agroforestry, where food plants are grown at the foot of trees. This soil would otherwise have been barren (and is becoming so again today as whole swathes of the Amazon rainforest are cut down to cultivate genetically modified soybeans), but enriched with *terra preta* it became an agronomist's utopia. Human impact went further still, through what can be described as selective horticulture. During their daily movements, Indigenous tribes selected the trees that were of use to them and broke or removed the saplings of undesirable species. Up until the sixteenth century, the Amazon rainforest was a well-tended garden—and then it went wild. The vast zones of *terra preta* are the graveyards of a civilization.

Monkey had an impact on the forest as a whole, perhaps from the very beginning. It had just begun to grow when the first humans moved there in the wake of the last major ice age.

So, can we say the Amazon is an artificial forest? That would be too extreme. This selective horticulture was low intensity. The Amazon is an old-growth forest, but it is also anthropized—it grew under the distant supervision of its gardeners. But if it was home to such a flourishing society when Capitán Francisco de Orellana arrived, why is it now home to only a few tens of thousands of hunter-gatherers? Why did the Spanish and Portuguese explorers who set out to find these societies never find anyone? Why did they find only a jungle that had become impenetrable, hostile, in which lakes silted up, paths became overgrown with vegetation, and the far-dispersed villages avoided all contact? Why did Europeans long believe, and some continue to claim, that this is "virgin rainforest," or at least almost uninhabited?

The answer lies in a single word: pandemic.

War against the Invisible

Tenochtitlan was a radiant city. In the early sixteenth century, the Aztec capital was a masterpiece of urbanism. Like all the major cities in Mesoamerica, it was built around a ceremonial center including religious pyramids and various buildings embodying the power of the elites who lived in them. Around this center, as far as the eye could see, stretched large avenues full of gardeners and artisans tending all sorts of food gardens and orchards. At that time, Lake Texcoco was 5,000 square kilometers (about 1,930 square miles) of water covered with *chinampas* (floating gardens) constructed from rafts made of reeds or built into the river mud among a dense network of canals and drains that transferred water to all the nearby cities.[2] This conglomeration was possibly home to three hundred thousand inhabitants, the Mexica, the rulers of the Aztec Empire. When the Spanish arrived in late 1519, they were stupefied by the wealth of the city. They wrote of how stunned they were by the diversity and abundance of the produce sold in the markets. The people wore embroidered clothes and intricate jewelry, the quality of which surpassed the conquistadors' dreams. All free men in Tenochtitlan received an education, particularly in mathematics. In the sixteenth century, the Spanish were still using Roman numerals instead of the decimal system, which had been invented by the Indians and spread by the Arab world in previous centuries. Like the Arab, Indian, and Italian merchants of the time, the Mexica had mastered the use of zero. In fact, the Mesoamerican calendar was then the most precise in the world.

The Triple Alliance, led by the Aztecs, had reigned as the unchallenged power over Mesoamerica for a century. Its territory covered most of the cultivated land of what is today southern Mexico and Costa Rica. It had a system of writing inherited from the Olmecs, paper (based on bark fibers whitened using lime), and even bound books (codices), invented in Mesoamerica independently from what was happening in Eurasia. It was a highly organized state able to wage permanent war and lay down the law to some fifteen or twenty million subjects. And the conquistador Herman Cortés took control of it with only six hundred soldiers! The Spanish saw this as a miracle. In their eyes, an invisible hand had intervened to assist them; they saw it as Providence, or God. In reality, the assistance came from creatures so small that they were indeed invisible: bacteria, viruses, protozoa, and other pathogenic microorganisms that fall into the category of "microbes" (from the Greek "small life").

At the time, the Americas were inhabited by approximately sixty million people, from Alaska to Patagonia, according to the most reliable estimates. Some authors stretch this figure to one hundred million or more. Others remain faithful to the colonial estimates of a century ago, putting the number of Indigenous peoples at around ten million, sometimes less.

During the sixteenth century, more than 90% of the Indigenous populations of the Americas were wiped out—more than one eighth of the world's population. It was the only time since the Agricultural Revolution, ten thousand years ago, when the slow but progressive increase in the density of carbon gases in the atmosphere, measured in polar ice cores, declined significantly. This was temporary, but it is symptomatic. Tens of millions of Indigenous people were dying and therefore no longer burning plants. Their death would only be legible much later, written over the long term, marked by the thousand-year-long curves of increases in atmospheric carbon. From the end of the seventeenth century, the rate of carbon dioxide in the atmosphere increased again, a phenomenon that contemporary research correlates with the modification of American ecosystems.[3] All over North America, the disappearance of Indigenous people and the invasion of earthworms (see chap. 9) transformed the forest understory into a jumble of scrub.

The proliferation of opportunistic plants absorbed carbon from the atmosphere. It probably also helped make the seventeenth century the peak of the Little Ice Age,[4] the coldest period the world has known in the last three thousand years (chap. 11). New plants brought from Europe proved to be formidably invasive in American biotopes. They were also able to spread so successfully because humans were not around to weed them out. This reached the point where in the late seventeenth century, this dense forest began to burn with a spate of gigantic natural bushfires. The fires pushed the carbon rate back up, contributing—along with the Industrial Revolution—to the beginnings of the new era known as the Anthropocene.

It is difficult to imagine what it might mean for a society to see 90% of its population wiped out in the space of a human lifetime. Spanish records tell of large numbers of suicides among the Indigenous populations. The survivors had seen their civilization crumble, their chiefs and their elders perish at a time when authority and experience were more necessary than ever to maintain social cohesion. They had seen the agony of their families, lovers, and children. They lost the knowledge they had and forgot the techniques they once used. Confronted with violence they could not have imagined, they tried to escape forced labor. It seems that many of them let themselves die from despair. Apathy goes some way to explaining the success of Spanish evangelism among the Indigenous peoples of the Americas. They had lost everything. Their gods had shown themselves powerless, and the promise of their new masters' future paradise may have been the only glimmer of hope they could see.

Let us return to late 1519. Naturally, Cortés had clear military superiority. He even had cannons. But these were like the elephants in the armies of ancient times, unwieldy, dangerous, and rarely efficient, primarily serving to terrorize the enemy in the first encounter and not thereafter. The Spanish soldiers were armed with steel weapons; they had horses and powerful, ferocious war dogs, but these animals were vulnerable. The conquistadors' advantage faded when outnumbered by fierce and determined warriors.[5]

Taking advantage of their opponents' weaknesses, Cortés proved himself to be a skilled diplomat. The Mexica lived in a state of permanent predation over their neighbors. Like all Mesoamericans, they believed that the sun would only rise each day if it were fed with blood. They obtained gallons of this sacred hemoglobin in so-called flower wars in which individual value was exalted, the objective being to capture an adversary in hand-to-hand combat in order to sacrifice him to the gods, who welcomed this offering "like hot bread out of the oven" according to the expression used in sacred chants. During great celebrations, priests, wearing the skin of their victims, would take the captives to the alters at the top of the great pyramids and cut out their hearts. The Spanish were repulsed by these sacrifices even though they were themselves planning to implement a system of genocidal slavery (chap. 10).

This curious geopolitical situation explains how Cortés, helped by information provided by his Indigenous lover La Malinche, managed to federate an army of auxiliaries who wanted to throw off the Aztec yoke. But these accumulated benefits, even along with a healthy dose of luck, are not enough to explain the outcome of this war.

Having entered Tenochtitlan, the Spanish protected themselves by taking the *tlatoani* (literally "he who speaks") hostage. This was Moctezuma II, the supreme chief of the city-state of Tenochtitlan and its vassal cities, who combined both religious and military leadership. But he died even as the Aztecs rose up against the Spanish. The latter executed their prisoners and managed to retreat, fighting one against a hundred. It is commonly said that their escape was a miracle. The outcome of this "night of sorrows" (la noche triste)—July 1, 1520—can be better understood once we know that Tenochtitlan had fallen prey to smallpox and that the epidemic had already killed between one quarter and half the population in less than two weeks. The young people, those who fought, were the most affected, the most physically weakened. Their strong bodies fought most energetically against the pathogens, forcing them to produce antibodies and exhausting themselves in the process. Cortés conquered a town in which not one of the three hundred thousand Mexica who had lived there two years earlier was left alive. In Mexico, as elsewhere, the Spanish built the cities of the New World on the smoking ruins of the Old.

A Feeling of Apocalypse

If the Spanish saw the invisible hand of God helping them in their plundering and conquering of towns in which the population had been decimated, the Aztecs must have thought the gods had deserted them. Their chief, the *tlatoani*, was as clever as Cortés and probably better informed. His network of spies was as sophisticated as that of any European state in the same period, and like all those of his time, he used astrology and magic. To justify their defeat at the hands of the Spanish, his sorcerers told him they were powerless against these creatures from another world. He probably wondered whether the end of the world was nigh, whether Cortés was an avatar of Quetzalcoatl, a divinity whose return from exile was prophesized as leading to the ravaging of Aztec lands while the world descended into chaos.

The *tlatoani* was no different from other rulers around the world at the time. All put great faith in what we would now call the occult. This feeling of apocalypse was present in most cultures of the sixteenth and seventeenth centuries. The Chinese saw dragons foretelling the destruction of their empire, European Christians burned witches and heretics by the tens of thousands and would shortly slaughter each other in interminable wars over religion. Tension and conflict were constantly reignited because these cultures were eschatological; they were fighting over the end of the world. Both Catholics and Protestants were profoundly convinced that God wanted them to destroy each other in order to purify the world to prepare for Christ's second coming, believing Him to have promised one thousand years of happiness for the righteous.

Christopher Columbus himself, who bought the Old and the New Worlds into contact in 1492, is a good illustration of this kind of pivotal figure. He had one foot in the beginnings of modern time, which was individualistic and humanist (chap. 11), and the other in the past, which was dominated by the great universalisms of the Moral Revolution (chap. 5). His exploration was funded by the Spanish monarchs who had just expelled Jews and Muslims from their kingdoms. Over previous decades, the Spanish elites had aspired to *limpieza de sangre*, aiming for a social body only made up of supposedly "pure-blood" Christians (at least third-generation Christians). The year 1492 therefore saw Jews and Muslims faced with a choice: exile or conversion. Some left, others stayed. Among those who stayed, some sought to preserve the faith of their ancestors. Marranos (crypto-Jews) and Moriscos (crypto-Muslims) publicly converted to Christianity but practiced their former faith in private. They were hunted down by the Inquisition, which in Spain had passed from the hands of the papacy to the monarchy.

It was in this context that Columbus set sail toward the west. Inspired by Marco Polo and other travelers who had "discovered" the lavishly wealthy Chinese Empire two hundred years earlier, he would likewise "discover" not what he set out to find but the highly populated Americas. The Spanish monarch had entrusted Columbus with the task of converting the "Great Khan" to Catholicism, the Europeans imagining he was still the leader of China.[6] In his copy of Marco Polo's *Travels* he made notes next to the passage where the author tells of how the roofs in "Cipangu" (Japan) were covered in gold. He hoped to bring back large amounts of this precious metal. This wealth would allow him to fund a crusade to take back Jerusalem from the Muslims, an event that he thought would provoke the second coming of Christ. Somewhat narcissistically, Columbus apparently saw himself as a biblical prophet, a "new Elijah."[7]

But his project was based on scientific calculations from antiquity. Columbus initially asked the King of Portugal to fund his exploration of the vast unknown of the Atlantic. The Portuguese refused. They were pursuing their own explorations down the African coast, which they hoped would come to fruition. Soon they would reach the southernmost tip, which would allow them to go around the continent and sail unimpeded to the twin hubs of the global economy, India and China. For the moment, the powerful Ottoman Empire controlled transcontinental trade. By land or sea, the silk roads necessarily passed through lands it controlled: the

Levant, Turkey, and, from 1517, Egypt and Arabia. The city-state of Venice was a compulsory intermediary for Europeans hungry for Oriental goods, particularly pepper, and it built its fortune on its privileged connections with Muslim states.

As they were already the masters of an alternative route, the Portuguese declined Columbus' offer. But the disagreement was also a scientific one. This was the beginning of modernity. That the earth was round was no longer a subject of debate among the cultivated European elites of the Renaissance. The question was its circumference. Columbus was convinced that the geographer Claudius Ptolemy had overestimated when he calculated, in the second century CE, that circumnavigating the earth in a straight line meant traveling 34,000 kilometers (about 21,126 miles) before returning to one's starting point. Columbus believed the planet was smaller, somewhere around 20,000 kilometers (about 12,427 miles) in circumference. The scholarly advisors to the Portuguese monarch based their advice on the measurements of the mathematician Eratosthenes, who measured the circumference of the globe in the third century CE using a clever geometrical calculation. He estimated it was around 39,400 kilometers (about 24,482 miles), which is surprisingly accurate, because we now know the figure is somewhere around 40,075 kilometers (about 24,901 miles). The Portuguese did not deny that China could be reached by the Atlantic, they just said it was too far, inaccessible. Columbus believed it was much closer. He wanted it to be very close, to the point of contradicting classical scholars. He was convinced that he would be able to reach the Orient in less than forty days sailing toward the west.

Although Columbus was wrong in this scientific debate, his reasoning was nonetheless modern. He spent his whole life obstinately arguing—despite increasing evidence to the contrary—that the land he had "found" to the west was India. The Spanish conquered these islands that he had stumbled on by chance. When they arrived, the Antilles were inhabited in the southeast by the Caribs and in the northwest by the Taíno, who were then experiencing a demographic boom. After the Spanish arrived, these Indigenous peoples began to die, victims of a cocktail of lethal germs to which they had never been exposed (smallpox, typhus, and pneumonia, along with plague, flu, hepatitis, measles, scarlet fever, tuberculosis, diphtheria, rubella, and mumps). Those who survived the first epidemic were weakened and soon had to confront a second, then a third. The chances of escape were slim. The Spanish took the rare survivors as slaves and killed them through forced labor.

The first impact of this microbial shockwave was the annihilation of the Taíno of the Greater Antilles followed by the Caribs of the Lesser Antilles, which deprived the Spanish of their workforce. It also led to the collapse of the powerful political structures of the Aztecs and the Incas and left their people under the yoke of the conquistadors. It reverberated throughout the Amazon, provoking depopulation both there and in the seminomadic lands in the north of Mexico. In the space of a century, tens of millions of Indigenous people died.

But why were these people so vulnerable to disease? Why had they not already been exposed to them like most Europeans had?

What Rabbits Can Teach Us about Microbes

There is little evidence of illness in the Americas before the arrival of Columbus. Only syphilis may have gone from the New World to the Old, and this is debated. This was despite the fact that there was significant population density in Mesoamerica, in the Andean plateau (where the Inca Empire consisted of twelve to fifteen million subjects), and in Amazonia (where settled civilizations had developed systems of intensive farming). This difference between the New and Old Worlds can be explained by two things. The first was the sterilization zone of Siberia. When people originally arrived on the North American continent, they had to go through Siberia during the Ice Age. Most of the germs they inherited from our prehistoric ancestors did not survive the cold.

The second, which was also decisive, was that there were hardly any animals left to domesticate after the massacre of the megafauna when humans first arrived (see chap. 2). The only domestic animals in the pre-Columbian Americas were llamas, alpacas, guinea pigs, and chickens in the Andes (probably brought by travelers from Polynesia). Dogs (from Asia, possibly arriving with the first wave of human settlement), turkeys, and even rabbits (for a time) were found in Mesoamerica. But there were so few of these creatures that the close proximity of people and animals observed in the Old World was not an issue in the New. By contrast, people in Afro-Eurasia lived in constant interaction with pigs, bovines, chickens, and other potential hosts for pathogens. The transmission of germs to humans was facilitated by frequent mutations, and humans in turn transmitted them to animals. This process created specific ecosystems, even communities of disease. Mirko Grmek developed the notion of pathocenosis to refer to such communities, but this concept can be expanded to include the situation of equilibrium not only between diseases but between diseases and their host organisms.[8] The people living in such pathocenosis and repeatedly exposed to a cohort of germs adapted and developed resistances.

This can be illustrated by rabbits. A dozen pairs of these animals, tamed in Western Europe during the Middle Ages, were introduced into Australia in 1859 as sport for hunting. Half a century later, a lack of natural predators meant that their six hundred million descendants had invaded the island continent. Because they ate the pastures and left them full of holes (potentially catastrophic for grazing animals) they became a threat to the survival of the sheep and cows that were so fundamental to the country's economy. They were even more disastrous for indigenous species, which were weakened through competition. The introduction of foxes, which was supposed to combine the pleasure of sport (fox hunting) and utility (regulation of rabbit populations) only made the biological catastrophe worse. The foxes preferred to prey on native species, which were not familiar with them and therefore easier to catch. So a virus was suggested as an alternative predator for the long-eared nuisance.

Myxomatosis was a disease that appeared in Uruguay at the end of the nineteenth century. For the most part, rabbits all over the world had adapted to it except for in Australia, where they had had no genetic contact with other populations. In 1950, biological warfare was declared. Blinded, weakened, feverish, 99% of contaminated rabbits died. Their agony lasted less than two weeks; the disease should have wiped out the whole species. But rabbits breed like rabbits. The second generation exposed to myxomatosis was more resistant, and only 60% died. In the third generation only 20% died. The rabbits who survived had genes selected to resist the disease, but that alone was not enough to explain the phenomenon. For the first time in history, we saw a virus that adapted its virulence to its hosts. The most powerful germs killed their victims before they had time to reproduce and were therefore eliminated by natural selection. It might seem counterintuitive, but in a context in which victims are defenseless, only the weakest microbes survive.

Many illnesses developed over several thousand years in the Old World. Measles, mumps, and scarlet fever; they were mass killers. But natural selection led to the survival of genetically resistant human populations as well as the least virulent microbes, those that could happily reproduce without killing their hosts. These epidemics became childhood illnesses, striking young humans as a kind of prophylactic. This equilibrium allowed these children to grow into adults who would survive contact with the illness while allowing germs to continue to comfortably reproduce.

Despite himself, Monkey had also made a pact with the invisible. This guaranteed us a tense but lasting coexistence with microbes as long as they did not cross certain boundaries.

These states of equilibrium, in which pathogenic agents prosper—weakening but not killing their hosts—constitute various pathocenoses. When an epidemic occurs, it means a disease has moved out of pathocenosis, the zone of equilibrium in which it can prosper alongside the population it lives off. When they are exposed to a new disease against which they have no defense, humans die en masse. Justinian's plague for example, which came from either Asia or Africa, was the first outbreak of plague to ravage the Old World. As we saw in chapter 6, it wiped out one quarter of the Chinese population and one third of the population of the Mediterranean basin in the sixth century.

Remarkably similar conditions were in place in the sixteenth century when the two colonizing forces of the Iberian Peninsula set their sights on the Americas. Cortés put down the Aztec confederation for the Spanish, and between 1531 and 1537 Francisco Pizarro, adopting the same strategy, did the same to the Inca Empire, which was ravaged by smallpox.

As for the Portuguese, they brutally inserted themselves into the secular trade networks operating in the Indian Ocean. They created a line of fortified trading posts along the African and Indian coasts, and from there, taking advantage of the technological superiority of their armored ships, imposed their law over the seas. In 1511, they captured Malacca (near today's Singapore), an immensely prosperous city-state and a hub of world trade. However, in 1519–1520 Tomé Pires, an apothecary-turned-diplomat, failed when he tried to conquer China.

Pires's project was not as foolhardy as it may seem at first glance. The Aztec confederation was hardly less formidable than the Chinese Empire, and Pires's methods were closely modeled on those of Cortés. But China was adept in Western combat techniques and had in fact invented most of the technology on which they relied. Most importantly, the Chinese had long since been exposed to and inoculated against all the illnesses circulating in the Old World. The initial attempts at conquest were so short lived that they went unnoticed. The Chinese pushed back the Portuguese apparently without even having registered that this was an attempt at annexation.[9]

The Great Plague of Humanity

Justinian's plague was a terrible microbial shock for Europe. It started around 541 CE and lingered in the Mediterranean basin until 767 CE, when it disappeared. Over the centuries people lost their hard-won physiological resistance once they were no longer regularly exposed to the disease. During this time the plague fell back to its centers of origin—one in sub-Saharan Africa, the other in northern India. The bacteria slumbered in fleas, infesting rodents' nests. Local people developed superstitions that protected them from contamination: if an animal looks weak or sick, it is cursed and must not be hunted or even touched.

In the thirteenth century, the Mongols set out to take Burma and North India. We can imagine that they burst into this ecosystem in full ignorance of the prophylactic traditions. The plague was carried out of this zone by Mongol riders and spread steadily through the steppes of Asia, from the Pacific coast to Ukraine. It passed from rodent to rodent, spreading outbreaks of the disease.

In the 1330s it struck China. One third of the 130 million strong population was carried away in the space of fifty years, leaving only ninety million people. This was also a time of climate problems, famines, and the Chinese rebellion against the Yuan Mongolian dynasty (chap. 11).

In 1346 a Mongolian army laid siege to Caffa on the banks of the Black Sea. This merchant town was a trading post for the Italian city-state of Genoa. Plague broke out among the Mongols, who abandoned their prize.[10] But the fleas had infested the city. Seven ships sailed from the city, and their crews were soon in their death throes. The harbors that took them in quickly saw death disembark. In 1347 the disease took root in Marseille. In 1352 it reached Poland. In five years, the Black Death struck down around a third of the European population, so that by the end of the fourteenth century the demographic impact had reduced the population to half the size it had been in 1346. The French population dropped from seventeen to ten million, and England lost 60% of its inhabitants. In some places, like Mallorca, as few as 20% survived. The genetic diversity of European people was seriously affected. For reasons that are still unknown, people with O blood type were more affected than others. If, like most Europeans, you do not have O blood type (in other words, you are A or B or AB), you probably owe that to your ancestors who survived the plague.

What happened in a town infested with the plague? The authorities—where they survived—had to implement emergency measures and convince people to collect the dead and bury them in enormous mass graves. They had to arm militia to defend empty shops and houses. They had to try and control people's movements to limit the spread of the disease even as people were trying to escape. Local authorities increased their power—they financed hospitals and appropriated the assets of the dead or of Jews (who were often victims of lynching, accused of having provoked the wrath of God). The decline of the labor force obliged them to abolish serfdom and replace it with free peasants and rented lands.

The Italian city-states were at the cutting edge of economic and administrative techniques, and they perfected certain emergency measures to confront the spread of the disease. They quarantined ships, which proved effective, and this technique was emulated all over Europe. They created passports that certified that the holder had remained outside infected areas. And they also entrusted their fate to the great merchant dynasties such as the Medici in Florence, who shook off the ideological yoke of the church. The time was right. Italy's economic growth was due to these merchants who had adapted Islamic commercial techniques, bills of exchange, trade associations, specific market institutions, and the use of so-called arabic numerals (which in fact came from India), including the previously unknown zero, which enabled the calculation of interest rates.

The plague returned over the centuries that followed. It was less indiscriminately deadly, but mortality rates were still high. The Great Plague of London in 1665–1666 killed one in five of the population. The last great continental outbreak occurred in Marseille in 1720–1722, striking down forty thousand of the city's ninety thousand inhabitants, and in Provence it killed 120,000, nearly one third of the population. Up until now, history has presented the Marseillaise elites as the guilty parties in this epidemic, painting them as selfish and quick to neglect quarantine measures, anxious to sell on a textile cargo from Syria. But genetic analysis appears to show (although this still needs confirmation) that the strain of bacillus was the same as that in 1347.[11] It had been lying in wait for four hundred years, hidden in an as yet unidentified rodent.

The preventive precautions taken by Christian states nevertheless enabled them to control and then reduce the impact of the epidemics. The European population experienced substantial demographic growth that allowed it to colonize the Americas and then impose its rule on the world. Islam experienced the opposite demographic trajectory. The Qur'an recommends not fleeing before epidemics, which slows down the spread of disease but also leads to fatalism: it would displease God to take action against these scourges. Islamic leaders were anxious to prove themselves worthy in the eyes of the *ulema*, the guardians of Islamic tradition. They would not have adopted the voluntarist measures on which European states were beginning to build their power, with the exception of the sporadic measures adopted in the Ottoman Empire. In the West, public authorities demanded that potential carriers of disease be quarantined according to international law, whereas the Islamic world seems to have generally been unconcerned with this. Whatever the reasons, the cycles of plague in the Islamic world were more violent and more persistent, lasting up until the end of the nineteenth century, whereas the disease had disappeared from Western Europe by 1722. A demographic chasm rapidly emerged between the northern and southern shores of the Mediterranean even though the European dynamic was partly rooted in what it had borrowed from the Arab world (agricultural, trade, and social techniques, including the invention of universities).

Fleas and Mosquitoes: Taking Down Empires

In the late sixteenth century, South America was subject to a second microbial shockwave, described by John R. McNeill.[12] To replace the Indigenous workforce that had been wiped out by epidemics, the Spanish and the Portuguese set up the Atlantic slave trade. Up until the fifteenth century they had obtained African slaves via the Arab world, which sold them along the caravan trails of the Sahara, buying this human "merchandise" from their trade partners, Muslim sub-Saharan kingdoms such as the Mali Empire. Henceforth, Portuguese slave traders would be able to collect whole cargoes of slaves directly from the source. Epidemics in Africa prevented Europeans going farther inland, so they bought their prisoners from the local authorities even if it meant arming the latter with steel blades and muskets so they could more easily round up their neighbors. Captured and sold by fellow Africans, bought and deported by Europeans, African slaves flooded into the New World. This trade, initially monopolized by the Portuguese, was quickly adopted by the other European colonial powers.

The deported African captives brought two new pathogens to

the New World: malaria in its most deadly form (*Plasmodium falciparum*) and yellow fever. These mosquito-borne diseases would play an essential role in geopolitics. They are extremely violent tropical diseases, but a child exposed to yellow fever has a much greater chance of surviving than an adult and as a result is immunized for life. For malaria, elements of resistance emerge through contact with the parasite; it is impossible to be fully immunized, but human organisms that are repeatedly exposed are less vulnerable. Mosquitoes were rife in the vast sugar plantations. A particularly robust pathocenosis developed among African slaves who were resistant to the disease but who often died from harsh treatment. Criollos, descendants from Spanish colonizers, and mestizos, of mixed race, were also resistant. For those who arrived from the Old World, not having contracting malaria or yellow fever in childhood proved to be a death sentence.

From 1655, the English dreamed of annexing the Caribbean, and they had the military means to do so. An expedition easily took control of Spanish Jamaica, which was defended by only a handful of fighters. Two thirds of the seven thousand British soldiers sent on this mission died during the wet season. Jamaica remained British, but it would be the only conquest of this kind. The Spanish learned their lesson. Henceforth, they would defend their possessions with a fort and a local militia. The military strength of the English was of no use here. They were forced to lay siege to the fort while the mosquitoes devoured their solders. The fevers finished them off in a matter of months. In assault after assault, the English sacrificed their armies. In 1741, a twenty-nine-thousand-strong British force tried to take the Spanish port of Cartagena in Colombia. It was defended by four thousand soldiers, sailors, and militia entrenched behind the walls. The dry season ended, the first rains came, and with it the mosquitoes. The English persisted, attacking another port. The endeavor claimed twenty-two thousand soldiers, of which 96% died from fever.

Spain lost its American Empire little by little over the nineteenth century. In Venezuela, Colombia, and Cuba, local populations rose up to claim their independence. Thousands of soldiers were sent from Spain to suppress the insurrections, and they suffered the same fate as their British predecessors. In this specific ecosystem, political power fell to those who were the most well equipped to survive malaria and yellow fever.

In the 1890s scholars studied the mechanisms for the transmission of these diseases and identified mosquitoes as the vectors. The first programs for eradication began around 1900. By 1915, there wasn't much left of yellow fever and malaria in the Caribbean. Mosquitoes would never again be a decisive factor in geopolitics in Latin America.

Let us go back one hundred years to the cause of the weakening of the Spanish Empire. The colonies began to rebel when Napoleon invaded Spain, which prevented reinforcements being sent from the continent. On June 24, 1812, acting against advice from his closest advisors, the French emperor decided to invade Russia with 660,000 troops and fifty thousand horses, a *grande armée* of the like Europe had never seen. The Russians fell back using scorched-earth strategies as they went. They burned villages and harvests and filled in wells. Napoleon's soldiers were hungry, and illness was rampant. Covered with flea bites, weakened by hunger and fatigue, almost two thirds of the soldiers died an atrocious death—a veil of mucus covered their faces, their blackened tongues hung out, legs and arms stiffened at improbable angles. Those who could still do so begged their comrades to put them out of their misery. It seems typhus was on the side of the Russians.[13]

When Napoleon entered Moscow in mid-September, he had only one hundred thousand soldiers left. The Russian commandos had left the (essentially wooden) city at the last moment and set it alight in several places. The Russian capital burned for five days. On October 19, having triumphed over the ruins of the city, Napoleon ordered a retreat. But winter caught up with him, with −22°C (−7.6°F) blizzards. On November 28, crossing the Berezina river, the emperor abandoned the last forty thousand survivors of the *grande armée* and returned to Paris. His downfall had begun with a flea bite somewhere in Poland.

Can the Killers Be Stopped?

We tend to think that epidemics were eliminated in the nineteenth and twentieth centuries. This legend has its cast of heroes. The English doctor Edward Jenner (1749–1823) developed the first vaccine against the terrifying smallpox, inspired by inoculation practices that were developed somewhere between India and China around 1000 CE and then spread throughout Asia and the Islamic world from the sixteenth century. Napoleon used this to his advantage by having all his troops vaccinated against smallpox. The Hungarian surgeon Ignace Philippe Semmelweis (1818–1865) demonstrated the importance of hand washing with soap before assisting childbirth, particularly when the doctor had previously conducted an autopsy. The French doctor Louis Pasteur (1822–1895) developed the vaccine against rabies. His Franco-Swiss student Alexandre Yersin (1863–1943) identified the bacillus responsible for plague and prepared the first serum against it.

In 1904–1905, the Japanese army simultaneously achieved two major victories. Much to the surprise of the rest of the world, the first was a military victory against the Russian army in Manchuria—it was the first time in a century that a Western power had lost a war to an Asian one. Japan's second victory went unnoticed, but its consequences were just as important. The Japanese army took an important step in the fight against illness. It became the first army in the world to implement large-scale multiple vaccinations for its troops. For the first time in history, death in war was more to do with enemy combat than germ attack.

And then came the last great epidemic onslaught, the so-called Spanish flu. Spanish? A terrible illness has to come from somewhere else, but this one probably came from North America, France, or China.[14] This extremely virulent and contagious flu, probably associated with pneumonia, spread throughout the world between 1918 and 1920 and carried off between sixty million and one hundred million people. Probably avian in origin, this pandemic killed three to five times as many people as the First World War, especially given that during the latter, famine and disease, particularly typhus in the trenches, were often far more deadly than enemy fire. The Spanish flu also increased awareness that the poor are more vulnerable than the wealthy because mortality rates were very clearly correlated to revenue. The higher one's income, the less deadly the disease. This is an iron rule that stands for all epidemics: being rich always ensures greater chances of survival.

The Spanish flu taught us another lesson too: the more interconnected the world is, the more rapidly epidemics spread. In the fourteenth century the plague took eight years to spread from China to Europe, which were connected by Mongolian trade routes. In 1918 in a world in which all paths of communication were mobilized as part of the logistics of war, the Spanish flu contaminated the whole world in less than a year.

The death knell for the power of epidemics was sounded in 1928 by the British biologist Alexander Fleming (1881–1955) who identified and isolated the first known antibiotic, penicillin. The World Health Organization was created after the Second World War to provide global coordination for states' attempts to control disease. An extraordinary goal was achieved in 1977 with the eradication of the smallpox virus, perhaps the biggest mass killer in history. It was also one of the most vulnerable germs because its reservoir was purely human. It was hoped that other pathogens would be similarly wiped out, but in vain. The failure to eradicate the polio virus in recent decades can also be explained by political and social factors. A trio of failed states—Nigeria, Afghanistan, and Pakistan —were not able to impose the vaccination on their populations. Since the victory against smallpox, there has in fact been only one other eradicated, the rinderpest (a disease affecting cattle) in 2010.

Since the 1980s we have seen the emergence of germs that resist antibiotics, such as certain strains of tuberculosis, as well as new diseases like AIDS. This has decisively buried any lingering hope of wiping out all epidemics; we must content ourselves with keeping them at bay. The intensification of human contact all over the planet has pushed more and more illnesses to spring up out of their pathocenosis with alarming potential consequences even though for the moment they remain under control. The Ebola epidemic that struck Africa in 2015–2016 was thus linked to the extension of human habitat to the detriment of natural environments where the virus was found. It was nonetheless circumscribed by the action of the African states concerned. Increasing inequalities have simultaneously led to the expansion of zones affected by illness linked to poverty. Among many possible examples of this is the spread of cholera cases in Haiti, which was left ravaged by natural disasters. Similarly, although AIDS did not have a major impact on the demography of wealthy countries, it has devastated a large part of Africa, leading to high mortality rates among young people. To date, the virus has left twenty-five million people dead and thirty-four million sick worldwide and has revealed glaring injustices in access to treatment.

Another threat lies in the potential use of viruses as biological weapons. This could be accidental, because a certain number of laboratories around the world contain strains of the plague along with an extremely long list of other pathogens; some even have smallpox. It could also be deliberate in the form of conventional warfare or terrorist attacks. Or it may be "natural," as in the summer of 2016, when freed by global warming melting the permafrost in which it had been trapped, the anthrax bacteria reemerged in Siberia.

Perhaps we should pay heed to the warnings of the past. Lend an ear to the Italian poet Petrarch, who wrote in 1348, "in what annals has it ever been read that houses were left vacant, cities deserted, the country neglected, the cemeteries too small for the dead and a fearful and universal solitude over the whole Earth? . . . Oh, happy people of the future, who have not known these miseries and perchance will class our testimony with the fables."[15]

The meeting of the two shores of the Atlantic produced two major biological consequences. The first was the contamination of the Indigenous people by epidemics, leading to the near total depopulation of the Americas and facilitating their annexation by the Europeans. The second consequence was similar in scale: the plants of the New World totally changed the way humanity fed itself.

9

DEMOGRAPHIC HAZARDS

The conquest of the Americas fueled the growth of the Old World while Africa bore the burden of the slave trade. Monkey relished tobacco, sugar, and corn, and always found good reason to ignore the suffering of others.

There are no reliable portraits of John Rolfe. You might recognize his name if only because he married Pocahontas. Her birth name was Matoaka or Amonute, and she was baptized Rebecca when she was captured by the colonizers, but she was nicknamed Pocahontas, "little wanton one," in her native language. Was this a reference to her reported tendency to rebel against patriarchal authority? Or the fact that she was a captive of the English? It is difficult to know. The ambiguous nickname gives us an idea of the destiny of many Indigenous "wives" of European colonizers who set out to make their fortunes in the immensities of the New World. And the trajectory of her husband Rolfe is a good reflection of the sixteenth-century European mindset: a smuggler, an ambitious trader, and an involuntary contributor to one of the most massive biotope alterations ever carried out on Earth.

Rolfe is suspected of being the man who introduced earthworms to North America.[1]

Worms and Tobacco

Virginia has the pompous honor of being "His Majesty's Most Ancient Colony and Dominion." Jamestown was the first permanent settlement there. It was apparently in this port that the fate of the world shifted. The British had discovered a new way of colonizing distant lands: joint-stock companies (chap. 12). It was thus in 1607 that the Virginia Company of London sent four hundred English settlers to found Jamestown.

Not knowing how to feed themselves in these unfamiliar lands and at the mercy of the biting winters of the Little Ice Age (chap. 11), epidemics, and the mistrust of the Indigenous people, 85% of these first settlers died within three years. In 1610, a rescue expedition was shipwrecked off the coast of Bermuda. The survivors eventually arrived at Jamestown on makeshift rafts. Among them was Rolfe, an adventurer who was firmly set on making his fortune. The time was right. The settlers had developed a good relationship with local Indigenous tribes, who saved them from famine with their gifts of turkeys and corn. Rolfe, having lost his first wife in the shipwreck, married Pocahontas, the daughter of Powhatan, the chief of a confederacy of tribes in the region. This woman of both worlds who converted to Christianity when she was forcibly captured by the English ultimately chose to stay with them.

Rolfe convinced a captain to bring him a few seeds of the *Nicotiana tabacum* plant from Venezuela. Although the tobacco *Nicotiana rustica* grew in Virginia, it was not to the liking of the English, who found it "poor, weak and of a biting taste."[2] True fragrant tobacco was monopolized by the Spanish crown, to which it provided a substantial fortune in trade. Stealing even a single seed to acclimatize it to lands beyond the control of His Most Catholic Majesty was punishable by death.

In the early seventeenth century, London had opened more than seven thousand tobacco houses where clients could "drink" the heady, addictive substance imported from the Amazon. Back then the British drank it like tea, as the Indigenous peoples of the Americas did. Some dandies used it as snuff, keeping a handkerchief handy for the ensuing sneezes. Tobacco houses were found all the way to Japan, Ottoman Turkey, and India. At the end of the sixteenth century, a poet in southern China related his astonishment that everyone, including twelve-year-old children, had begun smoking. The smoke seems to have been effective at repelling mosquitoes, and thus it served as a protection against malaria. But it had already become a way for soldiers to kill time, a consumer status symbol for the ladies of the elite, a little taste of luxury for ordinary people, and a subject of poetry for intellectuals. Indeed, "hymns to tobacco" had become a full-fledged literary genre.

In 1616 Rolfe returned to England with his new wife, their son, and a cargo of his new Virginia tobacco. One of his friends advertised the new product as "pleasant, sweet and strong."[3] Rolfe immediately became a trader. His fortune was made. True patriots preferred his product to that of the Spanish enemy. Fleets of ships left England loaded with ballast, earth, and rocks, which would be replaced in New England with barrels of rolled tobacco leaves, containers 1.3 meters (4.25 feet) high and 80 centimeters (2.6 feet) in diameter and carrying half a ton each! The stony ballast was then thrown overboard.

At the time, there were no earthworms in the northern half of North America. They had disappeared during the Ice Age and had not returned. Perhaps this was because worms move very slowly when they are not helped along by human intervention. But there were probably common earthworms and red worms in all the dirt mixed up in the swapping of ballast for tobacco. Europeans long believed that these creatures were harmful, but today agronomists know that they are essential. Charles Darwin (1809–1882) was one of the first to sing the praises of the worm, surmising that the total mass of earthworms exceeded that of any other animal species. Above all, he understood that their incessant activity consisted in eating soil to feed on the microscopic organic debris it contains, resulting in the creation of an infinite number of labyrinthine tunnels. These tunnels carry air and water that decompose the excrement of worms into fertile contributions to the soil.

The northern forests of North America went through two simultaneous transformations. The first involved the disappearance of the native societies that had practiced mass burning to facilitate both agriculture and hunting, and the second was the introduction of earthworms. Without worms, the fallen leaves from trees made up a thick permanent mulch that changed the undergrowth of the forests into endless shady prairies. With the arrival of worms, the leaves were transformed into compost. Low shrubby vegetation benefited from this even though the intense burning off (which regularly fertilized the soil with wood ash) practiced by Indigenous people had stopped. The Europeans arrived unknowingly carrying seeds that would colonize the Americas. As a result, more than 70% of the plant species that today cover the northern half of North America originated in Europe. Indigenous plant species went much the same way as the First Peoples: they were progressively replaced by the invaders.

This is also true for the Caribbean. Goats and sheep from Europe ate the native plants, and African grasses grew in their place from seeds carried in the bedding on slave ships. The local vegetation suffocated. Palm trees, mahogany trees, and Kapok trees were pushed back by acacias and other African shrubs. Mongooses, imported from India, exterminated nearly all the snakes. Today it is difficult to find an acre of forest in the Caribbean that is not contaminated by the invasion of nonnative plants. Yet although it was widespread, it went essentially unnoticed. It was not until the 1970s and the rise of environmentalist concerns that people became aware of it. The historian Alfred W. Crosby Jr. (1931) showed spectacular intuition on this point. His book *The Columbian Exchange* initiated a history of the biological consequences of the European discovery of the Americas, which progressively broke down the skepticism of his colleagues.[4]

The Columbian Exchange

Swapping germs for food. That is the basis of the Columbian exchange, an expression invented by Crosby, who stresses that Christopher Columbus remains the only human to be able to claim to have brought the world into a new biological era. The Americas were the source of the world's tobacco but also of tomatoes, maize, potatoes, beans, squashes, sweet potatoes, cassavas, pineapples, vanilla, chocolate, peanuts—and possibly syphilis. In exchange, the Old World provided cows, pigs, sheep, horses, and honeybees. And also wheat, rice, barley, oats, oranges and lemons, grapes, bananas, coconuts, peaches, pears, apples, turnips, yams, onions, olives, and finally coffee and sugar. But stowaways were exchanged as well:

smallpox, plague, flu, hepatitis, measles, encephalitis, rubella, mumps, pneumonia, and yellow fever as well as the bacteria that causes typhus, scarlet fever, tuberculosis, and diphtheria and the parasite responsible for malaria *Plasmodium falciparum* (chap. 8). An inestimable number of plants, wildflowers, insects, rats, and so forth were also inadvertently imported. As were large populations of nonnative humans, mostly from Africa. Between the sixteenth and nineteenth centuries, the slave trade contributed more than 80% of the demographic growth in the New World. Later, from the nineteenth century, human imports came from Europe and Asia.

By making connections between the four corners of the world, Monkey became an inadvertent actor in the decisive alteration of the world's biotopes.

The first wave of globalization was the work of the Spanish and Portuguese. From the 1520s the Spanish and Portuguese met in the Philippines—the former having come from the Americas and the latter via Africa and India. They were all competing for access to Chinese riches. Behind geopolitical globalization was economic and religious globalization. The military, the merchants, and the missionaries always worked together. But the effects of a fourth factor would eclipse the three other processes: biological globalization increased the availability of food all over the planet. American tubers were extremely productive. Potatoes, yams, and sweet potatoes provided bumper crops in Africa, Asia, and Europe. This led to a period of demographic growth between the sixteenth and eighteenth centuries, although this was somewhat tempered by the fury of the elements (chap. 11). Maize was a favorite on all continents. It was the dream of a land of plenty made real. A single ear of maize could hold more than one hundred kernels, much more than any other grain crop; only rice could compete in terms of productivity per hectare.

Some agricultural techniques were lost in this exchange, others improved. Squashes, maize, and beans were no longer systematically grown together as they were in the Mesoamerican *milpa* system (chap. 2). Methods for growing African rice, however, crossed the ocean toward Mexico, where they were improved by the Arab expertise in hydraulics that had been preserved by the Spanish.

Biological globalization went hand in hand with a hybridization of agricultural systems.

Biological Empires

Alfred W. Crosby Jr. pursued his analysis with a second book titled *Ecological Imperialism* (1986).[5] The concept of imperialism reformulated an old question and provided a new response. The initial question is in fact the basis for global history—why does the West rule? But Crosby put the question this way—why do we find populations of European origin on all continents, from South Africa to Australia, as well as all over the Americas? The answer was no longer based on the idea of military superiority (which came later, as we will see in chap. 13) or on moral, social, or scientific superiority. The cocktail that gave Europeans the advantage, to the point where they came to dominate the world in the nineteenth century, was originally due to a hazard of biology. As soon as Columbus crossed the Atlantic with his ships carrying animals, plants, and germs as well as humans, whole systems that had been isolated for millions of years came into contact with each other. Thousands of invasive species discovered new lands to conquer, all the more easily given that there were no longer any gardeners to tend them.

Having devastated the populations of the Americas, the microbial shockwave resounded elsewhere. For example, in Hawaii in 1778, the explorer James Cook (1728–1779) encountered a prosperous population of three hundred thousand Polynesians. That number had dwindled to sixty thousand survivors by 1850. In Australia, entire Aboriginal tribes were wiped out by illnesses in the years that followed their first contact with Europeans. There are many examples of depopulation around the world. The deadliness of these unfamiliar germs for Indigenous populations, who had remained isolated from the melting pot of pathogens in the Old World, was combined with the brutality of European settlers. Survivors from these weakened populations were enslaved to provide labor for plantations. In five hundred years, more than half the documented languages in the world died out. In homogenizing the earth, humanity itself became less diverse.

Thousands of plants and insects from Europe colonized these new habitats. Europeans transformed most of the Americas—and then Oceania, a large part of Asia, and southern Africa—into biotopes similar to those of Western Europe. Although the spread of plants was partly involuntary, it was also partly intentional. European colonial policies aimed to transform indigenous landscapes into familiar ones in order to better exploit them. Europe exported its plants, its crops, its ways of managing the living world. For example, it imposed monoculture in colonized Africa from the nineteenth century. When the British got their hands on India in the eighteenth century, they adapted their form of forestry planning to it. India lost large swathes of the forests that its royal elites had preserved for the safekeeping of the prestigious elephants.

Strengthened by demographic growth, bolstered by the arrival of food plants from the New World, and favored by the biological invasions from the Old World to the New, Europeans terraformed their colonies to fit their tastes. Canada, Australia, New Zealand, and the United States would all become overseas scions of England. One expression was particularly powerful: *terra nullius*, empty land, or so the colonizers said, as they built their homeland overseas. In the eyes of the Europeans, the way Indigenous peoples used the land showed that they did not value it, and therefore they did not possess it. The complex interactions these traditional societies had with their environment—selective hunting, burning off, and agroforestry—were completely beyond the comprehension of Europeans for whom cultivated land could only be cleared and plowed. Canada, Australia, and New Zealand became dominions, settlement colonies as identical as possible to the metropolis. Indigenous people were considered extraneous.

In South America, the surviving Mapuche people later called eucalyptus trees "planted soldiers" because they were green, lined up in rows, and it seemed impossible to stand in the way of their progression. The eucalyptus tree is the only Australian plant that has found a dominant place in the globalized ecosystem established by Europe. But it provides a nice metaphor for this environmental imperialism. Born in the fifteenth century from Europeans' desire to reach Asia and its wealth, the dynamic of commercial and biological exchanges ended up transforming the planet into a unified ecosystem, which some biologists have called the homogenocene,* a new biological era dictated by humankind. It was only in the second half of the twentieth century that we began to analyze this process on a global level. Globalization generates colossal economic gains, but it also produces extensive ecological and social turmoil.

The Horse, a Comanche Conquest

Wild horses disappeared from the Americas around ten thousand years ago, along with so many other large mammals. Europeans brought them back domesticated. Some escaped, others were stolen. The depopulation of the Americas also led to a temporary increase in bison populations from the sixteenth century following the development of many human activities. A new economy emerged in the Great Plains based on seasonal migration, a new way of living off the land, a new nomadic dynamic enabled by the introduction of a formidable form of transport: the horse. Riding on horseback meant that the tribes living in the Great Plains could follow the herds of bison and hunt them more effectively. Horses also provided a degree of mobility that encouraged guerrilla tactics, which had a profound effect on geopolitics.

New powers emerged. American Indian tribes learned to temper the microbial shockwave. The most prestigious sedentary civilizations of the time—the Aztecs and Incas—all disappeared. The density of their populations made them prime targets for germs. The societies that resisted were the least prestigious. The nomads in the Mesoamerican jungle, for example, the descendants of the Maya who abandoned the great cities of their ancestors for a hunter-gatherer seasonal farming society resisted for 170 years after the conquistadors, whereas the Aztecs lasted only three. The nomadic groups in northern Mexico adopted a radical preventive strategy: anyone who was sick was immediately abandoned. In this way, they managed to control the spread of sickness. From the seventeenth century, these groups experienced demographic growth, and a new tribal confederation emerged in the south of what is today the United States, the Comanche Empire,[6] which became dominant from the early eighteenth century. To a certain extent it was the offshoot of the Pueblo revolt (1680–1692), when the Indigenous societies of New Mexico temporarily pushed out their Spanish colonizers, confiscated their horses, and sold them to other tribes in order to bolster their short-lived war of independence.

Like nomadic structures in Asia, the power of the Comanche Empire, or Comancheria, was not initially based on the control of territory, but on mobility. Its control of horses as a resource allowed it to federate many tribes. Warriors essentially ate meat, although maize was also eaten when it could be bought or extorted from sedentary societies, and this led to rapid demographic growth. This society was quickly able to sporadically bring together warriors to form a substantial army with a specific objective: for example, pillaging or racketeering the Spanish towns in Mexico and Texas, threats that they did not hesitate to carry out given their strength in numbers. Then the Comanche Empire broke up and absorbed its rival, the Apache Nation. It drew an increasing number of American Indian nations into its orbit. Its model was very attractive; its trade relations with the Spanish, whether forced or voluntary, guaranteed its prosperity. Its mastery of war and coercion reinforced its power and influence.

For one long century this Indigenous political entity built diplomatic ties with British colonizers to the east, French colonizers to the north, and Spanish colonizers to the south. It managed to create a subtle balance among these actors. By pitting the different Spanish communities against each other—encouraging some through trade and undermining others by raids—it prevented any union that might threaten its hegemony and discouraged all attempts to fortify the border. The Comanche Empire deliberately cultivated the weaknesses of the Spanish that it dominated, guaranteeing itself access to the metal tools, weapons, fire, maize, and horses that the American Indian nations needed so badly. The Comanches preferred to get their horses from the Europeans by stealing and violence rather than by breeding them themselves. Comancheria was an intermittent military machine. But it was also a common market, both permanent and central, connecting Texas to Canada and the Rocky Mountains in the west with the Appalachians of the young United States. Although violence was integral to this process, the heart of the empire was peaceful.

But the Comanche Empire was a problem for the Spanish officers who had trouble understanding that "savages" were capable of sophisticated political action. The same situation unfolded in New Zealand in the nineteenth century when a Maori kingdom managed to resist the British invasion for seventy years while exercising the prerogatives of a modern state, including international diplomacy and a regular army. But these societies only left written traces in the archives of their enemies, which often skews the perspective of researchers consulting these documents.

The Comanche Empire had neither a capital nor a permanent army nor a supreme leader. It can be understood more as a coordinated political entity structured around great annual assemblies of warriors who set up in semipermanent camps and who generally led raids on Mexico. In the 1830s they raided all over Mexico, from one side to the other. They stole thousands of horses and thousands of slaves, who made up around a quarter of the Comanche population at the time.

Slavery, practiced by American Indian tribes well before the arrival of Europeans, was modified to become an element of power. Comanches made their captives, whether Mexican or from other Indigenous tribes, undergo a dehumanizing ritual aimed to strip them of their names. Tattooings, beatings, mutilations, food deprivation; these ordeals were a prelude to a regime of servitude different to that of the Europeans, whose slave trade was ravaging Indigenous societies elsewhere in the Americas.[7] Once the initiation ritual was completed, however, the Comanches' slaves were entitled to certain rights, such as not being subject to harsh treatment or not being resold once they had been assigned a master. The women were able to marry free men, which led to their own emancipation. The children of slaves were considered full Comanches.

After throwing off their status as a British colony from 1776, the United States experienced significant growth. This led to the annexation of half of Mexico in 1848, which was greatly facilitated by the actions of the Comanches. But around that time, a three-year-long drought led to the radical decline of the bison populations.

What had been the basis of the strength of the Comanche Empire brought about its almost immediate downfall. Famine destroyed political unity. American firearms, which had just benefited from unprecedented technological progress, did the rest. In less than three years, the Comanche Empire had fallen apart. It was replaced by a new empire of white colonizers, black slaves, railways, and ranches.

Chinese Indigestion

China has always been faced with a near impossible equation. It is home to between 20% and 30% of the world's population but only has around 8% of the world's arable land. That is why the Columbian exchange was such a boon for the country. In the sixteenth century it imported tobacco but also sweet potatoes, maize, peanuts, chilies, pineapples, cashew nuts, and yams. Each of these plants has its own story. For example, in the early 1590s, the Chinese trader Chen Zhenlong smuggled some sweet potato plants out of Manila and acclimatized them in his garden. Around that time, floods devastated the rice fields and famine broke out. Chen's son had the idea to have the governor of Fujian Province taste the tuber. The latter then had a whole field of them planted and eventually distributed cuttings to farmers. Sweet potato farming spread and saved millions of lives. As for maize, it would soon come to be known as "Jade Rice," which established it as a food that even the immortals would relish. This was a blessing for the populations like the Hakkas that had been pushed onto higher, less fertile ground. The land that their more powerful neighbors had allotted them had never been fertile enough for traditional grain crops, which obliged them to burn vegetation as a temporary fertilizer for their fields. But maize was happy in this environment, and it also matured faster than rice, barley, or wheat. Even better, the Hakkas discovered that sweet potatoes like all kinds of soil, even those lacking sun and nutrients. Sweet potatoes and corn were rapidly associated with a third crop from the New World: potatoes.

The demographic growth that resulted from this contributed to an immense wave of settlement toward the west, which was encouraged by the government. In the eighteenth century, tens of millions of Chinese set out to settle in the western territories like Sichuan, Shaanxi, and Xinjiang. They outnumbered the Indigenous populations of Tibetans, Yao, Miao, and Uyghurs. And they increased the fertile land by raising new crops. Without the nutritional contributions from the New World, the Han's demographic conquest of western China would not have been so extensive. Although the plants provided a means of development, it was the agrarian policy of the Qing dynasty (1644–1911) that was the primary factor here. The Qing came to power with the help of a cold period and a peasant revolt. They had one primary obsession: that the masses were protected from natural disasters to prevent them from rebelling. They established the first immunization program in the world, inoculating most of their population with weakened strains of the terrifying smallpox. They consolidated the national network of collective granaries, which allowed the state to buy excess grain during the harvests to be resold at reduced cost during times of shortage.

But the storage program, although conducted on a massive scale, proved insufficient for the environmental disaster that was looming. The settlement in the west led to widespread deforestation. Every uprooted tree made way for a few stalks of maize. But crop roots do not retain humus like tree roots do. Floods became progressively more violent and more frequent in the river basins that were still home to most of the country's population. Speculation was also a factor; farmers preferred to plant tobacco rather than rice because it could be sold for more. The imperial administration could not overcome this trend. In 1727 Emperor Yongzheng proclaimed that "tobacco is not healthy for the people," "because cultivating tobacco requires rich soil it is detrimental to growing grain."[8]

In the nineteenth century, the Chinese began a downward spiral not knowing how to manage the environmental impact of a population of 450 million (chap. 12).

Sugar and Suffering

Sugar is one of those plants that contains the seed of the whole history of globalization.[9] Its etymology comes from old French *sukere*, Italian *zucchero*, borrowed from the Arabic *sukkar*, which was in turn derived from the Sanskrit *sharkara*. It seems that the plant was domesticated in Papua New Guinea possibly around eight thousand years ago or perhaps a little later in what is now Myanmar. The ancient Romans imported it as a medicinal spice. Possibly, frustrated at not being able to trade sufficiently, they were already burning lead to produce a toxic substitute sweetener, the equivalent to today's saccharin. The desire to harvest ever more of the precious substance, which only grows in tropical climates, meant slave plantations emerged progressively at least fifteen centuries ago in China, India, and Persia. Nearchus, an admiral under Alexander the Great, wrote of these sugar canes from which honey is made without the intervention of bees. Armies of enslaved workers were driven to exhaustion in these cane fields. The juice had to be extracted, dried into loaves, and chemically treated to remove impurities. Sugar could be preserved indefinitely in this way, even over the longest journeys. The fact that it could be sold far away meant it could serve to build great fortunes.

In the eighth century, Allah's conquerors decided to dedicate the hot expanses of Mesopotamia to cultivating this plant as part of the Arab Agricultural Revolution (chap. 7). "Sugar, we are told, followed the Qur'an," wrote the historian Sidney W. Mintz.[10] The Muslim world was addicted to sugar as was the Chinese world. Both developed a production model based on slavery. The Abbasid Caliphate's empire paid for this with several millenarian rebellions, when the slaves converted to the Shiite ideas of revolting against injustice. The Zanj Rebellion of black slaves deported from West Africa inflamed Bahrain in 869 CE before being suppressed in 883.[11] So the Muslims bought Slavic slaves (which is where the word *slave* originates) from Byzantine traders to have forced laborers from different origins. Slave masters always made sure they broke down any connections between slaves to avoid feeding dissent.

In the eleventh and twelfth centuries, the Normans took Sicily from the Arabs, and the crusaders discovered the sweet life of the Orient with its subtly perfumed sorbets and pastries and Damascus rose perfumes. The race was on. From Crete to Cyprus to the Balearic Islands, the Europeans seized the islands of the Mediterranean and covered them with sugar plantations. African and Slavic slavery continued here even as it decreased in continental Christendom, when the plague made the workforce more valuable. In the fourteenth century, technical progress in mechanical presses enabled manufacturers to extract even more juice from the sugarcane.

Trade had become transnational. Even though the Christians were producing sugar, they still wanted more. They imported it from the Maghreb and bought slaves at the same time. The Islamized empires in sub-Saharan Africa, such as the Empire of Mali, used the steel weapons and horses procured from the Arabs to raid the populations to the south for slaves. Sold to trans-Saharan caravanners, then sold again to the colonizers from the Mediterranean islands, these Africans were joined by Slavic captives trafficked by Venetian and Genovese traders. The latter bought their slaves from the Eastern European powers, extending their trade to the Mongol Khanate of the Golden Horde, who sold their captives from central Asia. A common market of sugar and suffering was thus established, dominated by powerful Italian city-states like Genoa and Venice. Precapitalism was taking its first steps.

In the fifteenth century, the Spanish took control of the Canary Islands and cleared them to plant this precious sugar cane. The Indigenous population, the Guanches, were wiped out in less than a century by the violence of sugar slavery. In the 1450s the Portuguese cleared Madeira to grow the wheat that they sorely needed before changing their minds and growing sugarcane, which was more profitable. Demand was constantly increasing, production was never fast enough, prices were rising relentlessly and arousing envy. Sugar was supposed to cure everything. It overcame colds, protected against the plague, increased strength, and it was even claimed to boost sperm count—the Viagra of the Renaissance! It had to be plentiful at any banquet in the form of jams or sugared almonds. Genovese and Venetian bankers funded the Spanish expansion as well as the expenditures of the elites of the French kingdom and the Holy Roman Empire. The sugar capitals, which were under state control in the Middle Ages, slipped into the hands of major entrepreneurs.

In 1478 Christopher Columbus was the commissioner for a Genovese merchant house with substantial interests in Portuguese sugar. So it was not surprising that in 1492 he thought of measuring the temperature of the lands he annexed. Tropical? *Bueno!* The second expedition brought the sugarcane plants that would devastate the Caribbean a century later much as they had destroyed the ecosystems of the Canary Islands. This Moloch of the sugar plantations damaged biodiversity, created a perfect habitat for the mosquitoes that carried malaria and yellow fever, swallowed up the Indigenous people, and was reliant on the slave trade. The Canaries were a trial run. The Caribbean reproduced the model on a larger scale. As for the Portuguese, they transformed the Brazilian coastal jungle into an ocean of sugar.

The plantations were governed by a division of labor later theorized by Adam Smith (1723–1790). Tasks were centralized, specialized, and supervised. Slaves were considered blind, docile machines. But the system was not hugely profitable. In the nineteenth century, it was dismantled after the prohibition of slavery (chap. 13). Political and social evolutions along with the production of beet sugar on the mainland meant that the cursed equation between slavery and sugar could be undone. Work would henceforth be paid, and workers were often (not always) freer than before. Sugar production became massively industrialized. Control was concentrated in the Anglo-Saxon capitals. In 1800 the world produced 245,000 tons of sugar. By 1890 it was six million tons. This is the largest growth of a single foodstuff ever recorded. It demonstrates to what extent sugar was and remains the fuel of the food industry and the globalization of taste.

In the New World, the exploitation of slaves used on the sugar plantations was applied to other crops and industries. Tobacco and cotton, in particular, were farmed using the same methods. Slavery was quickly legitimized by the church, which borrowed the moral justification from Arab slavers in the form of the myth of Ham—an obscure curse handed down by Noah to one of his three sons (according to an overhasty reading of the Bible). His motivation? Ham had mocked his father for his naked drunkenness at the end of the great flood. Resentful before God, the father of humanity thus cursed the descendants of Ham to serve those of his other sons, Sem and Japhet. We only have to imagine that Ham had black skin and was the ancestor of Africans, that Japhet was white, and that Sem had olive skin, as Renaissance Europeans imagined Semites should, and the world order appeared right and justified.

It is difficult to calculate the impact of these centuries of slavery. Traditionally, a distinction is made between the Western slave trade—conducted by Europeans from the end of the fifteenth century to the first half of the nineteenth century and primarily destined for the Americas and secondarily for their Asian colonies such as Dutch Indonesia—and the Eastern trade conducted by Muslims between the eighth century and the beginning of the twentieth century. This latter trade was carried out via the Indian Ocean, the Red Sea, and the Sahara and was primarily destined for the Maghreb and the Levant.

The Western slave trade can be measured by studying slave ships registers. Around 12.5 million Africans were deported to the Americas between 1500 and 1865. If we suppose that nearly half of the captives died between being captured and being sold onto the ships, we arrive at a total of 25 million victims. Most of them were male because the West wanted strong laborers for the plantations and often relied on new captives rather than new generations. African slavers preferred to keep the women for farm labor. Fifteen percent of deportees died on the crossing and the bodies were thrown overboard. Forty percent of survivors ended up in Brazil, and as many again in the Caribbean, with 5% landing in the future United States. The remainder were distributed between the Spanish, French, and Dutch colonies. From the eighteenth century, the trade toward the Americas targeted the West African coast from Senegal to Angola, but also Mozambique. Human cargo was paid for in textiles, metal tools, rum, tobacco, glass jewelry, and guns, the millions of firearms that contributed to the power of the slavers. The Western slave trade reached its peak in the 1780s with eighty thousand deportations per year.

The demand for merchandise fueled industrialization, particularly in Great Britain, and also stoked the banking and insurance sectors. Precapitalism developed in the trade networks of the

Atlantic around the slave and sugar routes up until 1834, when the British Empire decided to abolish slavery in the name of free trade (chap. 13). Slaveholders were compensated, and the victims were forgotten.

The Arab slave trade is much more difficult to quantify because of a lack of documentation. It covered nearly thirteen centuries. Most historians consider that it was roughly equivalent in size to the Atlantic slave trade, with approximately twelve million people deported. Some suggest it was half as much or less; others double this estimate. With a probable but very approximate figure of 50 million victims, the two slave trades combined must have had a devastating impact on demography, although this is a matter of debate among historians. Some consider that these criminal practices led Africa into a demographic chasm, causing its proportion of the world population to drop from 20% in the fifteenth century to only 8% in the nineteenth. Others persist in arguing that its impact was more limited.

Today sugarcane is the most widely grown crop in the world, well before maize, wheat, and rice. This sweet cane represents more than 5% of global agricultural production, a little less than 10% of revenues. Unfortunately, aside from feeding immense commercial profits, its main contribution is a global public health crisis in the form of obesity, cancer, heart disease, and diabetes. These conditions develop everywhere that sugar-based carbohydrates are imposed as indispensable additives for the taste of processed foods. Sugar represents 20% of the world's calorie intake.

Monkey has been genetically programmed to be addicted to sugar since we came down out of the trees ten million years ago looking for overripe fruit (chap. 1). Our organism has learned to do without, but once we have had a taste of it, our brain wants more and more. Over thousands of years, Monkey has become a slave to sugar, and we introduced the precious cane anywhere it would grow.

The history of sugar and tobacco are interconnected. Both were produced by slaves and sold as optimizing the vitality of those who consumed them. Promoted as mass consumer goods since the sixteenth century, they have led to health risks that humanity only discovered the extent of in the late twentieth century. They also gave rise to vast profits on which industrial empires were built.

Why We Ate Mummies

Sugar is good to eat, but as we have seen, it was also considered "magical." It was claimed to cure the plague or impotence and to guarantee a strong, healthy body. Many spices, and even certain ceramics or textiles, were considered to have similar properties. The historian Christopher A. Bayly (1945–2015) described this type of product as "bio-moral." He argued that archaic globalization was constructed around bio-moral goods.[12]

From the eleventh century, a strange substance began to make its way down the digestive tract of those who sought exceptional longevity: powdered mummies. In Egypt, this became a flourishing trade. It involved extracting millions of ancient Egyptian human and animal mummies from their tombs and exposing them to the sun so that they spontaneously disintegrated into a powder in a process that was then considered similar to combustion. This powder was then sold as a magic ingredient to extend life, because mummies are everlasting. This ritual of sympathetic magic, operating through analogy, took the form of pills or teas. This bio-moral product was found in Europe as early as the twelfth century and widely commercialized in the sixteenth. Very quickly the powder became prohibitively expensive, resulting in a counterfeit industry. An anecdotal example of this are relics said to be from Joan of Arc that turned out to contain the remains of charred Egyptian mummies. In the South American Andes, where Indigenous peoples traditionally mummified the bodies of their ruling elites on the mountaintops, the Spanish apparently drew freely on this source to assure themselves long life. This is highly ironic because we can suppose that these acts of cannibalism were more likely to encourage the spread of certain diseases.

But there was another product that the world was then coveting: precious metals. China, at the heart of world trade, was now within reach. In the sixteenth century the Spanish and Portuguese navigated around its coasts. They had direct access to delicate porcelain and the finest silks. Although silk had been made in India for several centuries and in Europe for a few decades, the quality of Chinese silk remained unsurpassed. The problem was that the Chinese believed there was nothing they needed. Except perhaps a little precious metal, gold or silver? So much the better. The Spanish had plundered the Aztecs' and Incas' gold, and they had found silver in the Andes; an entire mountain, the mining of which would soon swallow up hundreds of thousands of lives.

The philosopher Peter Sloterdijk has described modernity as a leap into the monstrous.[13] This concise but regrettably accurate expression reflects the extent to which modernity extends the past into a violence that is identical in form but radically different in scale, shifting it into a global dimension.

Monkey would spread the horrors committed on the world's surface to the subterranean world.

PART III
Monkey Transforms the Earth

10

THE PROMISES OF QUICKSILVER

Monkey poisoned Latin America to get at the silver deep in the Andes. The Spanish were dazzled by the precious metal, the British claimed it, and the Indians and Chinese hoarded it.

In late January 1757 an elderly priest by the name of Juan Antonio De Los Santos arrived in the village of Laguna (in today's Bolivia) on the back of a mule. When he saw that there was no welcoming committee assembled on the parade grounds outside the church, as befitted the arrival of a new priest, he became scarlet with rage. Spluttering with indignation, the priest hurled out his fury, frenetically excommunicating the members of his flock. People began to gather. The lieutenant governor Mateo Padilla hurriedly implored the cleric to calm himself. Juan Antonio alighted from his mule and proceeded to unleash a stream of stones and insults on the officer, who was forced to retreat shamefaced.

De Los Santos had not bothered to ascertain whether his constituents had found out about his affectation, but he cared little. According to the rumors, his last six years of service had been horrendous. All his waking hours seemed dedicated to screaming curses, insulting saints, and attacking passersby on trivial pretexts. This was why his superiors had chosen to remove him from Potosi, the prestigious town in which he had been officiating.

The decision was a good one, but it was late coming. The priest was probably suffering from severe mercury poisoning. Potosi was drowning in this volatile metal.

The Curse of El Tio

In 1545 the Spanish got their hands on a mountain of silver 4,000 meters (about 13,123 feet) above sea level. The Indigenous man who discovered it was immediately relieved of his find. This was a freak geological occurrence, a vein of very pure native silver several meters thick breaking through the surface of the ground. The town of Potosi was built at the foot of this "Rich Mountain," the Cerro Rico. It was a dense city, bringing together Spanish fortune hunters, administrators defending royal interests, evangelizing priests, and the Indigenous workers requisitioned to extract the precious metal.

Twenty-five years later, the vein was exhausted. The mountain had become little more than a pile of rocks. The residual silver was trapped in worthless gangue (admixtures of valueless minerals) and was inaccessible. Intoxicated by this short-lived surge of wealth, the Spanish state wanted more. But the new Viceroy of Peru, Francisco de Toledo, knew of new extraction methods.

The technique was simple, and it worked with gold too. Extract the ore, crush it, add salt and mercury (which amalgamates with silver), and mix vigorously. The amalgam, heavier than the rest, falls to the bottom of the funnel. Collect this blackened paste, which is five parts mercury and one part silver. Beat it to remove some of the mercury, and then burn it to evaporate the rest. When all the mercury (also known as quicksilver) is removed, what remains is purified silver.

But without mercury, this process is impossible, as noted by Luis de Valasco, Francisco de Toledo's successor to the post of viceroy of Peru. No problem. God granted Cerro Rico to the Spanish, but he also added cinnabar deposits (the ore that is used to make mercury) some distance away in Huancavelica in today's Peru. This treasure was also discovered by an Indigenous man, and he, too, was quickly divested of it. This would be the "remedy" that would cure Cerro Rico of its barrenness. It is worth noting that mercury had long been associated with good health. Fascinated by its eternal brilliance, the Chinese Taoists had their patients ingest droplets of it to prolong their lives, and the Swiss doctor Paracelsus prescribed it as a remedy for syphilis in 1527.

Between 1570 and 1800, the Huancavelica mine produced a total of sixty-eight thousand metric tons of mercury of which seventeen thousand leached into the environment during the refining process. Potosi used forty-five thousand tons, and the remainder was distributed elsewhere for use in smaller silver mines all the way to Mexico. Potosi burned thirty-nine thousand tons of mercury, and six thousand tons ended up flowing through the town's streets, fountains, and streams.

Miners were initially taken from among the local populations. During the Inca period, adult males had to dedicate one year of labor to the empire every seven years, a practice known as *mit'a*. It was with this labor force that the Inca—and the civilizations that preceded them—were able to build 25,000 kilometers (about 15,534 miles) of paved roads through some of the roughest terrain on the planet. This was a feat of engineering surpassed in virtuosity only by the gigantic Chinese constructions, such as the Great Wall, and especially the Grand Canal and its many tributaries. But the procession of pathogens had done its work. The population was down to a twentieth of its former numbers. The Spanish decreed that the *mit'a* (work that was theoretically paid, thus distinguishing it from slavery) would be carried out every second year. There would be one year of hard labor for one year of freedom for as long as the workers survived the mines. But hacking at rock all day long in the narrow tunnels under the mountain inevitably led to silicosis, a disease where dust accumulates in the workers' lungs until they eventually die.

Those who returned to their villages after a year of hell at the bottom of the mines would often discover that they had been given up for dead, that their wives had left, that their neighbors had shared out their meager goods. Banished from their communities, they huddled in the slums around Potosi. They became *forasteros* (foreigners). They were now free from the *mit'a* because they had left their birthplace, but they were forced to earn their living the Spanish way: through paid work, for example, such as purifying ore. They had to tip cartloads of rocks into the grinders of hydraulic mills to crush the ore into gravel. This guaranteed silicosis from the dust as a result. It also meant poisoning for all those who drank from the fountains fed by the grayish water from the mills. The workers then had to tip the gravel, along with its mercury, into stone basins and trample it with their bare feet, their legs burning up to their knees in the caustic blackish sludge, for twelve hours a day and weeks at a time. Finally, they had to burn the amalgam. The mercury vapor released was insidious, omnipresent. It contaminated everything: air, food, running water, the rammed earth used to build houses, clothes, and eventually even bodies and minds.

The colonizers' version of the *mit'a* finished off Andean societies. Remaining in your village meant spending half your life a slave. In the seventeenth century, most Andean people preferred to become *forasteros*.

Quicksilver exacerbated colonial violence. At its height in the seventeenth century, Potosi had 160,000 inhabitants. At the time, it was the most densely populated urban area in South America. It was a world of its own, a world of extreme poverty and extraordinary wealth. A world built in a day. A writer at the time described the town as "unique in opulence, first in majesty and ultimate aim of greed."[1] In this city, where everyone got drunk on Sundays, there were apparently as many dogs as there were humans. They fed on the mountains of rubbish that accumulated in the slums on the outskirts of town. Armies of black slaves were imported as servants for wealthy Spaniards, but some also worked as foremen in the mines and around the town. That slaves gave orders to Indigenous workers who were supposedly free says a lot about the status of these *Indios*. And when they rebelled, the Indigenous workers reserved a particularly atrocious fate for African foremen.

In the mines, out of sight, the Indigenous workers worshiped El Tio, the god of the underworld and the husband of the earth mother Pachamama. They gave him the attributes of the devil so reviled by the Spanish. To endure their fate, the cursed of the earth drank *chicha* (a sort of fermented corn beer) and chewed coca leaves,[2] products previously reserved for the elites of Inca society.

The exceptional high-altitude forest, which used to cover the surrounding areas, had long since been burned in the forges of the mine or cut down to prop up the tunnels of what had become an infernal anthill under the Cerro Rico.

Since the beginning of modern times, there have been enlightened voices, both Inca and European, who have warned that mercury is highly toxic. It makes teeth fall out, changes people's moods, and provokes violence or excessive apathy. Later it became clear that it also causes infertility, birth deformations, asthma, balance disorders, chronic fatigue, arthritis, autoimmune disorders, insomnia, and vertigo as well as audio and visual difficulties. It accumulates in the brain, the kidneys, and the liver, causing other pathologies specific to these organs. Like the irascible priest Juan Antonio De Los Santos, the inhabitants of Potosi and Huancavelica overwhelmingly paid the price.

And yet mercury has been in constant use ever since in paint and in electronics, as a laxative, a whitener for paper and toothpaste, or even to promote virility. It has been used as a spray to repel insects, in makeup and skin lotions, in dental fillings, and in thermometers. Today, the fish at the top of the food chain, particularly tuna and swordfish, sometimes have large quantities of this poison accumulated in their flesh. And mercury emissions continue: 80% of them are linked to the burning of coal, which remains the first source of thermal energy in the world. Humanity continues to eat, drink, and breathe mercury. This process began three thousand years ago in mines in Greece and China and grew exponentially in the modern era, when Spanish colonizers combined the quicksilver from Huancavelica with silver from Potosi, just as Japanese miners were doing in their mountains.

Twenty-two centuries ago, the first emperor of China, Qin Shi Huang, poisoned himself with mercury, drop of long life by drop of long life. He believed this would make his body invulnerable to decay, much like Monkey's was preserved by the peaches from the garden of the immortals. But the emperor was depressed and megalomaniac and died of his quicksilver drops; in his colossal tomb he hoarded thousands of statues and an entire lake of mercury. In a way, all of humanity has inherited his dream. We all have a little mercury in us. But now we know that unfortunately it will not guarantee us a long life.

All the Money in the World

Between 1550 and 1800 Latin America produced a minimum of 136,000 metric tons of silver, providing 80% of global production. This was an unprecedented deluge of precious metals, three quarters of which came from Cerro Rico alone. The sixteenth century would be forged from Spanish silver. With settlements in Mexico, Peru, and the Philippines, the Spanish colonial empire was the first to be able to proclaim that the sun never set on its territories. It was fed on flows of silver it assumed were inexhaustible. But this flood of metal provoked persistent inflation, which pushed the empire into ever more debt. As long as silver continued to flow from the mines, debts could be paid tomorrow. For the moment, interest payments drained a substantial portion of the precious metal revenues out of Spain. Minted into *pesos* of 36 grams of silver that could be divided into eight *reales*, Spanish dollars, or "pieces of eight," became the currency of Europe and conquered the markets of the Islamic world. The fledgling American dollar was even pegged to the Spanish currency for stability after the Coinage Act of 1792. In the meantime, merchandise flooded in from the whole world to satisfy the demands of the Spanish elite. The Spanish monarchs were the champions of the papacy. Between the 1570s and 1609, one quarter of the Spanish monarchy's budget was dedicated to supporting Catholic orthodoxy, funding sixty-three thousand soldiers to fight against Protestants in Flanders, for example.

In 1570, Spanish diplomat Miguel de Legazpi succeeded where Columbus had failed. He made contact with the Chinese. Luckily, they were traders; unluckily, things went very badly, and his soldiers attacked them. Fortunately, they were merchants, and the shared language of good economic sense smoothed over these initial difficulties. This was in the Philippines, on the threshold of China. Legazpi founded the town of Manila, which is still the capital today, and the Chinese flooded in to trade. Under the control of the Spanish state, Manila's galleon, heavily guarded to protect it from British corsairs, crossed the Pacific every year.[3] It bought silver from the Americas to the Chinese as direct payment for the precious cargoes of silk and porcelain. Thus, at least a third of the silver extracted from the Americas went directly to Asia, without first passing through Europe. The return journey crossed Mexico by land to continue on to Cadiz or Seville.

The French, the English, and the Dutch gained access to Chinese and Indian markets from the seventeenth century by taking the same route as the Portuguese via the African and Indian coasts. As they went, they plundered the Lusitanian ships and sank them, wiping out the competition and increasing their profits. Free trade developed in violence. It aimed to counter the global monopoly established by Spain and Portugal, the great powers of the sixteenth century. Indeed, in 1494 these two powers had agreed, with the blessing of the pope, that the whole world belonged to them and that they could each have half. The Treaty of Tordesillas basically drew a line down the middle of the Atlantic; everything to the west went to Spain (all of the Americas except Brazil), and everything to the east was for the Portuguese (Africa, India, and all the way to the Philippines). This system worked as long as the twin superpowers maintained control of the sea and even more so when the monarchies were briefly linked (1580–1640). But in the seventeenth century, cracks began to appear in their hegemony.

Thanks to their more efficient economies, the English, the Dutch, and the French progressively ousted the former superpowers from the European global monopoly. The Dutch legal scholar Hugo Grotius (1583–1645) was charged by the Dutch East India Company with finding a legal basis for the plundering of all Spanish and Portuguese ships. In 1609, he argued that any monopoly on world trade contravened the "natural" rights of every nation to trade freely. Along with his British alter ego John Seldon (1584–1654),[4] he engraved the founding principles of international law into the marble of modernity: freedom of the seas and freedom of trade, enforced with cannon fire if necessary.

The euphoria of these European powers, in view of the wealth of this world that they could now travel freely, led to a fundamental upheaval in thought. Economic theories became structured around mercantilism. This ideology made trade—which had previously been intellectually scorned—a primary source of wealth and power. This had two effects. First, it led to the development of ideas that were in no way linked to the influence of the church (whose literal reading of the Bible meant they shunned merchants), and second, it led to the contribution of philosophers to the development of a public and national power (the state became a nation when it undertook to plan economic life). Several national mercantile theories emerged as Europe experimented with these new approaches: Iberian bullionism was based on hoarding precious metals; French Colbertism relied on the state to ensure the promotion of economic activity; and English and Dutch commercialism considered that the wealth of a country was developed through private trade with the support of a public authority.

Silver from the Americas played a major role in the shift of economic power. Although Spanish theologians theorized the basis of the market economy very early on, Spain was not able to preserve its precious metal. The silver arrived in Madrid only to be immediately redistributed. It was absorbed by the church in the form of donations, indulgences (entry tickets to paradise), masses, and church furnishings in precious metals. It was snatched up by Italian and German bankers to pay the prohibitive interest rates on the loans they had granted the Spanish crown. Finally, it was also used to buy merchandise from the north (Europe) and the east (India, China, and the Spice Islands), as we will see.

Sheep, Herring, and Beavers: The Baby Steps of Capitalism

In England, the weakening of royal power in the thirteenth century had led to the establishment of a parliamentary system, which was in the hands of the owners of large fiefdoms. From this era, a tendency toward "enclosure" emerged. This consisted of making collective goods, which previously benefited whole villages or monastic communities, into private goods controlled by members of the same elite that held political power. *Commons*, the term that describes these goods that would today be called public,[5] were in general rather infertile lands. But they were home to a substantial amount of biodiversity because they were not intensively farmed. As a result, they provided a range of services to communities: as pasture for pigs, or a place to gather firewood and wild herbs, including the indispensable broom plant that was used for straw mattresses and thatch.

When the floodgates of Spanish silver opened on the world, the wealthy British elite wanted their share. They had to fund the maritime adventures that would soon become the glory of the empire, and for that they had to find merchandise that was of interest to Spain. The solution was wool. This precious textile required hundreds of thousands of sheep, which had to be fed on vast pastures. From the end of the sixteenth century, enclosures expanded as they never had before, marking the end of the commons. Fields became private and were fenced off, separated by fences or thorny hedges. The demise of the commons was marked by strategies combining aggressive lending, administrative redefinition of land registries, coercion, and an emphasis on the need for better land use.

Enclosures and fencing overturned ecosystems, increased the frequency of flooding, accentuated deforestation in certain regions, and accompanied the adaptation of new crops imported from the New World. They also shook up societies. Increasingly, peasants found themselves in dire poverty as their resources crumbled and the lands were monopolized and laid waste by sheep. When peasant rebellions failed, the exodus into the towns began. France experienced a similar situation in the eighteenth century.

English wool pushed the world onto a new trajectory. First, it imposed the idea at the heart of liberalism that economic goods are better managed by private interests than by collective ones. It also enabled the amassing of substantial capital. The wool trade ensured England enough Spanish silver to finance its expansion, and particularly the annexation of India (chap. 12). It also brought large numbers of the poor to the edges of towns. They had to be "assisted," found housing and food, to prevent them rebelling. This assistance was granted in exchange for various kinds of hard labor: working poor farmlands, crushing rocks to make roads, grinding animal bones to produce fertilizer, making rope and tar for caulking ships. The poor were considered responsible for their pitiful condition, and therefore they had to atone. Packed into poorhouses, or more exactly workhouses, they provided the cheap labor of the Industrial Revolution (chap. 13) that would allow Great Britain to dominate the world in the nineteenth century.

Let us move to the continent, to the Dutch port of Delft, one of the epicenters of the seventeenth-century world. European globalization, which initially wore the colors of Spain and Portugal, had shifted north. The Dutch Republic, also known as the United Provinces, was a territorial confederacy subject to the hostility of the great powers that encircled it (the Holy Roman Empire, France, and England). The paintings of Johannes Vermeer (1632–1675), closely analyzed by Canadian historian Timothy Brook,[6] provide a revealing perspective on the austere interiors of the Calvinist Dutch bourgeoisie. In these paintings, commercial success is subtly evoked by a Turkish tapestry, a map of Asia updated with the latest navigational secrets, a Chinese porcelain tea set in the latest fashion, or a young African manservant.

Delft was one of the homes of the Dutch East India Company (the VOC, for *Vereenigde Oostindische Compagnie*). This charter company was created in 1602 to negotiate with Asia, and its monogram became the first worldwide logo. As a commercial business, the VOC owed its existence to the Little Ice Age (chap. 11). Delft was the point of departure for fleets of herring buses, three-masted ships built for offshore fishing in the North Sea. Endless winters and short summers modified both the economy and ecology of Eurasia. Grains became expensive because the long winters hampered the harvests, forcing the Dutch Republic to find new sources of food. The herring migrated south because of the cold. These fish, found in large schools off the coast of Newfoundland, were extremely easy to catch and preserve in salt. For several hundred years, Scandinavians had dominated the Christian economy surrounding the eating of fish during Lent, but it was now in the hands of the United Provinces. The herring trade would fund expeditions to the East Indies, allowing one million Dutch people to leave for Asia over the course of the seventeenth century. In 1650, at the heart of the Dutch Golden Age, nearly sixteen thousand

Dutch ships sailed around the world. In the same period England had around four thousand and France only five hundred. As the British philosopher Francis Bacon (1561–1626) noted, the three inventions that provided the technological foundation of global expansion—the magnetic compass, paper, and gunpowder—were all Chinese.

In the seventeenth century, Europe was technologically old fashioned and tried hard to copy the Chinese. They desperately sought the secrets of the bewitching transparency of porcelain, for example. And although Delft was pivotal to the circulation of trade in Chinese ceramics in Europe, it also became the counterfeit capital of the world. For the first time, its workshops began to produce European imitations that could pass for Chinese workmanship, even decorated with fanciful mock-Chinese symbols.

Something else was traded in Delft too: beavers. This semi-aquatic animal has entirely waterproof skin, which made it highly sought after for hat making. No respectable gentleman or soldier would leave home without his beaver hat, which came in a variety of shapes and styles, including the famous top hat. The fabrication of these hats, however, made the hatters go mad because the fur was soaked in a solution of copper acetate, Arabic gum, and mercury. But intensive hunting had wiped out beavers in Europe. Even Russian traders, who now had to go to Siberia for skins, struggled to meet demand. The Great Lakes region of North America was home to a large number of these animals who were also bigger than their late European cousins. The beaver therefore became a strategic pawn in the race between the French and English to establish overseas empires. The American Indians laughed at the insatiable appetite of the white men, who were happy to exchange one beaver skin for twenty good steel knives. But their predation threatened the environmental equilibrium of the Great Lakes. The Europeans could sell the fur up to two hundred times its original price and were not unhappy with the bargain. They nevertheless kept in mind that the most attractive profits were to be made in China. Expeditions through Canada—such as that led by Samuel de Champlain (1570–1635), who founded Quebec City—were above all intended to find an alternative path to the Middle Kingdom.

China and India, Beneficiaries of an Unequal Trade

Christopher Columbus, Miguel de Legazpi, and Samuel de Champlain were all driven by the dream of reaching China. Its prosperity, even when undermined by demographic growth and recession, always surpassed that of the rest of the world. It produced the two most coveted goods, porcelain and silk. Although the latter had also been made in India for several centuries and in France and Italy since the late fifteenth century, the quality standard remained Chinese silk.

But a rival manufacturing power emerged: India. Or rather the *Indies*, a term that included today's India, Pakistan, Bangladesh, and the southern half of Afghanistan. Much like China, it was a quasi continent, heavily populated in the sixteenth century, with some 150 to two hundred million inhabitants. In this era, approximately one third of humanity was Chinese and roughly another third was Indian. Both centers had economic potential, which explains the fascination of merchants from the Levant and Europe. They were well governed (China by the Ming Empire, the Indies by the Mughal Empire and neighboring states) and had strong currencies (the Indian rupee, created in 1542, was the first endogenous solid currency standard in the history of the Indian subcontinent), safe networks of communication in excellent condition, and large, densely populated towns full of artisans and state workshops that structured demand. From all perspectives, China and the Indies were much more developed than Europe. From the fifteenth to the seventeenth centuries, these two giants went through phases of agricultural expansion linked to the development of new methods of irrigation, which allowed them to grow crops more intensively, and then to the adoption of plants from the New World. The urban elites copied the luxurious lifestyle of the aristocrats and created genuine fashion markets.

This was particularly visible in the Mughal Empire, home to the most extravagant court in the world. Cotton, a plant endemic to India, provided thread to weave industrial quantities of textiles worn all over Asia. Indian artisans developed fixatives and plant-based dyes, such as indigo, that were long lasting. Chintz (printed fabric) became very fashionable when techniques for printing on fabric were perfected. Merchants offered models from a catalog. Once the order had been made, they bought the required quantity of the printed fabric from the artisans and took the profits without reinvesting them. Their clients were women from the aristocracy or wives of traders who came from all over Asia and sometimes from the Italian city-states. In India the textile industry made enormous profits without requiring any further investment. The same was true of the wootz steel industry, which followed the same rules of production. Artisans worked in groups, and rich merchants bought their production on an as needed basis, feeding long-distance trade that produced specialization. Whole regions were dedicated to providing the textile industry with cotton or the steel mills with wood.

From the first century CE, North India had mastered the technology of blast furnaces, which then fell into the hands of the Muslim conquerors, who preserved it. From the ninth to the sixteenth centuries, ingots of Indian steel flooded into the forges of Western Asia and beyond. Damascus steel blades, along with the best Viking swords, were forged in wootz, which trade networks obtained in India and then sent to Central Asia, where Scandinavians were then venturing.

The military expenditure of the Mughal Empire also fueled the leather and shipbuilding industries. Since the Chinese renounced their maritime technology and officially prohibited the construction of seagoing vessels in the fifteenth and sixteenth centuries,[7] it was Indian junks that crisscrossed the Asian Mediterranean,[8] the long maritime corridor whose ports serve the East Indies, Southeast Asia, Southern China, and Japan.

In early modern times, the Indian economy was made more fluid by a system of endorsable bills of exchange, hundis, which enabled significant transfers of money from one place to another. Indian merchants could be immensely rich, but they were no less vulnerable to market fluctuations and royal whims. These traders—Jains from the northwest, Gujarati from the west, or Tamils from the south—carried out most of the maritime exchanges between China and Western Asia, because the collapse of the Mongol Empire in the fourteenth and fifteenth centuries had led to the abandonment of the overland silk roads. The Western Indian Ocean remained dominated by Egyptian, Persian, Armenian, and Yemeni traders whose networks extended all the way to China and the Philippines.

It was into this thousand-year-old commercial space, structured around very substantial exchanges, that the Portuguese intruded in the sixteenth century. In the seventeenth century they were followed by the Dutch, the English, and the French. What was their chief asset in the face of the two superpowers, the Mughal and the Ming? The indifference of these two giants to the foreigners off their coasts certainly helped. But there was also gunpowder.

Cannon Fire over Sea

Artillery was the magic ingredient that would allow the West to face off against the multiple maritime powers of India, Indonesia, and Southeast Asia, whether kingdoms, city-states, or merchant guilds. Yet gunpowder was not a European invention. It had been developed by alchemists in China around 850 CE at the latest. Europeans were able to identify its components—sulfur, carbon, and saltpeter—and determine the optimal proportions required for the perfect explosion. It is often said that this black powder was first used for fireworks, which is seen as illustrating a supposed lack of pragmatism among the Chinese. However, there is nothing to prove the artistic use of gunpowder before the end of the sixteenth century, when rockets for aesthetic explosions in the night sky were suddenly documented in both Europe and Asia. In its early days, gunpowder was used primarily to make grenades. A painting in the Dunhuang caves (in Gansu, in far west China), dated from the mid-ninth century, attests to this. It shows a demon throwing an explosive ball at Buddha to try and disturb his meditation.

Thrust into an arms race to try and maintain the respect of their turbulent neighbors on the steppes (the Jurchen, Turks, and Mongols), the Chinese improved the procedure. The flamethrower (a bamboo tube spitting fire) was designed in the tenth century. Cluster bombs (which release shards of metal when they explode, to mutilate as many people as possible) appeared along with explosive and incendiary munitions in the eleventh century. These were followed by the bacterial bomb (a cluster bomb whose submunitions were covered with excrement and therefore germs) along with the battery of rockets and mines, both land and sea, which were developed in the first half of the twelfth century at the latest. A century later, the Southern Song dynasty launched a desperate war against its Mongol invaders. There are records of veritable battleships propelled by waterwheels and even tanks covered with iron plates, equipped with artillery, and pulled by armored horses. Leonardo da Vinci had predecessors in Asia who were apparently able to test the (probably quite limited) efficiency of their inventions on the field.

In the second half of the thirteenth century, the Mongol Empire completed their conquest of all the lands between Korea and Russia. They shared aspects of their lethal knowledge and contributed to increasing the efficiency of these weapons. This led to a technological exchange. Persian engineers, specialized in siege warfare, improved the design of the mangonels—a kind of counterweight catapult—and contributed their mathematical science of ballistics. The Russian foundry workers, able to smelt enormous bells in a single piece, enriched the repertoire of metalwork techniques that would later allow them to smelt large artillery pieces. The premodern maturation of the "great opening up of the world"[9] was also a moment of technological effervescence. The first rifle appeared in Manchuria in 1288. The first cannon was drawn in identical forms in China and Italy around 1325. Traders from Italian city-states who traveled the Turco-Mongolian steppes were apparently the primary intermediaries for this technological transfer between East and West. In 1453, the fall of Constantinople revealed the progress the Ottoman Empire had made in terms of artillery. In the sixteenth century they dominated the Mediterranean metalwork techniques for the cannon but then found themselves technologically outperformed. In France in 1453, the end of the final sieges of the Hundred Years War was hastened by the introduction of gunpowder. The French soldiers used culverins (ancestors of the musket) and early cannons, which allowed them to put paid to the English military superiority.

From the sixteenth century, Europe began a "military revolution" and increased the efficiency of its artillery.[10] The Italians invented fortifications able to resist cannon fire. The Italian "Bastian Fort" was soon copied throughout Europe, combining nonorthogonal geometry for the walls (e.g., pentagonal) to reduce blind spots, known as "dead zones." These constructions were associated with bare landscapes, enlarged curtain walls, and partly buried redoubts (a kind of bunker to protect defensive lines). This race to fortification and firepower led European states further into debt and meant they dedicated increasing amounts of their budgets to defense. An investment cycle was established that encouraged innovation. The bankers who allowed this market to develop and the merchants who prospered from it were the main beneficiaries. Military evolutions were increasingly expensive, necessary, and fueled by affirmations of state power. The size of European armies increased dramatically. In less than two hundred years they grew to equal those of the most powerful (and much more densely populated) Asian states. The invention of conscription in revolutionary France, at war against the rest of Europe, would increase the number of available soldiers still further.

The first European military innovations resulting from the alliance between state and capital took shape on the sea. For example, the battle of Lepanto (1571), an early joint venture associating various Christian powers who wanted to challenge the Ottoman Empire's navy.[11] The city of Venice led the operation, presenting six galleasses, giant galleys carrying cannons, against the Ottomans. The use of these ships would be decisive.

But it was above all the Dutch Republic, fighting for independence both against their former Spanish masters and against their French and British neighbors, who saw themselves forced to innovate or perish. The Dutch invented the handheld telescope, which meant the enemy could be seen from far away. They considerably improved the arquebus (an early gun) and its rules of use, ending the two-century-long dominance of armies of Spanish pikemen on the battlefields of Europe. And they eventually had the idea to line cannons up along the side of ships, behind hatches that could be closed and opened. Henceforth, naval artillery no longer fired "on" enemy ships but rather aimed at the waterline. Naval battles became fast. With a responsive crew, a small, swift sailing ship could sink a large galleon.

At the same time, arquebuses were overturning power relations in the New World. With just one of these weapons, the Frenchman Samuel de Champlain helped his Indigenous allies to victory through the fear that the explosions inspired among the enemy. As luck would have it, he killed three enemy chiefs in the first encounter, undoubtedly adding to that fear. Aware that the European superiority lay in the monopoly of firearms, he avoided trading them. His Dutch and English rivals were not so prudent, however, and arquebuses were quickly added to the repertoire of weapons among Indigenous peoples.

In the sixteenth century, the Portuguese disembarked along the Asian coastlines with extreme violence. They annexed islands and peninsulas and built small forts on them. They forced their way into the Asian trading routes and obliged merchant ships to moor in their ports, confiscating a tenth of the cargo as payment for their "protection." In 1511, after several similar armored attacks, the conquistador Afonso de Albuquerque (1453–1515) seized the port city of Malacca, the central hub of trade in Asia. Portuguese superiority in terms of naval artillery was clear. It allowed them to take control of the seas of the Asian Mediterranean in conjunction with a system of alliances with local powers. But although the Europeans could dictate their laws to the merchant cities in Southeast Asia, they had much more difficulty with the great inland powers. The Ottoman, Persian Safavid, Mughal, and Ming empires all went about their business as usual. These powers also mastered gunpowder, which they used when they had to lay siege to a rebellious vassal or fortify their borders. They were known as the gunpowder empires, but they did not consider artillery as important as the Europeans did.[12] They did not see the point in arming their ships, and their land empires covered vast territories where cannons were much too heavy and insufficiently mobile to be really decisive in battle. Here wars were won on horseback by fighters brandishing sabers. The Russian Empire understood this in its steady expansion toward Central Asia. On its western front it fought European style, with infantry and firearms. But it used Cossack riders to expand its immense territory to the east.

Monkey had now become a master of fire, a specialist in artillery and munitions. His ability to dig into the earth and rip out its precious metals would be exponentially increased by this explosive power.

In the second third of the nineteenth century, a burst of innovations in weaponry, fed by the Industrial Revolution (chap. 13), allowed the West to take a decisive step forward. This technological advancement would henceforth enable a handful of colonial soldiers to take on an army.

A global network of colonial coastal fortifications was woven in response to the fierce competition and near-permanent state of war among European states. Initially, the Spanish and Portuguese—and then the French, British, and Dutch—did not venture far from the coasts, with the exception of the Americas, where populations were decimated by the microbial shockwave. Inland Africa (except South Africa) remained unexplored, for example. It was too well defended by endemic diseases such as malaria, and it was not until the second half of the nineteenth century that the continent was divvied up. Of course, the slave trade provided African kingdoms with muskets, a form of currency like any other. But the rare Western observers of the time considered that they were used inefficiently here, the objective apparently being to fire a maximum number of shots to demonstrate one's courage. In Europe, the objective was to wipe out the enemy, which is why the weapons were constantly being perfected. Thus, during the war for independence fought by the Dutch Republic against Spain (1580s to 1609) the counts of Nassau, at the head of a Dutch army confronted with far superior Spanish forces, trained their soldiers by breaking down every gesture involved in firing a rifle, drilled them endlessly in the repeated gestures, and published illustrated manuals for instructor officers. That way soldiers could repeat the most efficient moves for firing a gun or wielding a pike precisely and in unison, increasing the capacity of an army to deliver death on the battlefield. But one century later in Africa or a century earlier in Mesoamerica, fighting was not necessarily about killing. War was a way to capture future slaves in one case and to provide sacrifices for the gods in the other. Good fighters were precisely those who knew how to win without killing.

The usefulness of firearms was nothing without the cultural context that meant they were "optimized." Without the social and economic infrastructures that had given rise to the European war machine, using or copying the technology that resulted from it would never have achieved the desired efficiency. Western warfare was based on a colossal apparatus of mass mobilization and logistics.

Victor D. Hansen has claimed that European military superiority has its origins in the emergence of Greek phalanxes.[13] However, it was only later, with modernity and the construction of centralized states, that the West was able to win the dubious title of the "best killers" in the world. In this they had an asset that was key to their domination. And it would prove decisive in the seventeenth century when an intense wave of cold swept the earth.

11

COLD, COLD EARTH

The Little Ice Age left Monkey shivering. Devastating droughts and winters, famines and epidemics hampered the ambitions of rival potential hegemonies from China to France and from the Ottoman Empire to Japan.

"Massive drought. Locusts. The price of millet soared. The corpses of the starved lay in the streets."[1] This telegram-style article was published in 1641 in a Shanghai newspaper, a barebones testimony to a collapsing world. Three years later, the Jurchen nomads reached Pekin and brought an end to the Ming dynasty. This coincided with the beginning of an almost continual three-quarters-of-a-century-long period of the worst winters the earth had experienced for ten thousand years. Climate disruptions brought low the powerful Middle Empire along with others and shook all the major states in the Northern Hemisphere. Some came out stronger, others were permanently weakened. This was an era of uninterrupted war, epidemics, political tensions, and populations forced into cannibalism to survive. Some historians describe the period between 1640 and 1715 as a "global crisis."[2] It led to a worldwide shift in power relations.

Can History Freeze to Death?

There have been several times in the past when climate changes almost wiped out life on Earth. Around 445 million years ago, a crust of ice suddenly covered the earth, destroying 85% of animal species. Another cataclysm, around 370 million years ago, wiped out

75% of species. Continual eruptions from megavolcanoes in Siberia 250 million years ago produced such a large volume of aerosol particles that the earth was plunged into the darkness of perpetual winter and 90% of all life died. Around sixty-five million years ago, some kind of cataclysmic event (a gigantic meteorite? An erupting megavolcano in India? Probably a combination of the two) led to the extinction of half the world's life-forms—and all the dinosaurs. Finally, the earth suddenly cooled again thirty-four million years ago, and whole branches of the mammalian kingdom died out. Our hominid ancestors were among those who survived this climate shift, known as the "Eocene-Oligocene extinction event," and they went on to prosper.

All these apocalypses occurred in the far distant past. Closer to our time, the explosion of the Toba volcano around seventy-three thousand years ago may have almost wiped out humanity (chap. 2). The Younger Dryas (chap. 3) period around 14,500 years ago caused the planet to cool considerably and probably favored the emergence of agriculture. All these climate shifts are prehistoric, in the realm of paleontologists. We can see glimpses of their impact in archaeological traces, but the voices of those who lived through them are lost to us.

Not so for the Little Ice Age. It took place between the thirteenth and nineteenth centuries and is consequently within the realm of history. We should not think of it as a period of homogenous cold but rather as a succession of climatic variations. In each of these centuries, there were episodes lasting two or three decades in which temperatures were more than 1°C (1.8°F) colder than those of the twentieth-century average. But these periods of polar winters alternated with hot summers and droughts, all of which were sometimes punctuated by periods of incessant rain. The civilizations of the time depended on seasonal and local agriculture for food and were therefore extremely vulnerable to these shifts. When the winter was too cold, early plantings froze. When spring was too wet, the seedlings rotted. When summer was too dry, the maize shriveled up. Everyone's lives, whether monarchs or peasants, artisans or soldiers, were dependent on the abundance of the annual harvest.

During this time humans commonly believed that the fickle meteorology was a sign of God's wrath. This is how the cold wave was seen in many parts of the Northern Hemisphere from 1270 onward. A global cold period was documented between 1300 and 1380 and characterized by a widespread famine in Eurasia between 1315 and 1319 and an episode of extremely cold temperatures in the winters of the 1370s. And then there was a warmer period for about one hundred years. In 1470 a new cold wave swept over the planet. It lasted until the end of the sixteenth century, sometimes relieved by sporadic periods of warming. From 1630 temperatures plummeted again. They reached their lowest level around 1645 and stayed there until 1715. This was the coldest moment of the last ten thousand years, and in France it coincided with the reign of Louis XIV (1643–1715), who was known as the Sun King. This was a time in which the Seine river in Paris completely froze over several times. According to the historian Geoffrey Parker,[3] this most recent cold snap saw a third of humanity die. People were malnourished, and their weakness meant they were less able to resist disease. They were also all the more ready to rush into war. Although many historians minimize the impact of these climate fluctuations on history, Geoffrey Parker—as well as Brian Fagan, Mark Elvin, Sam White, and Timothy Brook—have chosen to explore their consequences.

There are several possible explanations for the Little Ice Age. The most coherent of these focuses on major volcanic explosions (chap. 14), which are visible in the polar ice in Greenland and Antarctica. According to the astronomers Annie Russell Maunder and her husband Edward Walter Maunder, the effects of these eruptions were amplified by periods of low solar activity. Publishing in the late nineteenth century, they theorized that the cold peak documented between 1645 and 1715 was aggravated, if not caused, by an abnormally low number of sunspots, which are correlated with the activity of our star. Based on compilations of astronomical observations from the period, they documented fewer than fifty sunspots per year where the typical rates before and after this period were closer to fifty thousand sunspots. This hypothesis of a temporary drop in the activity of the sun, the thermonuclear reactor that lights and heats the earth, is known as the "Maunder minimum." A third explanation is based on a combination of elements affecting changes to the earth's climate machine, such as alterations of marine currents and thermal regulation phenomena like El Niño. Of course the complex effects of these parameters may have been combined. The devastating climate of the years between 1640 and 1715 may therefore be linked to the cumulated effects of a 1641 volcanic eruption on Mindanao island (in the Philippines), a significant El Niño effect, and a substantial drop in solar activity.

A fourth explanation, documented by Simon Lewis and Mark A. Maslin,[4] is connected to the theory of the Columbian exchange (chap. 9). It is important to remember that in the sixteenth century, germs from the Old World had reduced the Indigenous population of the Americas to just one tenth of its former numbers, slashing it from sixty million to just five or six million. But a large portion of those who died were farmers who, because they lacked metal and draft animals, had cleared the land using repeated burning. In the seventeenth century the natural environment of the Americas was simultaneously subject to the disappearance of their gardeners and the biological invasion of organisms from the New World—numerous exogenous plants, earthworms, pigs, and so forth. A new forest covered the continent, capturing a large amount of carbon and thus contributing to the substantial drop in global temperatures in the seventeenth century.

Greenland, a Green Land?

Let us go back to the late tenth century CE, to the Medieval Warm Period, in which the Scandinavian peoples were experiencing a period of very strong demographic growth. This was a time when the Vikings, Varangians, or Rus (as they were known in different areas) traded with the Islamic world and even with Central Asia. They had annexed most of Great Britain, part of France, and islands in the Mediterranean, where they had integrated into the local nobility. In their exchanges with Christendom they were fishermen and traders, bringing large volumes of herring and cod all the way to Constantinople. Their ancestors had learned long ago how to catch enormous quantities of cod from flat-bottomed boats. They removed the fish from the nets, gutted them, and dried them in their thousands in the cold air of the great north. Once dried, they were flat and easy to transport, imperishable, and still tasty when cooked in boiling water.

From the ninth century the Vikings reached and colonized an island surrounded by fish where warm and cold currents met: Iceland. The warmer temperatures of this period had covered the island with prairies. Some Norwegian Vikings set up settlements there to escape the authority of the King; they brought sheep, a few cows, and formed a parliament, the first in the world, inaugurating a long local tradition of democracy. In 930 CE they proclaimed the independence of this new colony. Shortly before 1000 CE, a redheaded, hot-tempered man was condemned to exile for having slain one of his peers. Eric the Red assembled a crew and sailed to the northwest, where, legend has it, he discovered high green cliffs. Hence the name, Greenland. However, much like today, the only thing growing on these lands were rare summer prairies of blueberries and dwarf birch even though the average temperatures were then 1°C (1.8°F) warmer than the twentieth-century averages. The name Greenland should really be considered a slogan directed at other future migrants. The Spanish had the same marketing technique when they arrived in the New World in the wake of Christopher Columbus, inventing names like Costa Rica (Rich Coast), or Rio de la Plata (River of Silver).

Very quickly other Viking expeditions setting out from Greenland reached what would become Canada. Colonies were even set up at the mouth of the Saint Lawrence River, and the coasts of Greenland were home to a dozen villages. The Vikings primarily traded wood, fish, and seal fur. They also hunted walruses for their tusks, their ivory having become a valuable substitute for elephants' ivory since North Africa had been conquered by the Caliphate.

But by the fourteenth century, there was no longer any sign of Vikings in Greenland. Global cooling made these seas inhospitable from around 1270. The icebergs reached and sometimes encircled Iceland during the winters of the 1400s and lasted until the beginning of summer, enabling polar bears to venture onto the island.

Viking colonies in Greenland and the New World were abandoned. Contrary to a persistent legend taken up by the biologist and historian Jared Diamond based on old sources, Vikings in Greenland did not die of hunger because they refused to eat fish or marine mammals. All the literary studies and archaeological sources at our disposal today demonstrate that they relished these foods. They left because the cooling made it impossible to maintain the connections on which the survival of the colonies depended.[5] Those in Iceland, however, hung on to their land and managed to survive.

The fish moved offshore, away from the cold. Between 1315 and 1319, as Europe was struggling with polar temperatures and the famine they produced, the Atlantic seems to have been the site for unprecedentedly violent storms according to reports from the period. From the 1400s, however, other fishermen set out in pursuit of fish. The Dutch, the Basques, and the English all developed fishing vessels able to confront the fury of the ocean. The English version, known as a dogger, was a stocky trawler around 15 meters (about 50 feet) long. With a mast and three sails, it used nets or long baited lines that could be hauled in. In its hold, it carried a ton of worms for bait and three tons of salt. Fish were now salted on board. The dogger could move up to a dozen tons of cargo and was like a miniature mobile factory. These boats covered the fish-rich waters of the North Atlantic and contributed to its exploration, brushing up against the New World, particularly off the coast of Newfoundland, before Christopher Columbus. As the Little Ice Age persisted, along with its devastating effects on food supplies, the influx of fish from the North Atlantic became increasingly valuable. Fish could also be used to feed ships' crews, facilitating lengthy expeditions, and allowing the colonies in the New World to satisfy their need for protein and survive harsh winters.

The Last Slough of the Ming Dynasty

According to historians Timothy Brook and Mark Elvin, the Little Ice Age arrived early in East Asia. The first shivers were apparently felt from the beginning of the thirteenth century and were perhaps a factor in the migration of the people on the steppes toward the south. In 1261, the cold really set in. A dramatic winter followed by an unprecedented drought left the Asian steppes barren. Famine struck down both animals and humans. This led to a spectacular exodus. But it was only the first in a long series of calamities. Brooke argues that China went through nine critical climate phases during the Yuan (1279–1368) and Ming (1368–1644) dynasties. He describes these periods of climate stress as "sloughs." The analogy of a swamp in which one gets increasingly bogged down emphasizes that these long periods of meteorological disaster contributed to the progressive collapse of social, economic, and political structures. He named each of these episodes with the honorific title of the imperial regime in which it occurred, from the Yuanzhen to the Chongzhen (1637–1643).

The first climate setback, the Yuanzhen slough (1295–1297), struck twenty years after the founding of the Yuan Mongol dynasty, itself built on the ruins of the southern Song dynasty. The regime was brand new, expansionist, and had no trouble weathering the storm. The following crisis, the Taiding slough (1324–1330), broke out in a context of political instability shortly after the assassination of an emperor. The Mongolian elite were swept up in a whirlwind of conspiracies and military rivalries. A dozen years later, the Zhizheng crisis (1342–1345) saw a massive wave of rebellions that led to the end of the Yuan dynasty in 1368. The sloughs often paved the way for political collapse, although the environment seemed to have a significant impact. Poor decisions by the imperial elites led to the degeneration of already fragile situations.

The Yuan dynasty went through three sloughs. Their successors, the Ming, would have to face five. The cataclysmic crisis of the Chongzhen era (1637–1643), the sixth and final slough under the Ming, struck at the heart of the seventeenth century. Increasingly glacial temperatures and severe droughts led to a series of spectacular epidemics, famines, sandstorms, and swarms of locusts. The Chinese imperial records noted the daily catastrophes that ravaged the country: visions of dragons demonstrating the anger of Heaven, recurrent invasions of locusts, formerly tropical lakes overcome by ice, and malnourished populations struck down by smallpox. The regime could not withstand this uninterrupted succession of natural disasters. The archives of the Shaanxi Province in the northwest tell of how the population of starving peasants, who had been forced to boil tree bark and eat the bodies of the recent dead, began to migrate in a desperate search for food. The grotesque orders of the central government to its local public servants remained unchanged: increase taxes, reduce military spending. The peasants formed armies led by demobilized soldiers and plundered public buildings and state granaries, lynching any public servants who fell into their hands.

The same thing happened all around the empire. Rebel armies formed, directed by charismatic leaders. The most famous of these war chiefs was a native of Shaanxi Province named Li Zicheng (1606–1645), an apprentice blacksmith once beaten for his debts, who proclaimed himself the new emperor. In April 1644 his troops entered the capital Peking, whose 30 kilometers (about 18 miles) of walls were defended by only a handful of soldiers and a battalion of imperial eunuchs. A few hundred kilometers to the north, the Ming army defending the border found itself caught between the advancing rebels in the south and the pressure from the movement of the steppe peoples in the north. Its general-in-chief Wu Sangui chose to form an alliance with the "barbarians" to take Peking back from the rebels. The latter dispersed without a fight. This was how the Jurchen from the north, a nomadic group of around a million members, were able to take control of a weakened empire of between 150 and 200 million people, one third of the world population. They renamed themselves the Manchus and founded the Qing dynasty, which reigned in China until 1911.

The climate was a decisive factor in the seventeenth-century fall of the number-one world power. But another factor was silver from Japan and Potosi. The Ming had given up the paper money that had ensured the prosperity of the Song dynasty and therefore perpetually needed silver to feed their economy. But they did not produce silver. They had to pay for it in silk and porcelain. Their appetite for silver was such that the south of the country was soon entirely covered in mulberry trees to provide food for silkworms. Various measures combining commercial incentives and tax coercion pushed landowners to prioritize silk production. The result of this was that Chinese silk, once it had crossed the Pacific and the Atlantic, arrived in Spain in such volumes that it was cheaper than the silk produced there! In China, silver encouraged the production of nonfood crops. Accumulation of the precious metal lead to inflation that soon ravaged the economy, and the state showed itself ultimately unable to finance the military actions required to defend its immense territories.

Three Ottoman Winters

Let us go back a little in time and cross to the other side of Asia. Anatolia, 1590, the heart of the powerful Ottoman Empire. Founded at the dawn of the fourteenth century, it was by then at its height, reigning supreme from Mecca to Budapest. Then a glacial winter struck. For thirty years, freezing winters kept western Asia under a blanket of snow. Even the Bosporus froze. For ten consecutive years, the cold destroyed the winter wheat planting, and the torrid droughts reduced barley seedlings to dust. These two grain crops provided most of the food for the empire. In this powerful, organized state there were grain stocks, which were intended to address the worst effects of a bad harvest but not an uninterrupted decade of scarcity.

With the harvests destroyed, tax revenue dwindled. A series of uprisings, known as the Celali rebellions, shook the Sultan's authority and emptied the countryside of its population hand in hand with the plague. People fled the slaughter, disappearing into the hills. Whole regions of Turkey and Greece were depopulated. This thirty-year crisis saw the Ottoman Empire lose its footing even though it was in full expansion. It was only much later that it was able to regain control of the regions that had rebelled, and it proved unable to repopulate them. This weakness extended throughout the seventeenth century and ultimately led to the loss of large areas of territory in confrontations with its two rivals, the Austrian Habsburg Empire, and empire of the Russian czars.

This was exacerbated by the fact that the climatic nightmare that occurred between 1590 and 1620 was repeated twice more, between 1640 and 1645 and then between 1680 and 1700. These crises transformed the Ottoman Empire, then the leading European power, into "the sick man of Europe." From his study of the available records, historian Sam White deduces that famine took half the rural population of the empire in the space of a century. This collapse was still being felt by the mid-nineteenth century. The Ottoman Empire's portion of the world population was reduced to a quarter of what it had been in the mid-sixteenth century.

Let us take the example of the year 1641. From analyzing village records in the Balkans, Crete, and Greece, we can see that since 1580 half the households had no heirs. Then there were eleven months without a drop of rain. In Safed, Palestine, the largest religious center for the teaching of Judaism was in trouble. Its revenues were based on the sale of textiles. But the rivers dried up. The waterwheels stopped working, and any kind of dyeing became impossible. The center was abandoned. Farther south, the great Nile shrank to a trickle. The fields lay barren. And Egypt was normally the main source of the empire's food. Its agricultural surplus benefited the army, fed workers, and guaranteed that towns were supplied with food, from the capital Constantinople to the great pilgrimage city Mecca. It allowed religious foundations to distribute soup to the needy throughout the empire. Famine made the masses everywhere restless. Only Christian winemakers in Anatolia could count themselves lucky—the grape harvest would be amazing. In September, however, a meteorological phenomenon (exactly what is a matter of debate; El Niño perhaps, or a volcanic eruption?) unleashed an uninterrupted downpour on the Mediterranean and on China. The much-awaited harvest went moldy on the vines, and many fields were simply washed away.

At the point where a state loses control of a significant part of its agricultural population, several events come into play. Peasants flee from the famine. Landowners no longer have the labor force required to maintain the fields. Those fleeing who do not become pirates, brigands, or rebels take refuge in the towns. This influx of poverty destabilizes economic circuits, feeds social tensions, and provides a rich seedbed for disease. Other peasants abandon sedentary life and become nomads, seeing it as providing a greater source of food security. But the lack of pasture systematically leads these communities to encroach on those of other peasants who are obstinately trying to grow crops. Intercommunity conflicts increase, the state is no longer able to ensure security, and armed gangs roam the countryside.

Although the Ottoman Empire, at its peak in the sixteenth century, had succeeded in encouraging many nomadic herdsmen to settle and ensuring that numerous populations could live together under its rule, the climate crises of the seventeenth century shook this edifice to its core. The communities felt themselves threatened, took up arms, and sometimes even opposed the state. One noble family of Albanian origin, the Köprülü, managed to gain control of the crucial executive responsibility of Grand Vizier and set about reforming the empire. At the end of the seventeenth century, a great surge of energy within the empire pushed it to try and take Vienna in 1683. Its defeat was followed by the empire's first loss of territory in Europe, particularly Hungary.

The Ottoman Empire then took the only path that seemed open to it—modernization. Its attempts to transform itself into a nation-state, following the model developed in Western Europe, did not prevent its destruction. By 1923 its fall was complete, but not before it had carried out genocide against its Christian populations, both Armenians and Chaldo-Assyrians, from 1915. The tolerance of the multiethnic empire, which had welcomed Jews expelled from Spain at the end of the fifteenth century, was no more than a distant memory.

The Revolutionary Trigger

In 1962 the government of the German state of Hesse carried out a survey. It asked respondents to rank seven historical events according to their solemnity, beginning with the worst. The Black Death, defeat in the Second World War, and the Nazi regime took places four, three and two respectively. But first place in this parade of horrors went to the Thirty Years War (1618–1648). Although this might be surprising for non-Germans, it was not a bad choice if we believe Geoffrey Parker, who recorded this anecdote. For this region, this conflict was more devastating than the Second World

War in terms of relative loss of population and economic consequences. It also had an impact on the European continent as a whole, drawing Spain, Denmark, the Dutch Republic, France, Poland, Sweden, the Swiss Confederation, and the northern states of Italy into its orbit.

In 1617 and 1618, a series of unusually cold winters and bad harvests led to the sparks that set light to the Holy Roman Empire, a giant puzzle of twenty million inhabitants federating today's Germany, Austria, Slovenia, Czech Republic, northern Italy, and eastern France. It covered 1,300 territories of various kinds, which although politically autonomous were united under the moral authority of the emperor. Religion fueled the fire. Catholic emperors reigned over a religious mosaic in which Protestant princes felt discriminated against. The war ended with the Peace of Westphalia of 1648 but left behind between three and four million victims. These peace treaties developed a new concept, the idea of a nation-state that is sovereign and absolute and that has a monopoly on legitimate violence to guarantee peace between the various components of society. This idea of the state is at the very basis of the foundation of the modern world and constituted a political revolution. In the centuries that followed, the reinforcement of European nation-states created an international law dominated by these entities that would come to replace empires as the dominant model of social organization. Where empires had governed very diverse populations by granting them different statuses in a pragmatic combination of tolerance and inequality, nation-states would defend the concept of free citizens who were progressively equal in rights, which meant smoothing over their differences.

It would be easy to list many other examples demonstrating the connections between climate disturbances and political conflict. In 1647, when the British Isles were plunged into civil war, King Charles I was forced to flee to the Isle of Wight. A local lord and member of parliament, John Oglander presented him with his services. This was summer, and a terrible storm was raging. The king asked whether this weather was typical on the island, and Oglander replied that it was the worst storm he had seen in his forty-year lifetime. The courtier also noted in his journals that such a hurricane could only be the expression of God's wrath for the blasphemy the rebels had committed by pursuing the king. This is one of the many anecdotes presented in extensive detail in *Global Crisis*, a dense, 850-page book describing the climactic and political uncertainties between 1640 and 1690. Its author, Geoffrey Parker, argues that sociopolitical and meteorological troubles are intricately connected. From London to Peking, from Moscow to Mexico, he describes a world in the throes of chaos and inhumanity. Starving people are heartless, and at times like this, fighters always prefer to burn a farm down and rape its occupants rather than test themselves on the battlefield. Parker argues that a drop of 1°C (1.8°F) made the world a hell, releasing all the negative impulses that the increasing prosperity of previous centuries had assuaged. The global crisis was the result of a fatal synergy between unfavorable meteorological conditions, which affected food resources, and human behavior that proved monstrous in times of crisis because it was focused on survival alone. He presents political and philosophical modernity, like the Westphalia treaty rules, as being born in this mire of blood, ice, and stress. In this, Parker proves himself close to the positions of a Chinese statistician and historian, David D. Zhang,[6] whose team demonstrated that the immense majority of the 1,727 military conflicts and rebellions that occurred in China between the years 1000 and 1911 were correlated to cold winters. The colder the winter, the more frequent the wars.

On the other hand, the French historian Emmanuel Le Roy Laduriein keeping with colleagues who would see a form of determinism in the idea that meteorology could have a decisive influence on history—saw the climate not as something that produced crises but rather as a "trigger,"[7] a final straw, a spark that sets fire to powder already in place. In the autumn of 1787, excessive rains meant that the crops rotted in the fields. Summer 1788 was particularly hot and marked by violent storms. That meant two consecutive significant drops in harvests. The price of grain skyrocketed. But in France it was not quite famine. Since the terrible shortages of 1710 and 1740, the country had begun to prepare for these things, particularly by importing food stuffs from the New World. Sugar was now readily available thanks to plantations in San Domingo, in the future Haiti, where 90% of the population was made up of black slaves. But on the continent, people in the towns and the countryside were restless. On July 14 food riots led to the symbolic capture of the Bastille prison. For Emmanuel Le Roy Ladurie, these demonstrations were above all the result of political tensions. On October 5 and 6, 1789, protesting the lack of bread, a large group of women marched on Versailles in order to fetch "the baker [Louis XVI], the baker's wife, and the little baker's boy," and this began a new phase in the French Revolution, shifting power away from the doomed monarchy.

Real famine returned in the summer of 1794, which was similar to that of 1788: burning hot, stormy, and with small harvests. The situation was aggravated by conflicts between revolutionaries (Robespierre had just been guillotined) and above all the maritime blockade by the Royal Navy, which prevented any grain imports. During the winter, the price of bread reached astronomical levels and provoked a series of riots. The revolution paved the way for Napoleon Bonaparte, having first brought forth several foundational elements of modernity: military conscription, which powerfully reinforced the state in its monopoly on violence; the emancipation of Jews and other minorities; and the assertion of equality and citizenship for all. However, this freedom was fleeting, very quickly circumscribed, and rapidly stripped from both "Negroes" in the colonies and from women.

In the Eye of the Storm

The seventeenth century was one of apparently constant catastrophe, but the eye of the storm is always peaceful. During this period, Japan was struck by numerous natural disasters, particularly a terrifying famine in 1641–1642. But the military government of Shogun Tokugawa Iemitsu was extremely diligent in its reaction, prohibiting lords from demanding hard labor and forcing them to reduce tax rates and distribute their rice stocks to populations in need. The archives of one village mention the implementation of an exceptional tax rate of 11% (the ordinary tax rate was 23%). When the famine persisted, taxes were lowered further (this village went down to 6%), and other measures were also taken. It was forbidden to use rice to produce alcohol or to plant tobacco instead of food products. Public servants roamed the countryside to educate peasants, demonstrate the solicitude of the government, and encourage farmers to not abandon their lands. This wise policy spared Japan the turmoil that swept over the rest of the world.

The political history of the world has clearly not been dictated by the fluctuations of the climate, whether in the seventeenth century or before. But the evolutions of human societies cannot be fully understood unless we also study their environmental context. It should be seen as a background against which human history becomes meaningful. The events of the Little Ice Age teach us that in the face of significant environmental convulsions, certain societies prove resistant, even dynamic in the face of adversity while others are weakened or collapse. The process also reveals the limits of our ability to learn from our experiences. In the seventeenth century, possibly a third of humanity died because of a mere 1°C (1.8°F) drop in temperature. The scenarios of today's global warming herald a minimum 2.5°C (4.5°F) increase and more probably 4°C (7.2°F), 5°C (9°F), or even 6°C (10.8°F) more by the end of the twenty-first century in comparison with the average temperatures of the late nineteenth century. Our children will live through an entirely unprecedented situation of threat, a climate revolution unparalleled in both size and speed.

In the early eighteenth century, Qing China was responsible for a third of global economic production.[8] The Ottoman Empire was already considerably weakened. Mughal India was also only a shadow of its former self. Their role in the global economy went from 25% in 1700 to 16% in 1800. Asia's thousand-year-long domination over the world economy was running out of steam. Europe would soon impose its law on the world. In 1950 the two demographic giants, India and China, found themselves reduced to economic dwarves, each producing less than 5% of global GDP, far behind the United States, the Soviet Union, and even Great Britain. How can we explain this gap, which historians have called the "great divergence?"[9]

12

DYING FOR THE FOREST

Monkey destroyed the forest that was his birthplace. China devoured its woods, barring the way to industrialization, while England exported its ecological impact to its colonies. This was the beginning of capitalism.

It is early morning. Amrita Devi comes out to sit on the doorstep of her house in the village of Khejarli, Rajasthan, in 1730. The sound of axes rings out from the two-centuries-old forest on the edge of her field. She rushes to confront the woodcutters, pleading with them to spare the trees. But they are deaf to her entreaties. She wraps herself around a tree. Her three young daughters run to join her. Weeping, each of them also takes a tree, putting themselves in the way of the axes. The woodcutters are soldiers of the Maharaja Ajit Singh of Jodhpur, a powerful Hindu king who dreams of breaking free of the Mughal yoke. The soldiers have weapons and orders. They must bring back fifty cartloads of wood to feed the lime kilns needed for the frescos of the royal palace in Jodhpur. But where do you get wood when you reign over a desert state? By plundering the rare communities who have planted and protected these oases.

Amrita Devi and her three girls belong to the Bishnoi community, the Hindu equivalent of the Amish. Faltering never crosses their minds. They hold tight to the trees. The axes bite into the wood and then into flesh. Four trees are felled. Four decapitated bodies fall with them. The murder of the father, who arrives late on the scene, completes the carnage.

Blood and Sap

The news runs through the villages. For two weeks the Bishnoi line up to save the forest, sacrificing one human life for every tree cut down. Children and old people, young men and pregnant women. The archives show that 363 people gave their lives there rather than accept that trees be cut down.

The forest always has a history.

This one grows in a region of India that its inhabitants have long referred to as the land of the dead, at the edge of the Thar desert. It is an arid place. In the fifteenth century, a drought that lasted several decades dried up the water sources, and most of the inhabitants left. Yet the Bishnoi community clung on at the edges of the desert. The word *Bishnoi* means "twenty-nine" in Hindi. This number refers to the twenty-nine commandments the community follows. Some of these are identical to the ideologies of the Moral Revolution (do not kill, do not lie, etc.; see chap. 5). Others are more specific to Indian traditions, particularly Jainism (absolute nonviolence, prohibition to kill animals or castrate them). Above all the Bishnoi consider that trees and gazelles are sacred to the point where a young orphaned gazelle will be breastfed by women in the village if need be. This community has persisted in living in this area and has repopulated hundreds of villages. They have fought tirelessly against the desert, planting trees and setting up immense chains of protective dunes with the sand that the storms swept onto their fields and stabilizing these sand dikes by planting blankets of scrub. In so doing they bought these areas back to life, encouraging the proliferation of birds and gazelles. They even paid a tithe to nature: 10% of the grains they harvested were set aside to feed wild animals.

Today there are less than a million Bishnoi left, spread out across northern India, making up 1% of the population in their original homeland, Rajasthan. History has seen many other rural communities who protect the environment around them, fight erosion, and foster biodiversity. We could say, somewhat anachronistically, that the Bishnoi exemplify a movement that today we would call environmentalism.[1]

Two weeks of cutting down trees and unarmed civilians ended up exhausting the soldiers' stock of violence, even drugged on opium as they were. Desertions in the ranks led to a visit by the Maharaja himself who had ordered the felling. This king had won his throne by patricide and was, like many Indians, influenced by astrological predictions. Probably as a result of this sensibility he issued a decree granting the Bishnoi the sacred and inalienable right to preserve their trees.

The forest was saved.

When we hear it today, this story sounds like a fairy tale. Its heroic narrative depicts Amrita Devi as the first eco-warrior and the Khejarli forest as a site of environmental action. It is also important to bear in mind the cultural context. The Bishnoi community remains marked by gender inequality; a wife must unquestioningly obey her husband. And yet, since its creation in the early sixteenth century, this society has improved the condition of women. For example, it forbids *sati* (the practice found elsewhere in India in which widows commit suicide after the death of their husbands) and establishes a period of thirty days rest and isolation for mothers after giving birth. This is the first of the twenty-nine rules. It was established to protect both mother and baby from potential microbial contamination. Although the community is Hindu, Vishnuite in fact, they bury rather than cremate their dead to save the trees that would be required for the fires. They outlaw all consumption of animal products other than milk in addition to alcohol and tobacco. They fast on each full moon and take a daily bath. Both water and milk are filtered to save any tiny living creatures they might contain. Any dead wood used for cooking fires is scrupulously cleaned of insects and arachnids before it is burned. In fact, the Bishnoi almost perfectly incarnate a form of traditional environmentalism, at least among those who remain in the countryside.

The Word for World Is No Longer Forest

At the end of the Ice Age, around twelve thousand years ago, the planet's temperate zones were covered with steppes and tundra.

The only dense forest was in the tropics. Then the megafauna died out, the world began to get warmer, and the increases in humidity generally helped trees to cover the planet. The agricultural boom and associated burning then steamrolled everything, destroying forests, reducing them to ashes, in order to better exploit them.

We can take the forest cycles in France as an example. Around ten thousand years ago the tundra made way for birch or pine forests. As the world got drier, between 6800 and 5500 BCE, hazelnut, oak, alder, linden, and ash trees appeared. They were followed by beech, fir, spruce, and yew when the climate again became cooler and more humid around 2500 BCE. In this period humanity shifted from a limited use of their environment based on hunting and gathering to a more systematic exploitation of it involving clearing areas for agriculture. Forests—in France and elsewhere—should therefore be considered as an environment that evolves and that may appear or disappear. In the third and second centuries BCE, the demographic expansion of the Gauls led to a significant decline in forests that can be seen through the increase in farms on all kinds of terrain. Julius Caesar even complained that there was not enough wood for his sieges. In Roman times, the forests seemed sparse, victims of the overexploitation of wood for industrial activities and the heating of large houses.

Forests have always gone through cycles of growth and decline. The end of the Roman Empire led to the expansion of the forests, which were still used for producing wood. There were many clearings, the results of burning required to produce charcoal, and this diversified ecosystem provided many domestic and wild animals with food. This was when the term *forest* first emerged, referring to the protection of an area of resources. In the seventh century the term referred to the *silva* (the original Latin), which were reserved exclusively for noble landowners or monasteries and protected from use by village communities.

The forests declined again with the demographic boom between the tenth and twelfth centuries and then made back their lost ground when the black plague and the Hundred Years War dramatically reduced the human population. From the end of the fifteenth century up until the mid to late nineteenth, the surface area of forests shrank continually. In France, like all Western Europe, forests were the keystone to an emerging national economy. The nobility's hunting reserves, along with state forests, were protected. Power was built on control of forest resources.[2] Modernity oversaw the use of wood to heat towns, to construct buildings and boats, to shore up mines, to extend pastureland and crops. In the first half of the nineteenth century, the wide-scale sale of logs was a means to reimburse the debts that the state had accumulated after the French Revolution and the Napoleonic wars. As a developed country, by the end of the nineteenth century France could allow itself a rare luxury: reforestation. But unfortunately this was too often overplanned, with species selected for their straight trunks and rapid growth and generally without considering the role of forests in biodiversity.

Archaeological evidence also clearly teaches us that in France, as in other places, there is no more a virgin forest than there is an original people. Located at the extreme west of Eurasia but along the path connecting the north to south, France has been home to many different peoples and cultures over the years. For as long as there have been forests, there have been people to alter them.

In the Levant, forests have declined continuously over the last ten thousand years, succumbing to the blows of the Agricultural Revolution (chap. 3). The *Epic of Gilgamesh*, written some 3,800 years ago, recounts the search for beautiful cedar trunks that led the masters of Mesopotamia to venture into Lebanon. The final blow to the forests of this region would come in the second half of the nineteenth century when the Ottoman Empire constructed a railway line connecting Istanbul to Mecca. Every sleeper on this track meant the destruction of one of the last trees of the Levant. And this happened all over the planet! If Europe was able to slow down its deforestation, it was because it took trees from everywhere else. From the seventeenth century the Dutch acquired large amounts of wood in Indonesia. From the early nineteenth century the British cut into Indian forests, tearing down most of the teak trees on the Malabar coast in southwest India. They then moved their felling to Burma. After the 1826 military conquest, the coastal province of Tanintharyi was stripped of all its teaks in less than two decades.

The Irrawaddy delta suffered the same fate after it was invaded in 1852. Metropolitan and international markets were hungry for this valuable wood that could be made into furniture. Industrialization also led to a need for railway sleepers. In the 1870s half a million trees were cut down every year around the world simply to build railway lines. In seventy-five years, the British razed four million hectares (about 9.9 million acres) of tropical forest in Burma alone. Neighboring Thailand also sacrificed its forests for sale to European powers and Japan.

In the eighteenth century, Enlightenment Europe came to believe that forests needed to be managed. Specialized trade in certain species, requiring planned felling, could prove very destructive. Sandalwood, for example, was fashionable throughout Asia and the West for use in perfumes and furniture. The progressive depletion of forests in southern and western Asia pushed Europeans to search for sandalwood in Oceania. The forests of perfumed wood in Fiji were wiped out between 1804 and 1809. The Marquise Islands were subject to the same treatment between 1814 and 1817, along with the Hawaiian archipelago between 1811 and 1825. In 1898, the United States took control of the Philippines, 80% of which was then covered with old-growth forests. By the time independence was won at the end of the Second World War, the Bureau of Forestry, one of the institutions established immediately after colonization, had "done its work": less than 40% of the territory was still covered with forest. But the felling continued. By 1980 less than a third of the old-growth forests remained. By the 2000s this had dropped to less than 3%, and the surviving trees were saved only by the creation of nature reserves.

We could repeat this example ad nauseam, because the same thing is happening today from Borneo to the Amazon. The last tropical forests have been reduced to ash, giving way to palm oil or soy plantations to feed the global food industry. Wide-scale reforestation policies in certain countries, often based on monocultures, are not enough to compensate for losses.[3] In two hundred years, more than a third of the world's forests have been wiped out. They used to cover nearly half of all land surfaces, some six billion hectares (about 15 billion acres). Today they cover around four billion. After reaching a peak in the 1990s, deforestation has slowed, but it continues. Between 2000 and 2010 the world lost around thirteen million hectares of forest every year and regained seven million through replanting policies, mostly in China, the United States, and Europe. These countries adopt strict reforestation policies while transferring their environmental impact onto others by importing large quantities of wood. Forests continue to shrink in Africa, Southeast Asia, and Latin America. Global warming also contributes to the spread of massive bushfires in heavily anthropized forest environments, for example, in Australia, Russia, and California. But let's not panic. Experts tell us that the trend will be reversed between now and 2030, and reforestation will overtake destruction. But can we wait that long?

Here is a quick reminder of how fundamental forests are for our existence. The air we breathe is produced by forests. The earth we farm is made fertile by humus from forests. The extreme climate events that we fear are lessened by the presence of trees. The fuel we burn in our engines is nothing more than energy created by photosynthesis stored in trees three hundred million years ago. The erosion that destroys farmland and habitat is too often the result of deforestation. Life on the shoreline relies on decomposing wood for survival. Happy societies would be unthinkable without healthy forests. The Bishnoi seem to have learned this lesson well.

Capitalism against Empire

By what feat of strength was the British Empire able to deforest the Indies? In the seventeenth century no one would have believed it possible. Back then, the Indies were an economic giant with 150 million inhabitants. England was a far-off little kingdom ravaged by civil war with a population of only five million.

In 1600 the East India Company (EIC) was founded in England. It would become a cradle for capitalism. Its Dutch cousin (VOC), which we have already discussed, was launched in the Dutch Republic in 1602. The latter had far more capital and was dominant throughout the seventeenth century, contributing to the prosperity of the "Dutch Golden Age" (chap. 10). Although other trading companies existed, such as in France, for example, they had less of an impact on history.

These trading companies were born out of the English state's debt. In the late sixteenth century, England understood the advantage of having colonies all over the world, like the Spanish and the Portuguese. But England did not have the means to acquire them. Spain and Portugal had resources with which to borrow or collect the money they needed to set up trading posts and colonial industries. Protestant England, on the other hand, was heavily in debt after its wars with Spain, the defender of the papacy. Unable to draw new loans with sustainable rates, it delegated the funding of colonial adventures to private entities—joint-stock companies. These businesses were the ancestors of modern companies, bringing together groups of wealthy investors who shared both risks and profits. Ventures were risky but they yielded substantial gains three times out of four. Groups of shareholders could therefore compensate for the disastrous effects of occasional losses.

The East India Company was an investor's dream. It had a state-guaranteed monopoly on trade in a particular region that covered the most lucrative of markets: Asia, including China and the Indies. As a result, it was protected from competition at least on a national level. The West India Company covered the Americas, which was much less lucrative. All these companies had substantial capital and insurance systems that allowed them to regulate profits and losses. The whole system was based on trust. Traders agreed to invest because they were sure their capital would not be confiscated by the crown. If it is to thrive, liberal ideology needs unwavering support from the state and a specific institutional regime able to stamp out authoritarian trends. From the cradle, Anglo-Saxon capitalism grew hand in hand with its twin, democracy. Back then they were both little more than babies.

Over the course of the seventeenth century, the Dutch East India Company (VOC) used its immense capital and Dutch superiority in maritime technology to progressively annex Indonesia. It took control over the very lucrative spice trade, granting itself privileged access to the Chinese market. But where the VOC focused on spices, the EIC preferred textiles and land conquest in India, the largest global producer of textiles and the original home of the famous painted "Indienne" fabric. This was a wager that paid off. First, because it was possible to adapt spices to grow in India (the Indians had been doing this for a long time, they were already producing large quantities of pepper, for example). Second, and above all, because the manufacturing of textiles served to feed industrialization.

In the eighteenth century the Dutch Republic declined, weakened by disastrous wars with France and England. The VOC was now outstripped by its rivals, the EIC and the French East India Company (Compagnie française pour le commerce des Indes orientales). France and Great Britain were now in direct competition for global hegemony over the oceans. Both had colonies in North America and trading posts in India. The Seven Years War (1756–1763) pit them against each other. Beaten on three continents, France ceded most of its colonies to England, including Quebec and its territories in India. It ceded Louisiana to Spain and retained only its Caribbean islands. In 1805 the defeat at Trafalgar, which sank the French fleet, was the last straw for France's maritime ambitions. Once Britannia ruled the waves, the EIC could focus on conquering the Indies.

In 1764 the Mughals gave up management of the prosperous Bengali province to the English, who appointed themselves lords and masters there. The EIC put substantial pressure on the province in order to pay off its debts to the Mughals as well as its shareholders. Then came a catastrophic rice harvest followed by a revolt. In 1770 famine killed millions of Bengalis. The company's stock prices plummeted. All over Europe the banks that had invested a substantial portion of their capital in these ventures went bankrupt. To avoid the system collapsing, His Majesty's Treasury chose to bail out the EIC, which was already "too big to fail." It introduced a tax on tea. This new tax pushed the American colonies onto the path of independence as protesters disguised as American Indians hurled tea chests into the sea during the Boston Tea Party. The company, now confident it would be backed by the state whatever happened, pulled itself together. It returned to whittling away the Indies.

Over the course of the eighteenth century, the company's ships methodically took control of the Indian Ocean. The EIC was thus able to appropriate the monopoly on the most lucrative of global markets: Indian cotton fabrics. They shipped them all over the Indian Ocean, from China to Africa and of course to Europe. As a bonus side trade, they also shipped large numbers of Muslim pilgrims from Asia to Mecca.

Around the same time, the Mughal Empire was beginning to crack. Persian armies ransacked Delhi twice. In the south, the Hindu kingdoms joined with the powerful Maratha Confederacy. Attacked from all sides the previously tolerant empire began a retreat into Muslim orthodoxy, which did nothing to limit its decadence. To survive, it had no choice but to implore the English to protect it against the Persians and the kingdoms of the south. The EIC chipped away at the Indies state by state, city by city. It took over factories and production, infiltrated the economy, and exploited its conquests as much as possible. In this way, a private company was able to completely cannibalize the second most productive country in the world (after China) and use its resources to build one of the most powerful armies on the planet. Transferring profits to its shareholders, it emptied India's coffers of the gold and silver that had poured in for the last three hundred years in exchange for printed fabrics. These riches were sent to China to pay for porcelain and tea. As for the British state, it continued to rule the waves on behalf of its precious company at the cost of increasing debt and taxes imposed on shareholders, which were often negotiated through political crisis. The Mughal emperors became puppets in the hands of the EIC, which locked them up in the fortress-palaces built by their ancestors, while the subcontinent sank into poverty.

Opium, a Currency Like Any Other

Printed Indian fabrics went out of fashion during the long reign of the austere Queen Victoria (1837–1901), who considered them far too frivolous. British owners abandoned the Indian cotton mills that they had bought for next to nothing. An unequal exchange was made. India now only produced cotton as a raw material. In this respect, it was in ferocious competition with the Americans. Bought dirt cheap, the cotton was shipped by the EIC to England, where it was then spun and printed in factory cities like Manchester, nicknamed "Cottonopolis." Cotton fabrics, now British rather than Indian, were sold all over the world. Great Britain would become the economic and military superpower of the nineteenth century thanks to the Industrial Revolution (chap. 13).

The absolute need to control access to India, the "jewel in the crown" of the British Empire, accounts for much of the global geopolitics of the nineteenth century. The British operated on two fronts in order to maintain their control over the subcontinent, first in Africa, where their intervention aimed to guarantee a path toward the Indian Ocean with the progressive control of Egypt, Sudan, Kenya, and South Africa, and second in the "Great Game," an intercontinental chess match between the imperialistic intentions of Britain and Russia fought over Asia.[4] Dozens of officers in the British Indian Army distinguished themselves over the course of the nineteenth century. They were perfectly fluent in a wide range of languages and disguised themselves as merchants to travel through Persia, Afghanistan, and Tibet. Among them was Richard Francis Burton (1820–1890), one of the first Europeans to enter Mecca (disguised, obviously). Along with his colleagues, he took note of everything, such as which local authority was ready to rebel against which other power, which oasis might be a site for invasion, and how far the Russians (who were also coveting and exploring these lands) had progressed.

The nineteenth century marked the end of the EIC. It had emptied the Indies of their riches. Now Indians had to buy their clothes from Great Britain. But the company's monopoly was dismantled, considered antiliberal. It survived by ensuring the production of tea for the mainland, acting as a substitute for the Chinese in this sector. It also produced opium, which it poured into China via smugglers to redress the trade imbalance. Without a doubt, the EIC was the biggest drug dealer in world history. China outlawed opium consumption as did Great Britain on its own soil. Millions of opium addicts were both a health burden and a moral threat. And from the 1830s, money began to flow the other way, from China to

England to pay for the drug. The Middle Kingdom took the necessary measures, burning one thousand tons of opium smuggled into the country by the company. England sent an expeditionary fleet that forced China to "open" itself to international trade and accept that opium was a currency like any other and that England therefore had the right to sell it. The two opium wars (1839–1842, and then 1856–1860) would be the decisive external cause of the collapse of China in second half of the nineteenth century. There were also internal causes, such as the gigantic messianic Taiping rebellion (1850–1864), the catastrophic environmental degradation resulting from deforestation and erosion of farmland, and the effects of colonial famines (chap. 14).

The secret that allowed a little island like England to dominate a territory thirty times more populated than itself was not a matter of superior military power. That was only achieved in the nineteenth century, after the Industrial Revolution. It was capitalism that allowed England to conquer the Indies, access China, and exploit their economic weaknesses.

The Great Divergence

In the eighteenth century, a Chinese scholar traveled all over the Middle Kingdom establishing a catalog of old trees. He consulted the archives, and his verdict was clear. In the space of a century, 90% of these big trees had disappeared. Deforestation was rampant, the consequence of intense demographic expansion, which led to the clearing of any and all potentially useful lands. The destruction of the Chinese forest tells the story of the world over the last three hundred years.

Why did the West come to dominate the world in the nineteenth and twentieth centuries? William H. McNeill made this question the central theme of his seminal book *The Rise of the West*.[5] Whether we use McNeill's phrase or speak of the "European miracle," like the economist Eric L. Jones,[6] or the "Great Divergence," like the political scientist Samuel Huntington or the historian Kenneth Pomeranz,[7] the idea is the same. How can we understand the fact that Europe, previously so peripheral to Asia, was able to become the dominant force over the rest of the world between the sixteenth and nineteenth centuries?

Several interconnected and mutually reinforcing explanations have been suggested. The authors working in this area often use the term *revolution* and observe a synergy between the following:

1. A scientific revolution from the Renaissance onward that led networks of scholars to challenge church dogma and progressively impose a desacralized vision of the world. This was associated with the rising power of humanist ideology (chap. 13) that emerged in the eighteenth century with the Enlightenment.
2. An economic revolution, or a trade revolution, that also emerged in the Renaissance. Up until this point the church had controlled a significant portion of inheritance, playing on the fear of the afterlife, and convincing people to donate their estates to pay for prayer to save their souls. From the thirteenth century Italian merchants managed to establish large amounts of capital that could be transmitted from generation to generation using accounting tools inspired by their Muslim counterparts. These merchants would become Europe's bankers. As ideologies such as the Protestant reform and Anglicanism allowed more and more European societies to progressively break away from the moral control of the church, capitalism progressed. Banks and joint-stock companies complexified their tools and became important partners for increasingly centralized states.
3. A political revolution that was fed by, and in turn fed, this economic revolution. The empire system gave way to absolute monarchy or democracy, where necessary brought about through armed revolutions, depending on national trajectories and periods. A new political figure emerged that would become the political reference of modernity: the nation-state.
4. A military revolution that began from the sixteenth century.
5. An industrial revolution (chap. 13).

How was this evolution expressed in economic terms?

From the nineteenth century, Great Britain began a race to economic development. It would be followed by France, the Netherlands, and Germany, and then by Italy, Spain, and Japan. The economist Angus Maddison estimates that around the year 1500, states all had a roughly equivalent GDP per capita, around €400 (US$500) per person (expressed in 1990 dollars), slightly higher for the Italian city-states. Over the course of the eighteenth century, the Dutch Republic and Great Britain began to take off, doubling and then tripling their GDP per capita around 1820. This is when the system really began to pick up steam. Great Britain reached a GDP per capita of roughly €2,400 (US$3,000) around 1850, €4,000 (US$5,000) around 1900, and €5,600 (US$7,000) in 1950. France and Germany followed the same trend, roughly three decades behind. In 1950 India and China's GDP fell below €400 (US$500) per capita. In the space of 150 years, the world had become profoundly unequal.

Beyond causes that were endogenous to Europe—or rival powers such as the Japanese, Chinese, Mughal, Persian, and Ottoman empires—authors have emphasized the role played by the availability of certain environmental elements and particularly coal. Kenneth Pomeranz argues, for example, that Great Britain was able to avoid a Malthusian check (see chap. 13) because it possessed colonies, a large labor force, and coal. The Northern Song dynasty in China went through a period of preindustrialization between the tenth and eleventh centuries that was interrupted by the Jurchen invasions (see chap. 7). The subsequent shift in China's demographic center of gravity toward the south moved the dynasty away from its coal resources. At a time when it was forced to overexploit available farmland, Europe took control of two continents that were "providentially" emptied of their inhabitants: the Americas, depopulated by imported diseases (see chap. 8). At the beginning of the nineteenth century, the English David was preparing to attack the Chinese Goliath. But David in fact had twenty million "invisible hectares" of colonial land that provided the resources it required, while Goliath was struggling under the weight of environmental limits. In Europe, the steam engine was being applied to something other than toys to distract the elites: the invention of engines. They provided energy that could be used to operate pumps and dry out mines to extract coal more efficiently. It is also important to note that England had rich bituminous coal deposits, but these are extremely humid, which meant they could not be easily mined until steam engines were invented.

Coal fed steam trains and steamboats. The labor force needed was already available in the workhouses (see chap. 10). The English forests would be spared thanks to the wood imported from India. Sugar, produced on the plantations of the Caribbean, was added to the tea massively imported from China and later grown in British India. The resulting drink, combined with a little milk, was nourishing, antibacterial, and invigorating. Back in the seventeenth century, tea was a luxury reserved for the Royal Court in England. But between 1750 and 1830 it became the appetite suppressant that sustained the British economy. The workers of the Industrial Revolution drank it in massive quantities as they worked the looms to weave the cotton now grown in India.

By hoarding resources from the entire world, England was able to take a decisive step forward in economic development.

Jesuits' Bark

In response to the Protestant Reformation, the Catholic church launched a Counter-Reformation in the sixteenth century. This revolved around two key strategies: using words and discourse to fight against a challenge that had successfully survived violent repression, and evangelizing the world through collaboration with the church's allies in the Iberian Peninsula and their territorial expansion. The new Jesuit order embodied these two objectives, emphasizing education, internal discipline, and intense evangelism. Very quickly, missionaries emerged in the Americas but also in the Indies, in China, and in Japan. There were Dominicans, Franciscans, or Augustinians, but the Jesuits were often the most visible.

Legend has it that it was the Jesuits who discovered the ultimate weapon against the influence of mosquitoes on history (see chap. 8). In the depths of the Amazon rainforest, there was a tree, *Cinchona*, whose bark can be used to make quinine. For a long time, this was the only remedy against malaria's deadly fevers. Without this discovery, the fate of the world would have been different; Europeans would have been hard pressed to conquer Africa in the nineteenth century.

The bark was brought back to Europe in the seventeenth century and widely used. At the heart of the Little Ice Age, malaria was rife all over the world, even as far north as Sweden, which demonstrates that the spread of this mosquito-borne disease is only secondarily linked to temperature. The first and foremost protection against malaria is the wealth of a given territory. The states that were able to educate their population about protection, drain marshlands and wetlands, and eradicate mosquitoes were able to conquer this disease starting in the nineteenth century.

The Jesuits proved jealous guardians of the secret of the tree with the golden bark. From 1609 the Spanish crown granted them the right to establish "reductions" (missions) in South America. Each of these autonomous principalities was governed by two or three priests and housed several thousand Indigenous people. At their height around 1700, there were about thirty reductions for 150,000 inhabitants. An unprecedented, even utopian society emerged. The local residents elected their leaders to manage their internal affairs, although government was also theocratic, with the priests maintaining actual power. The death penalty was abolished. Workdays were six hours long, half what was common in other parts of the world. Leisure time was spent together in prayer, reading, dance, music, and military activities. Medical care was free and accessible to all. The priests encouraged their charges to adapt their diet, eating more bread and growing more wheat and cotton in the fields carved out of the jungle. Each reduction was built according to an identical plan. In the center there was a main parade ground surrounded by a church, a house for the priests, a school, a dispensary, workshops, and finally, independent huts for each nuclear family. This carefully planned urban landscape also aimed to destroy the broader ancient societal structures and fight against traditional polygamy.

The reductions paid taxes to Spain. As well as the precious quinine, they also produced various artisanal crafts for export (printed books, wooden sculptures, textiles, ceramics, clockworks, and musical instruments). This enabled them to acquire staple goods such as wine, salt, and iron. But the Spanish and Portuguese colonizers were hostile to them; these reductions were full of Indigenous people who had fled the forced labor imposed on them by the *encomiendas* (land grants from the Spanish crown entailing labor from conquered tribes). To benefit from the protection of the Jesuits, one only had to convert, which was not a heavy price to pay to save one's life and recover a semblance of liberty. The colonial militia attacked the reductions. The Jesuits, who were often the younger sons of noble families experienced in war, organized the Indigenous residents into disciplined troops and somehow managed to hold the slavers at bay. But the papacy abandoned its charges and dissolved the Jesuit order in 1773 (it was later restored in 1814). The reductions were destroyed. After nearly a century of settlement, the survivors fell back to a forest that had become hostile. And it swallowed them up.

In 1795 the logging operation in Loja, formerly the heart of the Jesuits' production of quinine, received a visit from the geographer Alexander von Humboldt (1769–1859). He calculated that at the rhythm at which the trees were being cut down, some twenty-five thousand per year, this precious resource was overexploited. In the nineteenth century, as more and more Europeans were settling in the tropics, five hundred tons of bark had to be found to protect them from the consequences of mosquito bites. In the 1860s the English adventurer Clements Markham stole some seeds and adapted the tree to the British Indies. The world was getting used to such acts of biopiracy. Some fifteen years before, another adventurer, Robert Fortune, had secretly collected samples of the best Chinese tea plants to reinvigorate tea produced in India and thus better compete with Chinese tea production. Another British trader, Charles Ledger, was responsible for selling *Cinchona* plants to Dutch rivals.

Indian quinine turned out to be low quality, however, because the trees had trouble adapting. But thanks to Ledger, the Dutch, who were expert horticulturalists, were able to produce a hybrid plant three times richer in quinine than Peruvian trees and acclimatize it to their colonies in Java, Indonesia. As a result, they would soon have a near monopoly on the market. This situation did not change until the Second World War. In 1940, the Germans bombed Amsterdam and in so doing destroyed the world's stock of quinine. Shortly afterward, Japan conquered Indonesia along with its plantations of *Cinchona*. The Allies found themselves without access to antimalaria treatment even as the war was partly playing out in tropical regions. The Americans patented a synthetic substitute for quinine, Chloroquine, by copying synthetic molecules produced by the Germans. Cheaper to produce, the synthetic derivatives would progressively replace quinine obtained from bark. But *Cinchona* trees are still grown in Africa, Latin America, and Indonesia for the excellent reason that in some regions, malaria has become resistant to most treatments except for quinine, which is now the last resort.

Jesuits' bark is still saving lives.

13

UNLIMITED ENERGY

Monkey became a Titan. He freed the energy within coal. Steam powered the cotton looms and the new engines that allowed us to move faster. The West became the king of the world.

Fumifugium. This intriguing neologism is the title of a book published in 1661. The subtitle, with its Old English spelling, gives a clearer idea of the subject: *The Inconveniencie of the Aer and Smoak of London Dissipated Together With some Remedies humbly Proposed By J.E. Esq. to His Sacred Majestie, And To the Parliament now Assembled.*[1] Its author, John Evelyn Esq. (1620–1706), was what we might now call a whistle-blower.

No Fire without Smoke

John Evelyn was a philanthropist, a kind of Alfred Nobel. His vast fortune was based on the capital accumulated by his family business making gunpowder. He also proposed his services as a consultant. Wanting for nothing himself, he generously shared his opinion and advice. In 1664 he published a brochure encouraging landowners to plant as many trees as possible in order to provide wood to build the ships that would be required for the future power of the United Kingdom. His motto—"explore everything, keep the best"—is printed on the bindings of each of the thousands of books he collected. His second passion was gardening, for scientific purposes of course.

Fumifugium opens with an anecdote in the form of a letter addressed to the king. John wrote that he had recently taken a walk near His Majesty's uniquely magnificent palace but was most disappointed that he could not see it because of a thick blanket of smog that cloaked the street. It was so dense that John could barely make out the faces of the people he came across. He wrote, "this pestilent smoak, which corrodes the very iron, and spoils all the moveables, leaving a soot on all things that it lights: and so fatally seizing on the lungs of the inhabitants, that the cough and the consumption spares no man.[2]" To the irritated eyes of the writer, this London air evoked Mount Etna, Vulcano, or Stromboli, Italian volcanic sites that John had visited when he was younger and that he looked back on with horror both for the foul-smelling fumes and for the suffering of the workers extracting sulfur there. He expressed his concern for the health of the king and his sister, the Duchess of Orleans. It was well known that His Majesty complained of the smoke's effects on his lungs. These noxious fumes had to be dispelled as fast as possible. The king was invited to read the book that followed the letter, which laid out the origins of the problem, a possible solution, and then the implementation of that solution.

Let us trace this issue back to the very beginning.

Air pollution has long been a problem for humanity. We became human through cooking our food, among other things. This affected our genome, because Monkey is the only primate to carry a specific mutation making us less susceptible to the toxic compounds caused by combustion. But this does not mean we are fully protected. Any airborne particles are potential pollutants. It is just a question of exposure and concentration. Forest fires can be particularly virulent in this respect. Clouds of dust from deserts also make the air unbreathable in certain parts of the world. But combustion by humans has been the primary source of pollution since we tamed fire hundreds of thousands of years ago. Given how dispersed the population was, this pollution was initially only visible in closed environments, such as huts where meat was smoked, for example. But then urbanization and industrialization led to increased effects, such as the one described by John Evelyn. This even became the subject of a popular song:

> [Evelyn] shows that 'tis the sea-coal smoke
> That always London does environ,

Which does our lungs and spirits choke,
Our hanging spoil, and rust our iron.
Let none at Fumifuge be scoffing
Who heard at Church our Sunday's coughing.³

In the year 61 CE, the philosopher Seneca left Rome on the advice of his doctor. The vile air of the town was affecting his health, he wrote. He extolled the virtues of fresh country air, but is this a physiological truth or a political metaphor? Seneca had just fallen into disgrace. The paranoid Emperor Nero would eventually force his former advisor to commit suicide some years later, despite his withdrawal from public life. In antiquity, like in the Middle Ages, it was common belief that towns were a source of physical and moral pollution. Animal carcasses rotted in the streets, waters lay stagnant, and everyone complained about the smell. But the real problem was smoke.

During the Little Ice Age, the towns were all shivering. Paris had cut down its nearby forests with such enthusiasm that by the sixteenth century it had to send for wood from the Morvan mountains some 250 kilometers (about 155 miles) away. Luckily, tributaries of the Seine flowed nearby, which meant the wood could be sent down the river. The cities of the Dutch Republic, however, had neither wood nor coal. Their urban growth was based on peat. This is a mediocre fuel that produces large amounts of pollution for low energy returns, which possibly hampered industrial development in this country. The Dutch restricted its use to everyday purposes: brewing beer, refining sugar, and baking bricks. London turned toward brown coal, which was abundant in the English hills. The problem was that this coal contained a lot of sulfur, a corrosive substance that produces acid rain and toxic fumes.

John Evelyn did not live to see the end of London's air pollution. He would have had to wait a long time; it was an ongoing problem until the mid-twentieth century (chap. 15) because in the meantime, the combustion engine had appeared. Smoke from steam trains proved devastatingly destructive, especially when it was concentrated around terminal stations in the heart of major cities. The upper classes railed against the construction of such terminals near their homes. From the late nineteenth century, towns demanded electricity rather than steam on the train lines that served the heart of the urban areas.

If we look at the history of humanity's energy regimes, we can argue that there have been four major stages in which Monkey has managed to harness more energy than a plant can through photosynthesis. First there was fire, burning plants for cooking and heating. Then there was the Agricultural Revolution (chap. 3), which allowed us to condense the solar energy absorbed by plants into food. The third was the Industrial Revolution, which freed the energy from fossil fuels: first coal, then gas, then oil. The fourth was the reproduction of solar energy itself in the form of nuclear energy.

Choosing Fire

In eighteenth-century England, mechanization began to take shape progressively, but its foundations were not new. Coal had long been used as a substitute for wood, and both were still burned extensively. Metals were sought after for their mechanical qualities, and they remained expensive. Machines and their inner parts (cogs, wheels, levers, pulleys, screws) had all been around since antiquity. The waterwheel, for example, which is simply a way of capturing the solar energy that propels water forward through evaporation and rain, had spread throughout Asia and Europe from the very first centuries of the Common Era.

The history of the windmill is essentially the same as that of the watermill. It was invented in the Middle East three thousand years ago and spread widely over the course of the first millennium of the Common Era before benefiting from technological improvements during the Middle Ages. Energy from windmills allowed the Dutch Republic to carve land out of the sea in the seventeenth century at a time when their growing population was caught in a geopolitical trap and besieged by far superior land powers. In the 1650s, at least eight thousand windmills, providing the equivalent energy of forty-thousand horses, worked the pumps to keep the polders dry, protecting these artificial lands surrounded by dikes from the sea. For the Netherlands this technology was like a shield; they mastered the rules of this unusual environment that mired invading armies on many occasions.

Preindustrialization, where rural and urban populations provide for the needs of a specialized industry (textiles, ceramics) through labor that provides a complimentary but not primary revenue, had been a reality since at least the seventeenth century in many urbanized regions of China, India, the Dutch Republic, and Great Britain.[4] But in the latter, a particular combination of circumstances was forming. This dynamic would reach unprecedented proportions and reverberate around the world.

Why England? Why the eighteenth century? It would take more than one book to list all the hypotheses that researchers have developed to answer these questions at the origin of the debate on the Great Divergence (chap. 12). Let us begin by excluding, or at least minimizing, four explanatory elements. The idea that the Protestant entrepreneurial spirit and taste for economy were particularly favorable to capitalism does not stand up to historical analysis given that European Catholics were involved in similar processes.[5] The idea that the market was freer in England than it was elsewhere is also challenged by comparison with China. England may have benefited from an excellent transport network, particularly canals, but in that respect, too, it came far behind China. Finally, England did have a sophisticated banking system but one very similar to that in the Dutch Republic.

There is a simpler explanation: the availability of coal. The United Kingdom had it in abundance. But we cannot assume from this that coal was burned primarily because it was available. Indeed, the first impact of the steam engine was to increase the amount of coal extracted.[6] It is worth remembering that mines were not traditionally used for reasons related to climate. Given Great Britain's well-justified reputation for rain, digging a hole in the ground generally meant making a well rather than a mine. But now pumps could run continually and autonomously to dry out the mineshafts. And that was the spark for the Industrial Revolution. Using a little coal to be able to extract much more. This compensated for the short supply of wood. Between 1500 and 1630, Great Britain had increased its consumption of wood sevenfold, reducing forests to just a sixth of their surface in less than a century. After that, it had to import wood for fuel and construction from North America and Scandinavia, which placed increasing constraints on its budget.

The miracle of the combustion engine was based on something tiny: the discovery of the vacuum, the existence of which had long been refuted. In the 1650s, the German scientist Otto von Guericke proved that it was possible to empty the air out of a recipient using a piston. Around thirty years later, a Dutch mathematician, Christiaan Huygens, applied this principle to the practical improvement of the performance of firearms. And in the 1700s, an English ironmonger named Thomas Newcomen built the first engine. He burned coal to create steam, using this energy to move a piston to create a vacuum in a cylinder. Running at five horsepower, the machine pumped 350 liters (92.5 gallons) of water a minute without complaint.

Plant growth depends on the environment. A mill depends on natural elements like water or wind. But a motor will function for as long as its human assistants feed and maintain it. When Newcomen invented his machine, Britain produced less than three million tons of coal a year. By 1815 it was extracting twenty-three million, thanks in particular to the improvements made by Scottish engineer James Watt.

Monkey was now self-sufficient in energy. He released the power of fossil fuels, the condensed forms of thermal power that had been stored in the depths of the earth for hundreds of millions of years.

A Great Leap

Coal became ever more accessible, which launched a virtuous circle. The measures that economists rely on so heavily to gage a society's productivity began to grow exponentially. This trend really took off in the 1800s.

Over the course of the nineteenth century, more and more numerous and productive machines and tools significantly increased the production of manufactured goods. New metalwork and chemical processes improved the quality of products. The flagship textile industry, whose productivity was boosted by the invention of automatic looms known as "spinning mules," is the perfect symbol of this unprecedented growth. Steam-powered looms meant that a worker in the 1840s could produce forty times more fabric than an equivalent worker in the 1780s. Progress became systemic. Each improvement in a given sector influenced many other aspects of society. Average income began to grow exponentially, and Europe went through a period of substantial demographic growth. Women were still having large numbers of children, but there had been a change in the biological regime that had governed humanity since the beginning. Up until now, at least half of all children died before they reached the age of five. Not so anymore. London archives show that the percentage of children who died under five fell dramatically, from 75% in the 1730s–1740s, to 32% in the 1810s–1820s. Various factors were involved in this, particularly the substantial increase in average standard of living and the progress of medicine. Life expectancy increased dramatically. We died less often in the cradle, we lived longer, and we still had more babies. The demographic revolution was underway.

The figures for population growth in England and Wales reflect this explosion. From six million people in the first half of the eighteenth century, the population jumped to 8.3 million in 1800. It would quadruple over the next century: 16.8 million in 1850, 30.5 million in 1901. All of Europe was swept up in a similar momentum. The continent's demographic growth went from one hundred million to four hundred million between 1700 and 1900. The Industrial Revolution was the first time in human history in which both population and per capita income increased. Malnutrition had previously been standard in a large part of the world. Average life expectancy at birth was around thirty-five to forty years at the beginning of the eighteenth century, a little more in China and in Great Britain (thirty-seven years) than in Europe (thirty-five years). It was substantially longer in colonial regions where food was more abundant, such as the United States or Australia (forty-five to fifty years), where people were also an extra half a dozen centimeters (2.36 inches) taller than their European cousins. In the early twentieth century, life expectancy was over fifty years in developed countries but only twenty-seven years in the Indies. By the mid-twentieth century, life expectancy at birth for Western populations fluctuated around seventy years even though the Second World War had brought a battery of deprivations.

These demographic and technological changes were associated with a "little" agricultural revolution.[7] This considerably improved crop production by mobilizing more capital, labor, and technical means. The practice of leaving land fallow was replaced by crop rotation. This consists in alternating crops, such as clover and alfalfa, which draw nitrogen out of the atmosphere and fix it in the soil, after root vegetables such as turnips, which draw nutrients up. Clover and alfalfa provide food for large numbers of livestock, which in turn fertilize the ground with their manure. Progress and metalwork meant that steel plowshares—a two-thousand-year-old Chinese invention that had already contributed to more than one past demographic expansion—became widespread. It was relatively easy for Great Britain to turn toward intensive and diversified agriculture given that its colonies provided a platter of agricultural products, including sugar, a sought-after source of energy. For the first time in history, certain regions were now protected from famine. England had the guarantee of being able to import as much wheat, animal protein, and calories as it needed from the Indies, Australia, New Zealand, and the Americas. But it came at a price. With their best land used to feed the motherland, its colonies—particularly India and Ireland—were now more vulnerable to famine.

England became industrialized by deindustrializing the Indies. It transformed an active economy that produced finished products such as cotton fabric into a passive economy that only produced raw materials (cotton, tea). This phenomenon relied essentially on protectionism. Textiles imported from the Indies were subject to increasing taxes over the course of the eighteenth century, taxes that were initially adopted at the end of the seventeenth century to save the British wool industry from collapse. As a result the Indian economy was weakened while English revenues increased. China had fallen into the chaos of civil war and overexploitation of its farmland and was ravaged by the opium trade that the West imposed with their cannons. In the nineteenth century it became progressively deindustrialized, as its economic infrastructure crumbled.

The economist Joseph Schumpeter (1883–1950) dispassionately described capitalism as a continual process of creation and destruction. The unprecedented creation of England's industrialized economy (and Europe's more generally) in the nineteenth century was correlated with the destruction of preindustrial Chinese and Indian advances, which had for centuries provided a large majority of the manufactured products traded around the world.

Controlling Time and Space

The early Industrial Revolution had two distinct stages.

1. The first phase ran approximately from 1800 to 1860. It was built around stationary steam engines used in manufacturing. These did not become mobile until the 1830s when the locomotive was developed. After that, mobile engines on steam trains made it possible to create rapid connections between the major cities of the United Kingdom. The 1860s saw the emergence of a new application of this mobility: battleships. These steel-hulled military ships were propelled by steam. The waterwheels had been rapidly replaced by propellers, which were both more powerful and less fragile. This progress led to the growth of large civilian ships built of metal. But it did not make sailing ships redundant. There were still a large number of clippers on the seas up until the beginning of the twentieth century, optimizing the use of wind. However, the development of large ocean liners allowed Europe to send its excess population overseas. Millions of Europeans left for North America, Australia, and New Zealand.

After this first Industrial Revolution, mobile engines became the beating heart of modernity. They now irrigated the world with continual flows of people. Chinese, Indians, Europeans, and others were spread around the world to serve as labor. The primary physical law that had limited Monkey's movements up until now was removed. Fernand Braudel (1902–1985) expressed this superbly when he said that Napoleon moved with the same speed as Julius

Caesar, in other words, only as fast as his troops could walk. A few years after Napoleon's death, the steam train was finally able to break through this speed barrier. A century later, air transport made the vastness of our planet a distant memory. The time when sailing ships took three years to circumnavigate the globe was rapidly forgotten.

The steam train had another effect. It was a boon for market capitalism. From the 1840s, the City of London became the number-one financial market in the world. There was heavy speculation and investment on the growth of the railway network—first British, then European, then North American, then global. The City is still the number-one investor in the world, with some Commonwealth countries opportunely operating as tax havens to guarantee free-flowing trade. The introduction of the telegraph, also in the 1840s, helped accelerate the speed of the world. It was now possible to transfer information instantly. The business world was the first to benefit.

Among the industries boosted by this new steam energy was the printing sector. This contributed to intellectual life through the printing and distributing of newspapers and popular books. It also helped literacy in society as well as the spread of democratic ideas and the abolition of slavery. Everyday life was affected in other ways too. The production of gas from coal transformed London into a city of light from the 1820s. On nights when the smog, a mixture of natural fog and industrial smoke, was not too heavy, businesses could stay open later. People began to sleep less. Social rituals were profoundly changed. It is also important to mention the industrial production of large windows, which took architecture to new levels. The building of the Crystal Palace for the Great Exhibition of 1851, a gigantic building of steel and glass, was the symbol of the height of industrialized colonial England in the nineteenth century.

2. Beginning in the 1860s, the second Industrial Revolution had four major characteristics: steel, now produced in large homogenous quantities thanks to technical progress in blast furnaces; electricity, which had become omnipresent through a large-scale system of energy distribution; organic chemistry, based on coal and then oil, which increased both the palette of substances used

in industrial processes and their associated pollutants; and then oil itself, which allowed for the spread of small internal combustion engines.

The first stage of economic globalization led to the expansion of free trade between the 1860s and 1914. In financial terms it is illustrated by the large capitalization of British rail and maritime companies joined on the market by major companies such as Bayer AG (1863), Standard Oil (1870), General Electric (1892), and US Steel (1901). The second Industrial Revolution led to a shift in dynamic from Great Britain to the United States and Germany.

The growth of the oil industry occurred in conjunction with the success of personal cars, which became the essential pillar of urban development. This was the result of an industrial choice, in particular by the Ford company in the United States. Public transport was abandoned even as towns succumbed to urban sprawl. In 1800 towns and cities were home to only 3% of the world's population. By the year 2000, it was 50%. This urban concentration was made possible by the industrialization of agriculture (which meant producing more with fewer workers) and manufacturing (concentrating the labor force into factories). Over the course of the twentieth century, the large urban conglomerates of the rich English-speaking world established the model of the service-rich city center (with offices and businesses) surrounded by a residential periphery and serviced by a fleet of individual cars—the perfect symbol of modern consumerism. A commuting model, based on daily travel between home and work, was adapted nearly everywhere on Earth with disastrous environmental consequences.

This had an impact on the centralization of time itself. Up until the late nineteenth century, each major town had set the rhythms for the working day, establishing a time reference that was generally different from its neighbors by about fifteen minutes. But then London imposed its own time reference on the rest of the country, which was important for the rail companies from 1848 in coordinating the complicated network of trains over the national territory without accident. From the 1880s, British imperialism at the global level resulted in universal time being calculated from the meridian at Greenwich. French imperialism did not intervene. This was a deal between powers who divided up their influence over the world. France abandoned any reference to the Paris meridian in exchange the international adoption of its standard measure of differential distance: the meter.

Monkey had become master of time and space. But we also lost control of the technologies he created. In the 1830s, the physicist Antoine Becquerel discovered photovoltaics, and the chemist Michael Faraday developed the first dynamo (which converts mechanical energy into electricity). Windmills were still widely used, so we had everything we needed to convert them into wind driven electrical generators. But the industrial system demanded the concentration of capital resources, and this type of solution would have favored a dispersal of energy sources. The traditional industrial approach preferred large-scale interconnected systems like railroads and then networks for the distribution of electricity and oil. The nuclear power station, an immensely complex organism, is the epitome of this. Capitalism feeds on technocracy, and both favor opacity in the name of efficiency. Their complexity keeps them removed from all moral judgment. Business is business.

Fear of the Many

Hong Liangji (1746–1809) was a fervent moralist. Influenced by Confucianism, he was born in the most economically developed region of China, the Yangtze Delta, the heart of Chinese preindustrialization. He is remembered as being ruddy and good natured, famous for both his talents as a poet and his drunken excesses. His vulgar manners clashed with the cultivated society in which he lived. Working for the Qing dynasty, he was responsible for new administrative divisions within the country, he published a key geographic text, and he contributed to improving the many hydraulic mechanisms on which Chinese prosperity of the time relied. But his most important accomplishment went virtually unnoticed: in 1793 he was the first to consider the potential consequences of overpopulation in a finite world in a book titled *On the Governance and Well-Being of the Empire*.[8]

Twenty-three years earlier, having struggled to pass the administrative exams, he was appointed inspector of education in the Guizhou Province in southwest China. It was rainy, mountainous, difficult to access, and poverty stricken. The typical place a troublesome public servant might be sent. Strong demographic growth was already pushing Han Chinese toward these lands. Hong Liangji conscientiously noted that this Han exodus had multiplied the number of inhabitants of the province by five in just three decades. This meant that the traditional residents, the Miao and other tribes, had been pushed back to higher, more inhospitable lands. He turned to his abacus and calculated the growth of a typical family, then the possible progress that might be made in terms of crop densification and extension of farmland. He concluded that in a century, the population would be multiplied by ten or twenty, whereas food production would, at best, be doubled. The state would face a catastrophe. Living frugally would not be enough to avoid this fate. After much hesitation, he published his book. In it he wrote, "Question: Do Heaven-and-earth have a way of dealing with this situation? Answer: Heaven-and-earth's way of making adjustments lies in flood, droughts, and plagues."[9]

Hong Liangji indeed predicted what would happen in China. He just forgot to mention the fourth rider of the apocalypse. War would ride alongside the devastations of floods, droughts, and epidemics. The rest of Liangji's career was spent in taverns berating his incompetent colleagues and denouncing the corruption of the administration. He eventually resigned. Several years after it was published, his essay was noticed and considered (rightly) as a critique of the Manchus, an authoritarian and oversensitive regime. His punishment was announced in 1799: silence or death. He chose discretion. At the same time, on the other side of Eurasia, the Anglican pastor Thomas Malthus (1766–1834) became famous for expressing similar ideological positions. In his case, however, they would prove to be inexact.

In 1798 Malthus published an anonymous text titled *An Essay on the Principle of Population*. It was a reply to a debate he had with his father who, like many contemporaries, was convinced that humanity could transform the earth into paradise because that was the will of God. Malthus was not convinced. He argued that the population had a capacity for growth that was far superior to the earth's ability to feed that population. It followed that whenever food was readily available, the demographic growth resulting from that abundance led to the creation of more mouths to feed and thus a risk of famine. This was the trap. The Malthusian check. Humanity is perpetually on the brink of famine. The acclaimed author published several editions of his work under his own name. He was openly opposed to charity on the basis that feeding the poor would only give them the means to reproduce.

Malthus was both right and wrong. That is why his work has provoked such extreme reactions. He was right to the extent that his logic had governed the course of human history up until then. He was wrong because the British society in which he lived would be the first in the world to break out of this trap, as we will see below.

The Misinterpretation of the Minimal State

The year 1776 is a key date for the history of ideas. The United States Declaration of Independence was an illustration of the triumph of Enlightenment ideas. In the same year, Edward Gibbon published the first volume of his monumental history of the decline and fall of the Roman Empire, desacralizing the history of Christianity. He explained the spread of Christianity not as the accomplishment of God's plan but as the product of decisions by human actors. Finally, Adam Smith published his master work *The Wealth of Nations*, in which he clearly lays the basis for liberalism, which would accompany the development of capitalism.

The Industrial Revolution was associated with a strong mobilization of capital that enabled the intensification of the means of production in the hands of industrial leaders. Capitalism was the result of this, and it would progressively become more developed and more complex. In simple terms, Karl Marx (1818–1883) summarized the tension of this dynamic as an opposition between two classes: the bourgeoisie, those who have capital, and the proletariat, who only have their labor power. Liberalism accompanied the implementation of the system throughout the English-speaking world. But its symbolic father, Adam Smith, developed his theory before the emergence of industrial capitalism. His metaphor of the invisible hand, which imagined that trade would naturally harmonize the world, was primarily based on humanist morals inspired by his friend the rationalist philosopher David Hume (1711–1776). For Smith, the market economy is only possible if there is enough trust to concretize exchanges. He argued that the state had to intervene in economic life to guarantee this trust, without which there could be no growth. The state had to prevent companies, which are by nature predatory, from abusing their position. And it had to ensure a redistribution of revenue from the rich for the greater good.

That Adam Smith came to be associated with laissez-faire economics and unfettered economic liberalism is the result of a misinterpretation. Burgeoning capitalism in the eighteenth century was deeply protectionist. It was only when these early capitalists felt ready to conquer the world that they made free trade the prerequisite condition of smooth global operations. In the mid-nineteenth century, British, American, French, and Russian fleets forced China, Korea, and Japan to abandon state control of external trade under threat of cannon fire and to unconditionally open their national markets. Weaker countries were colonized in Southeast Asia as well as in Africa. At the same time, Great Britain and the United States were applying the world's most hefty customs taxes—50%!—on imported manufactured goods. The growth of global capitalism is indissociable from the affirmation of nation-states, and it is a symbiotic relationship that summarizes the last two hundred years of history. One feeds the other and vice versa.

History can teach us two lessons here that are in stark opposition to the fundamental precepts of neoliberal dogma that has been in fashion since the 1980s:

1. The rise of capitalism was only possible under the tutelage of a protectionist state. And sustained growth of any kind needs the help of the state. The Chinese have not forgotten this, having paid for it with a national humiliation at the heart of their history. Their economic policy today is in many ways similar to that of Great Britain in the early period of the Industrial Revolution.

2. Capitalism is independent of democracy. Nazi Germany was a totalitarian state but also an efficient capitalist one. Contemporary China is even more so, bringing banks and major corporations under the auspices of a single political party all at a time when new technical abilities increase the means available for social control.

Humanity at the Heart

At several different points in its history, China experienced the economic and social conditions that would have made the Industrial Revolution possible. Why did it not take this step before England? For some authors, it was because the West was able to create a grand narrative favorable to progress and freedom of thought: humanism.

Over the course of the European Middle Ages, humanity saw itself as subject to the whims of nature and the elements as the reflection of the will of an all-powerful God incarnated in the church. But from the thirteenth century, a new ideology began to develop. It started in Italy, which was then divided up into the dynamic merchant city-states of Genoa, Venice, Amalfi, and Milan. The Italian peninsula was the most advanced region of Europe. It was these merchant-citizens who first broke the rule that all wealth should be bequeathed to the church to pay priests to pray for one's soul. These merchants adopted the commercial techniques of the Muslim powers with whom they traded. Bills of exchange, merchant guilds, the rise of the banking sector, and the adoption of arabic numerals; these new innovations brought prosperity. Renaissance humanism was based around rereadings of ancient knowledge and philosophical texts. In the sixteenth century it went hand in hand with the Protestant Reformation, which did not hesitate to defy even the most deeply rooted certainties. Humanism can be summarized in a phrase attributed to the Greek philosopher Protagoras: "Man is the measure of all things."

The humanist ideal took off in the seventeenth century. From being the "measure" of the world, man became the dominant element in it. The philosopher and mathematician René Descartes

(1596–1650) encouraged us to be the "lords and masters of nature."[10] It was the beginning of the idea of progress. The advances European states had made in administration enabled better hygiene, which provided improved protection against epidemics. The last European plague struck in Marseille in 1720–1722 (chap. 8). It became possible to plan the economy and anticipate the future. The physical laws demonstrated by Isaac Newton (1643–1727) and other thinkers led to a scientific revolution that undermined the authority of the church.

The fight was on. The scientists ended up proclaiming loud and clear all the things that reason had proved. No, the earth is not at the center of the universe, it turns around the sun. Yes, animal species evolve and die out according to natural selection. One practical consequence of scientific reasoning was that fighting the plague was no longer a matter of chanting before holy relics but looking for the causes of contagion to eradicate them. And we were better off because of it.

Humanism was associated with the surge of technology. This was probably the ideology humanity needed to imagine a world of machines and ever more ways of producing energy. Humanism desacralized the world. Born in Christendom, it opened the door to the end of religion. The Industrial Revolution was also the period when the church lost its hold on people's souls, when European societies became secularized (which was not the case in the United States or the rest of the planet, except for postwar Japan). A product of a time when humans believed they were free, employment was presented as a contract. It replaced slavery and servitude. Signing this contract meant exchanging a little freedom for economic security.

The historian Yuval Noah Harari identifies three branches of humanism:[11] liberal humanism, whose ultimate avatar is neoliberalism, which has imposed its ideas and economic model on the whole world for the last three decades; socialist humanism, which largely collapsed, mired in its own contradictions, during the twentieth century; and evolutionist humanism, marked by the now disgraced nationalist ideology of Nazism.

Liberal humanism won. Although some religions are returning

to favor now, their influence is still relatively minor compared to how attractive humanism is. But this ideology is also a victim of its own success. It put science in charge of the world, but science has never responded to the ultimate questions of humanity. It made technology a kind of all-powerful magic, imbued with all the characteristics hitherto associated with the gods—the ability to change night into day, to fly through the air as fast as the wind, to speak across long distances, and to prolong life. But this technology also gave rise to the worst horrors imaginable: the atom bomb and nuclear radiation. This economic system brought most of humanity out of extreme poverty; famine has never touched so small a proportion of the human population. But our appetite is so great that soon we will need two planets to fulfill our needs. Our economic system is crushing nature. Humanism brought down God, but it has trouble outlining a desirable future. Moreover, its openness about this comes at a price. Humanism provides no more certainties. It pushes us to doubt the future. That is why God will not disappear. The embers of hope are still burning at the heart of religions, which can promise a future that reason cannot guarantee.

The question remains, however, Was the grand humanist narrative enough to explain why the West overtook China? The answer seems to be no. The Middle Kingdom had long had a secularized ideology, Confucianism, which was similar in many ways. Humanism probably helped Europe conceptualize the power it would acquire over the world through its increasing mastery of technology. But this potential also existed in China. The Great Divergence seems to result primarily from the exploitation of different environmental factors. Europe had easily accessible coal and arable land in its colonies. China lacked both.

Democracy: All about Energy?

As we have seen, morals were important to Adam Smith, who was steeped in the humanist teachings of David Hume. He was not the only one. His peers were also driven by moral sentiment, which was particularly inspired by puritanism. Two festering social wounds would be the focus of debate in this period: child labor and slavery.

Children born after the Industrial Revolution lived longer than they did before it. Initially there was no question that the progeny of the poor would go to school. They had to work, sometimes from as young as four years old. Their employers paid them less than they paid adults even though their productivity was essentially the same. In this newly mechanized world, labor now required precision rather than strength. Children were excellent at this. Their small size meant they could slip easily into the inner mechanical workings of the factories. In England, records show that the new machines of the textile mills of the 1780s meant the workforce was sometimes two thirds children! They worked twelve to sixteen hours a day with only Sunday as a day of rest. Coercive violence seems to have been the rule. Tuberculosis, scoliosis, rickets—diseases thrived.

Public opinion was moved by the plight of these children. Politicians began to defend their cause. They came up against their peers, factory owners, and industrial elites who were determined to preserve this cheap source of labor. In 1833 and 1844 laws were nevertheless passed to regulate work in factories and mines. It was now prohibited to employ children under nine years old, and those between nine and eighteen were not allowed to work night shifts or more than ten or twelve hours a day, depending on their age. Inspectors were appointed to police this, but there were not enough of them to be genuinely effective. In 1854 a radical decision was made prohibiting the employment of women and children in British mines. But it was not until the end of the nineteenth century that child labor was outlawed in the rest of Europe and in the United States.

This movement against child labor extended into a push for compulsory education and seems to have been the result of a moral awakening within public opinion, which led to a political mobilization of certain actors.

But what about workers' rights more generally?

From the very beginnings of the machine age, many artisans lost their jobs, particularly in the textile industry, where improvements to productivity meant that humans could not possibly compete with machines. An unskilled worker operating a factory machine produced the equivalent of ten specialized artisans. Reduced to unemployment and ruin, these skilled workers rose up. From the 1810s factories were attacked and machines destroyed. The probably fictional character of Ned Ludd, a sort of Robin Hood who defended workers forced into unemployment, gave his name to the English workers' movement, the Luddites. The army, sporadically reinforced with private militia paid for by the factory owners, was deployed to protect the factories. The rioters who were captured were tried, sometimes hung, and often deported to Australia to help populate and exploit that immense colony. But the movement lasted for several decades and spread to other economic sectors. Farmworkers rose up in a similar way in the 1830s, with a mysterious, fictional Captain Swing taking responsibility for the systematic destruction of the new horse-powered threshing machines.

It has been said that repression progressively stamped out this violent contestation. That is not necessarily the case. History is a battleground. We are at a point where workers' collectives, both on the farms and in the mines, would soon take up the proletarian identity proffered by socialist and communist thinkers and become unions determined to defend workers interests. These trade unions were modeled on older forms of mobilization, such as medieval guilds and corporations, or the British Freemasons lodges organized in the eighteenth century. Workers' collectives were outlawed between 1799 in 1824, which did not stop them gaining strength. The repression they were subject to goes some way to explaining the violence and secrecy of their actions.

When they were legalized, unions set themselves up as pressure groups and ensured the collective economic welfare of their members. Their financial abilities were maintained by membership dues that allowed them to pay out pensions to workers' widows and children. Building societies, mutual insurance companies, and other solidarity-based organizations were the fruit of the struggles of British unions.

In his book *Carbon Democracy*, the political scientist Timothy Mitchell argues that our deliberative democracies were forged in strikes that pitted workers' movements against those in political and economic power. For him, the nature of a society's energy regime conditioned the outcome of these workers' struggles. As a general rule, the countries that developed around energy from coal evolved toward representative democracy. But those that entered modernity through the exploitation of oil became authoritarian regimes or dictatorships.[12]

Black Gold: A Boon for Authoritarian Regimes

Could mineral energy resources condition political evolutions? This hypothesis appalls classical thinkers of liberal democracy, who prefer to see this system as the product of a confrontation of ideas, political crises, and social negotiations. But their theories do not take nature into account. In this respect they are in keeping with classical economics. According to this logic, nature is only valuable in political and economic terms as an externality. It provides free neutral resources enabling economic development. Fossil fuels are valuable only once they are extracted and processed. As a result, the ecological cost of their production and recycling is only taken into account after the balance sheet has been calculated. This reasoning persists because today it provides new markets for global finance, for example, in sectors to do with depollution or water treatment.

But Mitchell refutes this argument that energy is politically neutral. Let us look more closely at his thesis. In the nineteenth century, energy from coal was used on several levels. For example, in urbanization. Previously dependent on wood for heating, towns were geographically spread out. Coal meant that they could be denser and more concentrated. Democracy, which depends on the fact that the majority has access to the political center, became possible. Transport was reorganized and densified with the help of more efficient energy. This increased the possibility for individuals to access information; along with the development of the press, free information was one of the conditions for a functioning democracy. Political action, for example in the form of colonialism, became more efficient. Agricultural surplus increased along with the ability to trade these surpluses on the market. Coalminers could therefore demand better pay and better working conditions and defend the idea of universal suffrage—all through industrial action. This began with a general strike during the summer of 1842, which shut down most of the mines before spreading to the textile industry. The workers' collectives, future unions, were able to shut off the supply of energy to the nation. All they had to do was ensure that no one was bringing coal out of the mines. England's lifeblood was cut off. The whole system was under threat. The miners held the beating heart of the world in their hands. They could negotiate from a position of strength and demand representation in social and political negotiations. Westerners owe their enviable status as citizens today to the specific conjunction between the determination of nineteenth-century workers and their role in the means of energy production.

Then came oil. Extending the argument, Mitchell defends the idea that black gold is not a replacement for coal.[13] We still burn coal all over the planet to provide ever more electricity. Oil provides new energy possibilities but follows different regimes. Where democracy was already stable, as in the United States, oil was produced at the end of the nineteenth century through various operating structures. The liberal market encouraged the initiatives of thousands of pioneers who in turn created thousands of companies. Some managed to grow by absorbing their neighbors, becoming ultrarich cartels able to influence the political game but not change it. The United States remained a democracy. And the industrial fabric that developed here was specific to this environment. Its varied and responsive entrepreneurial ecosystem would provide an economically favorable context for the exploitation of shale oil in the twenty-first century.

But oil is also mined overseas, where the lessons of the "carbon democracy" were fully understood by the regimes where it would be mined. Let us look at the case of Saudi Arabia, an archetype of the "oil dictatorship." In 1924–1925 the House of Saoud—promoters of Wahhabism, a fundamentalist Sunni ideology—took control of most of the peninsula and gave it its name. Saudi Arabia is the only state in the world to bear the name of its leaders, for whom it functions as a deed of title. They control the holy cities of Islam: Mecca and Medina. Oil deposits are mainly located in Shiite majority zones. Exploitation of these deposits, which developed rapidly after the Second World War, were designed so that they could not be disrupted by workers' strikes. This was based on two principles. The first was division: ensure that workers were not bound to each other by any kind of solidarity by employing people from different backgrounds and instrumentalizing their divergences. The second was powerlessness: the extraction and distribution facilities were mechanized to the point where human action was limited to implementation and occasional maintenance. No strike could stop the pump connected to the pipes. A few armed men were all that it took to protect the oilfields from sabotage. For Timothy Mitchell, the conditions in which oil was extracted created a context in which representative democracy could not take hold.

From the 1870s unionism spread throughout all industrializing countries. The backdrop to this new social model was a shared dream: that the wealth produced could one day be shared equally among all. Capital owners did not necessarily agree, however. The struggles that followed would give rise to theories that would have a decisive impact on the history of the twentieth century. Henry Ford for example, decided to pay his workers more so that they could buy his cars. This principle would lead the United States to drain more natural resources than if it had chosen another economic path, for example, to encourage public transport, which was the development option chosen in the decades between 1900 and 1920. Ford's lobbying also involved taking financial control of the public transport system (along with his peers). The intentional decline of these infrastructures is what enabled the triumph of Ford's model. The progressive abandonment of buses, tramways, and trains was the necessary condition for the reign of the individual car. Fordism is based on increased productivity, which transforms humans into machines reproducing the same repetitive movements over and over on the production line. Workers are better paid and more productive; profits go up because although cars cost less, more of them are sold.

Nature necessarily lost out in this exchange. For humans, it all depended on where you stood, socially and geographically.

Moral Abolitionism, or the Correct Use of Slavery

Slaves are clearly in the most unenviable position of all workers: objectified, dehumanized, owned. The Industrial Revolution was associated with an abolitionist movement that would progressively spread across the globe. Today slavery still exists in different forms throughout the world. But it is considered shameful, concealed, and challenged by international bodies and nongovernment organizations.

The United Kingdom was the first country to permanently abolish slavery. Dissidents seem to have been the catalyst for this idea. In Great Britain, these groups were Protestant—Presbyterians, Anabaptists, Calvinists—who shared the refusal to submit to the Anglican Church. There were also moralists. Unlike most Catholics, they considered slavery to be immoral. If we accept that a Black man or woman is free to become Christian, then they must have all freedoms. For the plantation owners of the New World, this kind of thinking meant robbing them of their workforce. But the ideas of the dissidents gained ground nonetheless. The slave trade was prohibited in 1807. Slavery itself was abolished in 1838, preceded by a transition period of five years in which slaves who had been officially freed were forced to continue to serve their former masters by way of "compensation!"

Britain then managed to export the abolition of slavery. They did so through diplomatic negotiations with other European powers and through violence elsewhere. The colonization of Africa was, in part, justified by the fight against slavery. Former slaver kingdoms were invaded by their ex trade partners on the pretext that their slave industries, yesterday so banal, were now barbaric.

The United States was torn in two on the question of slavery. The feud ended in the Civil War (1861–1865) which was the first total war in the sense that both sides had unprecedented means at their disposal. Troops were deployed using railways, instant telecommunications were made possible by the telegraph, state propaganda was complacently spread in the press. There were repeating rifles, battleships, heavy artillery shelling, trenches, and widespread bloodshed. The Confederacy in the South wanted to preserve the plantation economy as it was, while the Union in the North defended the moral imperative of abolition. The Union won.

The global abolition of slavery now seemed unstoppable.

Given this success, some philosophers and historians of the environment have today taken up the idea of moral abolition. If moral awareness raising of the atrocities of slavery could bring an end to slavery as a system, could our dependency on fossil fuels be abolished in the same way? These thinkers defend the idea of "human engines" and "energy slaves."[14] It is easy to call for boycotts against a company that uses children to make basketballs. But it is much harder to mobilize people around environmental questions even though they raise a profound moral dilemma: what world will we leave to our children and grandchildren when we have robbed it of its mineral reserves and left it sweltering and polluted?

Thinking in terms of energy slaves means imagining the number of humans that would be required to produce the energy we rely on for our electricity and machines on a daily basis. On average, a human works at roughly 100 watts per hour. How many energy slaves do we each use every day to travel, to eat, to light our houses? One liter (about a quarter gallon) of gasoline, 10 kilowatt hours of energy, is equivalent to the work of ten humans (if we take into account an adjustment for the energy expenditure of the motor). This enlightening calculation was proposed by the French engineer Jean-Marc Jancovici.[15] In 2012 he reported that an average French person consumed an average of around 30,000 kilowatt hours of final energy per year, all energies and usages combined. This is the equivalent of around 50,000 kilowatt hours of primary energy. This works out to around four hundred "energy slaves" just to ensure a comfortable lifestyle. It does not include externalized energy (e.g., the energy cost of the Chinese manufacture of the smartphone this French person would buy that year). It also does not include gray energy, the transport energy required to bring the appliances to France.

The poignant analogy of energy slaves provides a magnificent way to raise awareness about the upheaval that the use of energies—from coal, oil, and uranium—represents for our existence. But just how effective its moral objective is in inducing guilt that will force us to limit our consumption or even make us change energy regimes is less clear.

The idea of energy slaves, which relies on the principle of a moral awakening,[16] should also be nuanced somewhat. Did industrializing countries in the nineteenth century abolish slavery purely for moral reasons? A second motivation seems to have been just as important: a good understanding of the interests of capital. From the eighteenth century, liberal economists had argued that slaves were less productive than paid workers. Free laborers worked better because they wanted to work.[17] The business world that became increasingly powerful in the industrial sphere came to have more influence than its rivals whose wealth was based on colonial and financial capital. Slaves were far away, in the colonies. But national factories operated with a paid, local workforce enabled by coal-burning motors that amply surpassed what could be produced by humans. Capital from the slave trade had perhaps enabled the means of industrial production to become more concentrated, but that became secondary. Financial flows were now based on manufacturing. If liberalism had actively participated in the abolition of slavery, this was also because economic conditions were favorable to it. Liberals worked alongside dissidents, those Protestants with such high moral values.

According to the "energy slaves" metaphor, we all live like kings today because the Industrial Revolution gave us the energy equivalent of four hundred people working exclusively to satisfy our individual needs. But is the image of ourselves as eighteenth-century plantation owners enough to make us feel guilty? Monkey has made energy his servant. Armies of (energy) slaves now bend to his every will. But the energy genie, appearing out of a gigantic Aladdin's lamp, has placed severe limits on our future. To tell the truth, Monkey does not care. He lies to himself to better ignore reality. Like the proverbial three "wise" monkeys, he sees no evil, hears no evil, and speaks no evil. But that does not mean he does no evil.

In the nineteenth century, three different catastrophes would, like sinister premonitions, sketch out visions of other possible futures.

14

THE COLD CHILL OF CATASTROPHE

Volcanic eruptions, solar storms, overexploited resources. Can natural disasters affect Monkey? Are we in fact the worst scourge of all?

Three columns of fire in the sky. Blocking out the stars. An incandescent mountain spitting out explosive volcanic rock, blinding ash, corrosive aerosols, and clouds of suffocating gas provoking violent tornadoes and repeated tsunamis. The Indonesian island of Sumbawa had become hell on Earth. Ten thousand people were buried under molten magma and thirty thousand more would die in the next forty-eight hours. A black cloud covered the center of Indonesia smothering the rice fields with a blanket of ash one meter (3.3 feet) thick. The month that followed would see another fifty thousand die, victims of starvation or fluoride poisoning from the ash that saturated the water. But in spite of everything, even at the very foot of the volcano, some people survived the apocalypse: the elites of the closest town. Perhaps they had the fastest horses to flee on? In any case, they lived to tell the tale.[1]

That was two hundred years ago. The eruption of the Tambora volcano ejected between 100 and 200 cubic kilometers (about 29 and 48 cubic miles) of ashes and gas into the atmosphere. Where twin peaks had once stood lost in the clouds, the mountain disintegrated and was replaced by an immense crater 6 kilometers (about 3.7 miles) in diameter and more than a kilometer (about 0.6 miles) deep.

On the fringes of the colonial empires, this event went largely unnoticed. The eyes of the world were fixed on Paris, which Napoleon had just reached after his extraordinary escape from Elba prison. He claimed to want to conquer Europe with an army of half a million men, before his final hopes were dashed at Waterloo on June 18, 1815.

But on April 10 of the same year, just two months before that bleak day of battle, the Tambora eruption plunged the entire world into three long years of turmoil.

The Long Winter

Perhaps the best portrayal of the mood during this period comes from romantic literature. In his aptly named poem *Darkness*, Lord Byron (1788–1824) put it like this:

> I had a dream which was not all a dream,
> The bright sun was extinguish'd, and the stars
> Did wander darkling in the eternal space,
> Rayless, and pathless and the icy earth
> Swung blind and blackening in the moonless air.
> Morn came and went, – and came, and brought no day
> And men forgot their passions in the dread
> Of this their desolation; and all hearts
> Were chill'd into a selfish prayer for light.[2]

These lines were written during the summer of 1816. Byron was living in Switzerland, where he spent his evenings with cultivated companions like Percy Shelley and his lover, who would soon achieve renown under her married name, Mary Shelley. The weather was abominable. There were one hundred and thirty days of rain between April and September. The lake broke its banks and flooded Geneva. Byron raved about the violence of the storms. Paintings by William Turner and others of the time depict skies tinted with an unnatural red glow. Recent analyses have confirmed the chromatic accuracy of these works. Between 1816 and 1818, these artists unknowingly recorded the optical effects of large quantities of particles and gases projected in the stratosphere by the Tambora explosion in 1815.

But it was not just the color of the sky that changed. The earth's climate was radically disrupted. Volcanic aerosols, airborne particles including sulfuric acid, altered the chemical composition of the atmosphere. They changed its physical characteristics, increasing its ability to filter the sun's rays. They blocked out the light and gave the sky a sinister yellow-orange hue. Average annual temperatures dropped by 1°C (1.8°F) during the summer of 1816. In some regions they dropped by up to 7°C (12.6°F). This was the case in London, for example, in 1816. The drop in temperatures continued until the autumn of 1818, with a global average of −0.5°C (−0.9°F), locally increased up to −3°C (−5.4°F).

In many European countries, the year 1816 was described as the "year without a summer," and often consequently "the year of hunger." Harvests were drowned under glacial rain in Great Britain or shriveled by drought in Scandinavia. Europe went through its last major famine. Even the colonies proved unable to supplement the food supply as they were also mired in similar meteorological changes. In the middle of August 1817, several severe storms covered a large part of North America and Europe in a blanket of snow 30 centimeters (about 1 foot) thick.

The deadliest effects of this catastrophe occurred in the world's most populated regions, southern and eastern Asia. The effect of the aerosols brought the monsoon to a halt. Yunnan Province, known as the rice fields of China, was struck by a famine so severe that mothers sold their children as slaves for a few liters of gruel in the hope that their new masters would be able to save them from starvation. This was not an option for Indonesians, however, because the extension of British abolitionist policy had brought slavery to an end in that part of the world. But debt bondage was one of the pillars of the Asian economy. With this option for relief of food insecurity removed, Javanese or Balinese families unable to sell their additional children preferred to put them out of their misery rather than watch them starve.

In Bengal, the showcase of the British East India Company, cholera mutated under the effects of changes in rainfall that warmed the water and reduced the flow of the rivers. Previously confined to Southeast Asia, the disease became an epidemic that conquered the world in the following decades. It followed in the footsteps of colonial powers, hungrily exploiting the globalized connections of Her Majesty's ships. But between 1815 and 1818 it was typhus that prospered from Indonesia to Ireland, accompanying the legions of starving peasants who converged on the towns. Throughout Europe, soldiers discharged from the army after the Napoleonic wars were unable to feed their families. Hunger riots spread, and many public buildings were burned down.

In this global disaster, there were some who got lucky, such as farmers in eastern Russia and in the Mississippi basin, for example. For them, the climate warmed slightly, providing spring rain and sunny summers. These peasants glimpsed wealth for a time, able to sell their momentous grain harvests at huge prices to a starving world. But they were struck with an economic crisis when global harvests returned to their previous levels after 1818.

Why did the Tambora eruption have such an impact? First, it occurred near the equator, which maximized the global dispersion of atmospheric debris. The same explosion farther to the north or to the south would have caused less damage, proportional to its distance from the equator. Second, it was violent enough to project immense amounts of particles into the stratosphere to an altitude of more than 45 kilometers (28 miles). Finally, the 100 to 200 cubic kilometers (about 29 and 48 cubic miles) of ash and gases released constituted a volume able to enduringly alter the planet's climate.

Recently, studies have suggested that the eruption of another Indonesian volcano may have been responsible for the beginning of the Little Ice Age.[3] On the island of Lombok, between Sumbawa and Bali, the Samalas volcano erupted in 1257. This explosion was comparable in intensity to that which destroyed Tambora in 1815, but it produced twice as much sulfuric acid. The effects of the volcano's anger were global, very similar to what occurred in 1815–1818. In London, the summer of 1258, which was meteorologically similar in every way to the summer of 1816 (continual torrential rain, glacial cold, permanently darkened sky), bought a famine that wiped out a third of the population. The analysis of a charnel house from that time, where more than fifteen thousand bodies were hastily piled, has shown that these people died of starvation. Some were even victims of cannibalism.

The Little Ice Age seems to have been prolonged by several other significant volcanic blasts with eruptive volumes exceeding 100 cubic kilometers (24 cubic miles). By way of comparison, the largest eruption of the twentieth century (that of Pinatubo in the Philippines in 1991) "only" produced 10 cubic kilometers (2.4 cubic miles) of matter. In 1453 a megaexplosion, probably exceeding 50 cubic kilometers (12 cubic miles) suffocated Kuwae island in Vanuatu and revived the Little Ice Age. This event affected skies and temperatures everywhere. It was documented in chronicles from Cairo, Peking, Moscow, and Paris.[4] In Constantinople, which was on the point of falling into the hands of the Ottomans, people were concerned about foreboding omens—blood red skies, glacial cold, or a comet that seemed unusually large. In 1600 it was the Peruvian volcano Huaynaputina's turn. This eruption would be largely responsible for the great famine that killed a third of the Russian population between 1601 and 1603.

The harmful effects of the Tambora eruption, visible in the polar ice, were amplified by another megaeruption six years earlier in 1809. The volcano responsible for that one has not yet been identified, but its aerosols had already had a considerable impact on the stratosphere. The cumulated effects of these two eruptions made the 1810s the coldest decade for a thousand years. The records of George Mackenzie, pioneer and meteorologist, show that in Great Britain the number of sunny days per month went from twenty between 1803 and 1810 to five between 1811 and 1820.

On August 27, 1883, the "small" eruption of Krakatoa (Indonesia), six to eight times smaller than that of Tambora, produced effects that Western science (which was now paying attention) was able to physically trace at the global level. The explosion could be heard 6,800 kilometers (1,631 miles) away, and the resulting tsunami, after devastating Asia, reached as far as London, where it petered out in the form of a 30-centimeter (about 1 foot) wave.[5]

Dr. Frankenstein's Moral Dilemma

Percy and Mary Shelley, like Lord Byron, are key figures in English romanticism. This literary school was an intellectual reflection of humanity's increasing power in the physical world. At a time when the earth was entering the Anthropocene and the Industrial Revolution was beginning the large-scale emission of human-generated greenhouse gases, romanticism evoked the power of nature and its destruction of former civilizations. The years between 1815 and 1818 marked the shift in power between nature and humanity. The echoes of Tambora's mighty belching would be the last sign of the geological supremacy of the elements. From this moment, Monkey became the primary actor in the transformations of the globe. The Anthropocene was the stage for this power shift. From 1818 humanity itself became the driving geological force of the new era. Telecommunications would soon concretize the idea of a shared humanity (chap. 16). Fiction would accompany, or even precede, humanity's awareness of the process. Certain horror, suspense, or science-fiction stories would paint premonitory visions of the new world in the making.

In this chilly, wet summer of 1816, Byron and the Shelleys talked a lot about the end of the world in their villa on the edge of Lake Geneva. A challenge was launched: which of them would be best able to depict the horror they saw behind the incessant rain and biting cold of this world "without a summer?" In response to this challenge, Byron described an apocalypse marked by leaden days, endless wars for food, and dogs devouring the bodies of their masters. It ends with a vision of a barren world. In passing, he sketched the modern figure of the "vampyre," undead and able to drain the energy of the living to prolong his earthly existence. For his part, Percy was convinced the planet would get colder. On the Swiss alps where the snow never melted, he imagined glaciers that would progress until they eventually covered the earth. But unexpectedly, it was Mary who won, with her novel *Frankenstein; or, The Modern Prometheus*.

Frankenstein is a multifaceted metaphor that operates on several levels thanks to the clever construction of the narrative. It raises the question of humanity's extinction as icebergs cover the earth. But it also explores the creation of a new, potentially immortal species brought to life by Dr. Frankenstein's experiments with electricity diverted from lightening. Shelley depicts a golem produced not by mysticism but by science, evoking the eighteenth-century experiences of the physicist Luigi Galvani, who demonstrated that electric charges could be used to animate dead tissue (frogs' legs). According to Ashkenazi myths, certain rabbis could bring an anthropomorphic clay statue to life by drawing the Hebrew word *emet* (truth) on its forehead. But the appropriation of divine power in conferring a soul on matter necessarily ended badly. Luckily there was a safety net; erasing part of the inscription on the monster's forehead transformed the spell into *met* (death) and returned the creature to its previous inert muddy state.

However, in Mary Shelley's story, Dr. Frankenstein is consumed by his ambition to equal the power of God and fails to factor in an off switch. His creature gets away from him. Initially the monster tries to integrate into society but is rejected because of the fear he inspires. He begins to loathe these humans who shun him, and his hatred becomes murderous. He then turns on his creator, demanding a companion to ease his loneliness. Frankenstein agrees before foreseeing a future in which the descendants of his creatures will eventually replace humanity—and do so all the more easily because they are indifferent to the biting cold sweeping the world. Frankenstein destroys his monster's bride and sets out to hunt his creation. The megalomaniac inventor dies in a world covered in ice. But Mary Shelley makes the creature ambiguous. His final words above all reveal his intrinsic humanity and his desire to do good.

In 1816 against a backdrop of climatic upheavals, Mary Shelley formulated the dilemma that our descendants will perhaps have to face in a world where androids and artificial intelligence may well be better adapted to our altered environment than we are.

The Threat of the Sun

On September 1, 1859, the English astronomer Richard Carrington (1826–1875) was observing several spots on the surface of the sun that had appeared a few days earlier.[6] They were getting bigger, to the point where they were visible to the naked eye. Then, at 11:18 a.m., a lightning bolt formed on the surface of the star joining the spots and soaking up their energy. It was a colossal sun storm followed by a blinding eruption of matter. Seventeen hours and forty minutes later, the earth's sky was lit by a flash of light so bright you could read the newspaper in the middle of the night. The aurora borealis stretched out to the tropics, and electric fires ran through ships' rigging while the sea went wild. Telegraph operators from North America to Australia, Norway to the United Kingdom, all reported the most worrying phenomena. The telegraph stations using the Bain system (named after its inventor Alexander Bain and based on giant chemical batteries producing electricity) were hit with substantial electric shocks. The operators had to disconnect themselves, fearing for their lives, and the paper on which the messages were printed caught fire. Several employees were wounded, burned, even electrocuted. The telegraph network was partially destroyed by this series of short circuits.

This was a geomagnetic storm, lucidly diagnosed by Carrington in the months that followed. As a result, it is now known as the "Carrington event."

Since then, the cycles of the sun have been studied very closely. A risk scale has been developed based on observations conducted since the late nineteenth century. But nothing of this magnitude has been seen since, and experts predict its probability at around one or two every thousand years.[7] Retrospectively, it seems that the Carrington event well exceeded all established scientific scales for measuring extreme geomagnetic events. It was gigantic in comparison with the solar storms that have struck Earth since.

Records from the past suggest, however, that the Carrington event may not have been the first of its kind. For example, the philosopher Seneca described a possibly similar event in 37 BCE. Around 2012 reports from NASA, the US Army, and insurance companies estimated that there was a 10% to 20% risk of a Carrington event reoccurring in the decade between 2012 and 2022. There was indeed an eruption in 2012 but on what is for us the "hidden face" of the sun. The energy was sent off in the opposite direction to the earth.

In 1859 the only widespread use of electricity was the telegraph system, which had a few thousand kilometers of lines. It was quickly repaired. If such an event were to occur today, the magnetic flow would pulverize the earth's magnetic field, sweeping away satellites. In all likelihood, it would affect most electrical and electronic circuits and destroy a portion of the transformers that ensure the conversion of current within electrical distribution networks. Certain parts of the world would be cut off from communication and left without transport, electricity, or heating and would possibly fall into social chaos. Imagine what would happen if nuclear power stations were affected. The memory of Chernobyl (1986) and Fukushima (2011) should be enough to alert us to the dangers of such a scenario, which would relegate these global catastrophes to the rank of minor preludes to disaster.

Sound like science fiction? Having been burned by a "little" solar storm that wiped out part of the North American and Scandinavian infrastructures for electricity distribution in 2003, preparations are now underway. The United States now has a duplicate of essential telecommunication networks, including a part of the internet, and have buried cables in deep underground bunkers. The objective of this is to allow the state and the army to remain operational in case of a cosmic cataclysm or nuclear fallout. The threat has been deemed sufficiently credible for decision makers to invest billions of dollars in it. In Canada, the Hydro-Québec electricity company has invested more than one billion dollars to protect its transformers from this kind of event.

Easter Island: A Metaphor for What?

A pinpoint lost in the immensity of the ocean. With a surface area of 163 square kilometers (63 square miles), barely one and a half times the size of Paris, Easter Island, now known as Rapa Nui, is 3,700 kilometers (2,299 miles) from its nearest neighbors and located between Chile and Tahiti. It is a mythical place. The Polynesian sailors who found it eight or nine centuries ago had both daring and maritime skill that demands respect. They conducted at least one colonization expedition using their immense catamarans, the same ones that took them to Hawaii and New Zealand in the same period. All this was accomplished without a knowledge of metals and by relying on subtle clues such as the color of the clouds, the reflections on the water, ephemeral maps made of shells and twigs laid out on the sand, or mnemonic chants that were transmitted between captains.

On Easter Sunday 1722, the Dutch explorer Jacob Roggeveen landed on the island with his men. They were there for reconnaissance, food collection, and to fire a few gunshots to scare away the island's inhabitants, who laughingly grabbed the curious objects that the new arrivals wore, particularly hats and pocket handkerchiefs. The native people seemed healthy; they raised chickens. And they had erected gigantic statues that the Dutch believed to be made from clay.

In 1770 the Spanish captain Felipe Gonzalez de Haedo landed. He brought together the native chiefs and had them sign a document by which Spain annexed their island. They were unable to understand the concept of colonization, but apparently they would remember that writing wielded great power to the point where they invented their own.

Four years later, in 1774 the British explorer Captain James Cook found the island in a state he considered to be disastrous. No wood, no source of fresh water. Everything that a ship's captain needed for supplies was lacking. The statues, which he identified as being sculpted from basalt, seemed to have been abandoned. Some lay in ruins.

In 1786 the French admiral Jean-François de Galaup, comte de Lapérouse, anchored off the island. His officers described a fertile island able to support its population. Like the previous visitors, they complained of incessant pilfering by the local inhabitants.

It was apparently during this period that the islanders abandoned the statues of their protector ancestors from the ocean. They adopted a new belief system based around the bird man. Every year an initiation ritual determined who would be king for that year. Each tribe selected an athlete to represent them in this competition. Climbing, running, swimming: the chosen ones competed in a deadly triathlon. The winner would be he who brought back a tern's egg, a sign of divine favor, after climbing down a vertical cliff 300 meters (985 feet) high, swimming 2 kilometers (about 1.25 miles) in turbulent shark-ridden waters to an isolated island, searching every nest, finding the treasure, and then returning by the same path, obviously without breaking his trophy.

From 1805 Western ships began raiding the island to collect slaves. The Indigenous people, who were formerly friendly, now received the arrivals with open hostility but were powerless in the face of guns. Deportations peaked between 1859 and 1862, a period during which Peruvian slavers rounded up 1,500 islanders to work in guano mines. Over thousands of years, this seabird manure had been deposited in layers several meters thick along the South American coast. It was an extraordinary fertilizer, which is why the Incas used it to improve their soil in the Andes plateaus. This treasure would be exported to Europe in such quantities that almost nothing is left of it today. Two thirds of the slaves from Easter Island who worked in this industry died of disease and poor treatment in the first months of servitude. Others met similar fates, having been exported to the plantations of French Polynesia and Australia. In the 1870s the island had no more than one hundred native residents. Having chased off the handful of missionaries who had set up on the island to evangelize the survivors, a French adventurer proclaimed himself king. He was assassinated in 1876 but not before introducing sheep farming onto the island and committing widespread abuse.

Chile annexed the territory in 1888. It would now become a sheep farm, entrusted into the care of a Chile-based British company. The Indigenous inhabitants were restricted to a village and had neither rights nor occupations. It was not until the 1960s that the situation improved and not until the 1970s that the island opened up to the world. NASA built a disproportionately large landing strip there so that the space shuttle could land if it had to come down urgently in this remote region.

The dramatic situation that unfolded here fueled of all kinds of fantasies, including those of extraterrestrials who helped to build the colossal statues.[8] The biologist Jared Diamond, among others, has popularized Easter Island as a metaphor for the possible destiny of our planet: a limited environment with no possibility for escape in which humanity exhausts all resources to the point of endangering our own existence.

According to Diamond, there are five factors that together may lead to societal collapse: the overexploitation of resources beyond their ability for renewal (such as deforestation); climate change;

the breakdown in trade with neighboring societies; military attacks; and inappropriate reactions in the face of change.[9] The historian Joseph A. Tainter put forward a complementary theory,[10] arguing that societies collapse out of inertia. They prove themselves unable to modify their relationship with the environment when that environment is changed.

For Diamond, the Easter Island civilization died out as a result of excessive competition for prestige between tribes. Driven by their chiefs, the Polynesians dedicated all the island's resources to the construction of the *mo'ai*, the monolithic stone statues of ancestor gods. They cut down the trees to make rollers to transport the statues to their finale sites. This deforestation resulted in erosion and famine, pushing the survivors to kill each other or even eat each other. The island's elites apparently destroyed their environment through lack of foresight. Diamond summarizes the process in a single word: ecocide. The murder of an ecosystem.

Experimental archaeology has proposed alternative scenarios imagining other methods of shifting these giants blocks of stone, some of which weigh dozens of tons, based on the means available to the islanders (reduced labor power, rope, and little if any wood). These alternative scenarios suggest that deforestation was probably the result of rats and the chickens brought by the Polynesians as food. It prevented them continuing their maritime activities (without wood, it was impossible to make other vessels). But the people knew how to transform the island into a garden with large walls and circular depressions to protect crops from the elements. Polynesian colonizers arrived with sweet potatoes, yams, taro, sugarcane, bananas, and mulberries (whose bark could be used to make clothes). They adapted to change.

Demographic collapse was plainly not linked to deforestation but to the arrival of Europeans and the microbial shockwave that hit in the eighteenth century. Possibly the decline in the population was exacerbated by periods of drought linked to El Niño, which we now know caused the inhabitants of the Pitcairn and other Pacific islands to migrate. Easter Island society was ultimately finished off in the nineteenth century by slave raids. Traditions were lost. The surviving islanders later sought inspiration from the culture of their Polynesian brothers in rebuilding their own. Easter Island was not the victim of an ecocide but rather a genocide that the French ethnologist Alfred Métraux described (in 1935) as "one of the most hideous atrocities committed by white men in the South Seas."[11]

Easter Island is a metaphor for Earth. It testifies to a destiny that was shared by many peoples and environments in the nineteenth century.

Outside Europe and the United States, which had begun industrialization, most peoples around the world still depended on local agricultural resources for survival. Yet colonial powers increasingly monopolized the best farmland for their crops, sugar, coffee, cocoa, and rubber plantations. Being forced to produce monocultures often broke down the transmission of local agronomic skills and knowledge. New diseases decimated food resources. Cattle plague, for example, wiped out the livestock of many herdsmen in East Africa and left famine in its wake. But this situation was also combined with substantial meteorological changes resulting from El Niño, which regulates much of the world's rainfall and which is particularly crucial for the monsoon.

In the second half of the nineteenth century, conditions were right for the "colonial genocides" denounced by American historian Mike Davis.[12] Three massive waves of drought struck the world in 1876–1879, 1889–1891, and 1896–1900. These "climates of hunger" led to an estimated total of between thirty and sixty million deaths spread over India, China, Southeast Asia, Africa, and Brazil.[13] Even as millions in India were dying of starvation, the colony continued to produce wheat for the British motherland.

This process was not new. It had just scaled up. The Irish example is particularly insightful in this respect. Ireland was the first English colony and had been a test for colonial exploitation since the seventeenth century. Colonizers seized the best farmland and reduced the initial occupants to the status of farmworkers. Mired in poverty, many of the Irish chose exile. If they did not, living alone with no dependents was the only way to have enough resources to survive. In the eighteenth century, the arrival of the potato from South America was providential for them. The poor-quality land that the colonizers had left them was well suited to these tubers, which provided bountiful harvests. The Indigenous population grew to 3.2 million in 1754 and to 8.2 million in 1845. But in 1840, a new arrival appeared from the homeland of the potato, *Phytophthora infestans*, or potato blight. This microorganism reduced the potato harvest to a pile of black rotten mush. Between 1846 and 1851, famine carried away one million people who were officially citizens of the most developed country in the world. One and a half million Irish people migrated, primarily to the United States. With a population of 6.4 million inhabitants today, three quarters of whom are in the Republic of Ireland and the rest in Northern Ireland, the country has never regained the population it had before the famine.

The Last Sea Cow

Here is a riddle: What am I? I am an animal. I weigh up to ten tons. I am a marine mammal related to the dugong, somewhat like a seal the size of a whale. Two thousand years ago, my habitat stretched from China to California and all the way up to Greenland. Four centuries ago I could still be found off the coast of Newfoundland. Today all that is left of me are three full museum specimen skeletons. My name? Steller's sea cow.

George Wilhelm Steller was both the "discoverer" of this strange creature and the cause of its extinction. He was a German biologist employed by the Danish explorer Vitus Bering. Working for the Russians, they set out in 1741 to explore the coast of Kamtchatka, the peninsula that extends Siberia into the frozen waters of the North Pacific. But they met with shipwreck and scurvy. The crew found themselves trapped on the Commander Islands, named Bering and Medny, more than 100 kilometers (62 miles) from any other land. During his exploration Steller noticed these strange animals. Gentle aquatic giants 8 meters (about 26 feet) long who lived in small groups around the coast and fed on kelp, large algae that formed genuine underwater forests. They were not afraid of humans. A sailor with a makeshift harpoon could easily get close enough to kill one. The most difficult thing was bringing the massive carcass back to shore. Steller also noted that these monogamous animals swam anxiously along with their dead partners and were essentially killed in pairs. The starving survivors of the expedition reveled in this good fortune, which allowed them to survive the winter. Stella describe this meat as similar to the best kind of beef—hence the name "sea cow"—and the fat as tasty as good butter.

Bering died of scurvy along with a third of the crew. The forty-six survivors managed to regain the mainland on rafts. The news of their feast spread. In less than thirty years, a series of expeditions in search of fur and meat had wiped out the last of these animals.

The sea cow went extinct at a time when Westerners had only just become aware of its existence and included it in their biological categorizations and museum collections. Gentle and easy to hunt, Steller's sea cow joined the ranks of the other aquatic giants that had once been the delight of the first humans to descend along the American coast (chap. 2).

Monkey has always hunted large animals. But from the nineteenth century, the conjunction of new forms of transport (steam) and improvements in artillery (in the form of the explosive harpoon) allowed us to attack marine giants without danger. Whale hunting had been practiced for several thousand years along the coasts of Korea, Japan, and in the Basque country. And inhabitants of the polar regions in the north—peoples related to the Inuit, Yupik, and Aleut—had developed skin-covered canoes and floating harpoons from the ninth century CE. This combination of technologies was the basis for a unique lifestyle involving seasonal whale hunting for meat, which was preserved to enable large numbers of people to survive winter on the ice.

This traditional hunting had a marginal impact. But the emergence of whalers decked out like battleships in the second half of the nineteenth century transformed an artisanal hunt into a systematic massacre. As soon as technology enabled us to attack sperm whales, hundreds of thousands of these creatures met the same fate. The carnage served to extract a few tons of blubber per whale to be used as oil for lamps, for lubricating machinery in factories, or for making soap. The bodies of these cetaceans were used to make fertilizer. This massacre reached its peak in the first half of the twentieth century.

In 1519–1520, the first white explorers to venture near the coast of Patagonia were the crews of Fernand Magellan's ships. They saw colonies of thousands of marine animals. Sea lions, elephant seals, and various kinds of penguin all thronged around the coast. At night the sailors also saw fires glowing in the dark, hence the name given to this Patagonian archipelago, Tierra del Fuego. Four ethnic groups lived here, Selk'nam, Haush, Yahgan, and Kawésqar. Those who lived farthest south had populated the barren islands of the archipelago, which were constantly swept by glacial winds and driving rain, all the way to the infernal Cape Horn. They spent their lives naked. Any kind of clothing would have quickly frozen into shrouds of ice. They kept themselves warm by lighting fires and covering their skin in the fat of marine mammals. They sporadically hunted sea lions. Their canoes were made of successive layers of bark which gave them unparalleled flexibility and meant they were able to navigate on the most unpredictable and hostile sea imaginable. The women fished naked in 4°C (39.2°F) water to collect the selfish that was abundant around the coast. They had lived like this for thousands of years.

White explorers arrived in the nineteenth century. To the north, with the support of the Chilean and Argentinian states, they enforced a genocidal policy to crush the Mapuche and Tehuelches who lived on the hillsides and steppes and survived by hunting the guanaco. These people were the source of myths about Patagonian giants. In the sixteenth century, the Spanish explorers who first encountered them found them huge. But these hunter-gatherers were simply physically active and well fed, and therefore healthy and athletic and around 1.8 meters (about 6 feet) tall. By contrast the European sailors barely reached an average 1.65 meters (5 feet 5 inches). But by the end of the nineteenth century, having been pushed onto less fertile land, and with the guanaco killed to leave grazing room for sheep, these "giant" Patagonian people were no taller than around 1.55 meters (about 5 feet). This can be seen in photos of missionaries, where tall, bearded figures are surrounded by crowds more than a head shorter than them. The height of populations remains extremely variable and dependent on available resources.

Today there are still Mapuche and Tehuelche people. The more southern groups, however, have almost entirely disappeared.

In the second half of the nineteenth century, these four southern ethnic groups were, like their counterparts in the North, expelled from their hunting grounds, which were turned to pasture. Public opinion in London took their side, and missions were set up to save them. But these Indigenous people lived naked, so the missionary's first priority was to clothe them. Used clothing was collected in London and distributed in Patagonia. Less than a year later the tens of thousands of those who had taken refuge in the missions were almost all dead, bitten by fleas and lice and contaminated by a wave of diseases against which they were not immunized. They were killed by those who had come to save them.

Their coastline met an identical fate. To process seal fat (which was of lower quality than whale but used in a similar way), it had to be heated. But these desolate lands seemed to have nothing to burn. Except for penguins. Factories were built, millions of penguins were slaughtered, and their carcasses dried in the sun before being fed into the factories' furnaces to melt fat from hundreds of millions of sea lions. Killing penguins and sea lions was seasonal work. The animals were caught at night when they were the most vulnerable and massacred at dawn with clubs.

It only took a few decades for Patagonia to become a desert. It was perhaps one of the last pristine places on the planet. The low level of technology among the people who lived there protected it from human depredation. But it was not a hospitable environment. That might give us a clue as to the wealth and biodiversity of the rest of the world before the arrival of Monkey.

Today, Patagonia is only partially inhabited. The hundreds of thousands of square kilometers of frozen islands where these naked people once lived are now completely deserted. Even top-level athletes covered head to toe in Gore-Tex and boasting the latest navigational technology have trouble surviving here more than a few days.

Only one immigrant has successfully adapted to this region. The

Canadian beaver. In the 1950s twelve pairs of this 30- kilogram (66 pound) animal were imported to begin a breeding program. But their fur proved unusable. Ethologists later suggested that the quality of the fur is linked to the presence of predators. Without wolves or lynxes to hunt them, quality seemed to decline. In any event, the market for beaver fur rapidly disappeared, and the unhappy owners of this fur farm let their animals escape into the Patagonian wild.

The beaver family was happy here. There were no predators, not even any humans to bother them. They changed their diet, switching from greenery to seafood. Their growing population left the coastline sterile and began to methodically deforest the interior. These beavers grew to an average 50 kilograms (110 pounds). They have become a pest that Chile and Argentina do not know how to eradicate.

This history of Patagonia is like a faded photo, stuck between Tasmania and Cameroon, picked among hundreds of other possible illustrations in the great album of the destruction of the world's biotopes over the nineteenth and twentieth centuries.

The Omens of Global Warming

The Industrial Revolution, which led humanity to burn ever-larger quantities of fossil fuels, was responsible for the beginnings of the global warming we are beginning to feel the effects of today. We often hear that humanity has only recently become aware of this very late in the day. That is untrue. And we can learn a lot from the way in which this inconvenient truth emerged.

The *Epic of Gilgamesh*, written 3,800 years ago, already demonstrated a certain awareness that deforestation can lead to the aridification of entire regions. Other early voices made similar observations over the centuries in Europe and in China. At the end of the eighteenth century, in Enlightenment Europe, the dominant consensus was that land clearing and agriculture moderated the climate and produced meteorological conditions favorable to the development of civilization. This was the source of the then-dominant belief that Europe was civilized while America was yet to be so. In 1780 the French naturalist Georges Louis Leclerc, comte de Buffon, explained that "the face of the earth now carries the imprint of the power of man."[14] He even speculated, as a precursor to geoengineering,* that humanity could "modify the effects of Nature ... to resist the intemperance of the climate."

From the nineteenth century on, various questions were raised for scientific debate, including whether or not cutting down trees is beneficial to the climate. In France the controversy was resolved in the negative. This allowed the state authorities in the 1820s to prohibit peasants from cutting down trees to avoid modifying the climate. The royalist state blamed the revolutionaries for the climate problems. Having confiscated the woods and forests that the aristocracy had managed up until then, and through allegedly savage deforestation, supporters of Danton and Robespierre were accused of having plunged France into an environmental chaos, and it needed saving. The same logic was extended to the colonies. It was implied that North Africa was too hot because it had been deforested and that it would be possible to make the climate more clement by replanting trees. In the 1860s the colonial authorities encouraged the planting of two million eucalyptus in Algeria with the deeply racist belief that this strong-smelling tree would block the "noxious odors" emanating from the local populations who were assumed to be dirty and thus potentially subject to epidemics. Like many other racist myths propagated by the colonial elites, this was above all a tool to gain control over environmental resources. Europeans were assumed to know best how to manage nature for the good of the local people whom they expelled from the best land.

These policies were applied extensively in sub-Saharan Africa, depriving Indigenous populations of resources in land and game and leading to the introduction of monocultures (e.g., coffee and rubber trees), often with catastrophic environmental consequences. Their social consequences were also often accentuated by the colonial states delegating management to private enterprises, which generally had no consideration for local populations. Worse, they did not hesitate to use strategies based on terror to force Indigenous people to work.[15]

The tropical islands were a specific case.[16] In these fragile environments the destruction of the biotopes was particularly visible. They were also the subject of fantasies, seen as reflecting a lost Eden. These islands became the laboratory for the testing of environmental protection policies. The agronomist Pierre Poivre was the administrator intendent of Mauritius from 1760 and set out to adapt various crops to the island. In so doing he was already aware that deforestation contributes to the deterioration of this environment—the ports filled up with silt, and landslides became more frequent. His program combined transplantation and conservation. He managed to slow down the rate of deforestation for sugar plantations and implemented a law enforcing the concepts of river and mountain reserves. His political approach was inspired by Dutch practices maintaining natural resources in South Africa and would in turn be partially adopted by the British in their colonial administration. During the nineteenth century these ideas spread in the philosophy of American conservationists such as the painter George Catlin (1796–1872) and the philosopher Henry David Thoreau (1817–1862). They would become one of the sources of inspiration for the creation of the first national parks, beginning with Yellowstone in 1872.

But over the course of the nineteenth century, Western elites progressively distanced themselves from the idea that humanity could influence something as enormous as the earth's climate. This was paradoxical because it was also around this time that humanity did in fact begin significantly altering the atmosphere by pouring large amounts of carbon dioxide into it. However, even as the elites were convinced such human-led alteration was impossible, some were aware that climate could change over time. In 1896, the Swedish chemist Svante Arrhenius was the first to theorize the greenhouse effect, extending the intuitions of the French physicist Joseph Fourier. In 1908 Arrhenius estimated that warming was probably underway as the amount of carbon dioxide in the atmosphere was slowly increasing. He predicted that it would double over the course of the three next millennia. However, Arrhenius considered this to be beneficial and rejoiced for the future descendants of humanity. According to his calculations this would lead to an average global increase of 5°C (9°F). He saw this warmer climate as bringing innumerable positive effects to the earth, particularly for agriculture.

The showdown between proponents and deniers of climate change played out in London. Over the course of the nineteenth century, dozens of British explorers ventured into Asia. They roamed widely, made maps, recorded atmospheric pressure, and measured the level of lakes, the flow of rivers, and the altitude of mountains. They exhumed the ruins of formerly prosperous towns. All this information was sent back to the highly select scientific club known as the Royal Society, open to only the most elite intellectuals of the age. The data available to these great minds seemed to suggest that Asia was becoming more arid. This caused a heated debate.

Two men, both geographers and explorers, personified the drama that would forge twentieth-century geography: Sven Hedin (1865–1952) and Ellsworth Huntington (1876–1947). Huntington, the rebel, formulated a theory of climate change in his book titled *The Pulse of Asia* (1907). He was convinced that the world was experiencing a "pulse" that he identified, for example, in Lob Nor. This swamp, in the Chinese desert of Taklamakan, was gradually drying up, and the layers of sediment that surrounded it demonstrated that this place was once a great lake that had progressively filled with sand. However, Hedin, the established specialist who had also studied Lob Nor, persisted in writing that the climate remained constant even though he had physical measurements and maps that were far superior to those collected by Huntington. In fact, it was Hedin who had discovered the site of the former trade city of Loulan on what was once the banks of the lake. Loulan had been deserted in the third century CE because of the advancing sand.

Ultimately the consensus within the Royal Society reinforced colonial prejudice. Like its populations, Asia was seen as an inert region, passive, indolent, resisting all change, even environmental change. Outnumbered and ostracized by his peers, Huntington remained true to his ideas. In 1915 he published a pioneering book titled *Civilization and Climate* in which he documented numerous connections between climate change and human history.

In 1938 the British engineer and amateur climatologist Guy Stewart Callendar (1897–1964) argued before the Royal Meteorological Society that global warming was underway and that humanity was responsible for the emission of greenhouse gases. Once again, he saw this as good news—humanity would avoid the return of the Ice Age. From the 1950s, several climatologists and physicists demonstrated on a local level that atmospheric carbon dioxide was increasing, that Iceland was getting warmer, or that certain glaciers were receding. The Cold War played a decisive role in this awareness. Scrutinizing the skies for signs of the USSR's nuclear and spatial progress, the United States invested massively in atmospheric research. The Vietnam War also led to attempts to artificially alter the climate. For example, the United States attempted to drown the Ho Chi Minh Trail under constant rain to cut the guerillas' logistical axis. That experiment failed. The alterations to the climate were not those that the models had predicted. The soldiers of the Việt Minh also adapted to this new situation, which only slightly aggravated the hell they were already experiencing under the bombs, napalm, and dioxins raining down as Agent Orange.

In 1957 two scientists, Roger Revelle and Hans E. Suess, having published several articles and conscious of the accumulation of carbon dioxide in the atmosphere, concluded that humanity was conducting unprecedented geophysical experiments. Their diagnosis was that "within a few centuries, we are returning to the atmosphere and oceans the concentrated organic carbon stored in sedimentary rock over hundreds of millions of years. This experiment, if adequately documented, may yield far-reaching insight into the processes determining weather and climate."[17] In 1958 the meteorologist Charles D. Keeling demonstrated from Big Sur in California that overall carbon dioxide levels were increasing in the atmosphere.

The idea that human-driven global warming is potentially dangerous matured over many years. In the United States it was swept under the rug by the same lobbies that persuaded the authorities and public opinion that tobacco was harmless.[18] It was only in 1988 that the debate would be launched on a global level when the climatologist James Hansen testified before Congress that he was

99% certain that the climate was getting warmer and that humanity was to blame.

This meant it was possible to construct the myth of scientists who had been innocently ignorant of global warming before it was "revealed." In reality there have long been scientists aware of the processes underway, but they were outnumbered in public debate because their ideas were not fashionable or favorable to political, industrial, or military interests. And their isolation meant they could not fully understand the consequences of the climate changes they had identified. Up until the end of the 1950s, those who saw that warming was underway generally saw it as a positive thing.

15

A TIME OF EXCESS

Industrial agriculture meant the world's population quadrupled in the space of a century. Monkey was now waging an all-out war on nature.

October 2016. Ensconced in the writing of this book, I take a walk outside to try and clear my mind. Unsuccessfully. Attacks on the environment are unavoidably visible everywhere. It is a sunny autumn day with an air quality reading of six, moderate. The horizon is a brown smear in the direction of Paris, some 200 kilometers (about 124 miles) away, a cloud of diesel particles, boiler smoke, and a cocktail of pesticides as vaporous as it is menacing. In the distance a tractor is spraying out products for "crop protection," a sinister oxymoron.[1] After a few hours walking I pass a farm. The farmer is bent over the machine hitched behind a tractor, a sower. With his hands, he smooths out the seeds. They are fat and red, covered in a pod that is supposed to help them grow. The farmer has black gloves on made of that thick synthetic material generally reserved for dealing with toxic hydrocarbons. He wears a gas mask.

There was a time when seeds symbolized life and hope. Today, those who handle them must dress as if they were soldiers in a chemical war. Which, in fact, they are. Agriculture has become a death machine and farmers are its first victims.

Since the end of the nineteenth century, humanity has learned the art of total war. And our number-one enemy is nature.

A War on the Living

In the Chinese opera that tells the story of Monkey, complete with breathtaking special effects, the insolent protagonist, the simian upstart we are now well acquainted with, believes himself invincible, stronger even than the gods. He gobbles the peaches of immortality before giving a good hiding to the gods supposed to be guarding the garden. He laughs in the face of all celestial and terrestrial power. But then he has the bad idea to challenge Buddha himself, He who has understood the nature of the universe. A gigantic hand emerges and closes around the imprudent Monkey who is caught in a moment of moral conflict. He is torn between his wickedness and the revelation of his powerlessness. Chastened, he promises to be a faithful servant of the Good, and to escort the monk who is charged with China's salvation.

This tale is an almost perfect parable for the twentieth century, when humanity abolished all the laws that previously constrained our actions. There is one important difference, however. We are still waiting for the giant hand that will close around our collective conscience to make us change our ways. Our planet, which was more or less disrupted by our activities before the energy revolution, has now become totally anthropized.

The clearest sign of this new power is the phenomenal demographic growth that broke through the Malthusian check. The human population increased from 1.6 billion in 1900 to 6 billion in 2000! But that is not even the most important thing. In his book *Something New Under the Sun*,[2] the environmental historian John R. McNeill explains that he spent the 1990s writing an environmental history of the twentieth century just as that century was ending. At the beginning of his investigation, he asked himself what the most important event of the century would be. His initial idea was perhaps influenced by previous research on the question, particularly Paul and Anne Ehrlich's book *The Population Bomb*[3] and the Meadows report of 1972.[4] Like the IPCC (Intergovernmental Panel on Climate Change) reports, the Meadows report extended current-day trends according to different possible scenarios. Over time, three conclusions became clear. (1) Carried by a wave of demographic expansion accompanied by unsustainable demands on the environment, humanity was heading for collapse in the long term. (2) It was impossible to determine when the collapse will happen. (3) The collapse was avoidable if strong political decisions were made.

The overarching conclusion McNeill was able to draw from the scientific literature available in the 1990s was that the unprecedented demographic growth of our species was the most important environmental event of the twentieth century. More people necessarily meant more farmland, mines, fishing, roads, and towns, all of which would eventually lead us to a tipping point. But this was still based on reasoning centered on the idea that each human being consumes a set portion of natural resources.

What McNeill realized as he advanced further into his research was that the twentieth century had fundamentally altered this relationship with nature. Hence the title, *Something New Under the Sun*, in reference to the quote from Ecclesiastes.[5] Up until the energy revolution, humanity had always lived in the same dynamic: much like animals, we used natural resources for our survival, occasionally produced surpluses, and modified our immediate environment. The twentieth century radically changed this paradigm. Extending the energy revolution of the nineteenth century, our industrial societies produced ever more substantial surpluses by burning more and more fossil fuels and continually increasing the power of money to artificially create wealth. We built a global civilization that suddenly altered the earth on a planetary level, both macroscopically and microscopically. As McNeill concludes, humanity gambled its future without really knowing the rules of the game. Humanity had long since burned wood, dug mines, destroyed fauna, managed flora, produced waste, and speculated on future profits systematically leading to the depletion of natural resources. But all this was only ever done on a small scale until the mid-nineteenth century. One thing has changed since then: the extent of our environmental impact.

According to opportunistic prophets of doom, the year 2000 was supposed to bring the end of the world. They predicted that the Y2K bug would cause computers to collapse and that we would be plunged into darkness because inventors had supposedly forgotten to program information systems to go from 1999 to 2000. In reality, of course, computers negotiated the transition smoothly, as did humans. The apocalypse was postponed. And yet if we look back, an assessment of the past century is enough to send shivers down anyone's spine.

Between 1900 and 2000, Monkey left one quarter of arable land on the planet sterile. We shifted vast volumes of earth (with our mines, constructions, dams, drains, leveling) thousands of times superior to those moved by natural forces. We increased the earth's temperature by nearly 1°C (1.8°F). We initiated a wave of major biological extinctions. In other words, we transformed what was a multitude of local environmental impacts into a global geological impact. This affected so many aspects of the world's equilibrium that it would be impossible to analyze all the changes in the space of a chapter.[6] Here I will focus on just a few key examples among countless others in order to illustrate some aspects of the global evolution.

Welcome to the Anthropocene, this new era of which Monkey is the unwitting hero. The term comes from the Greek *anthropos* and *kainos*, "new man." It was developed by the meteorologist Paul J. Crutzen in the year 2000 to describe this unprecedented extension of humanity's hold over the planet. He emphasized that what was new in this was that our domination over the environment was now unfolding at the geological level. Crutzen set the beginning of this period symbolically at the invention and early distribution of the steam engine by James Watt in the final decades of the eighteenth century.

The twentieth century was the century of humanity marked by a relentless war against nature, a drama that would play out largely in Germany and in the United States.

The Two Faces of Fritz Haber

If one man symbolizes the godlike rise of humanity during the twentieth century, it is Fritz Haber (1868–1934). He was a war criminal and yet probably saved the most lives in all human history because through his genius he also protected millions from starvation. He is a Janus-faced figure, a paradoxical image of a man who developed poison gas for warfare as well as inventing a procedure that would allow humanity to double and then quadruple agricultural productivity and then finally, irremediably, to alter the biosphere.[7]

Toward the end of the nineteenth century, Germany developed a colossal chemical industry based on the transformation of coal, which it has in abundance. Haber began his career with the invention of the procedure that allowed hydrocarbons to be refined to transform oil into petrol. His discovery, along with that of other inventors, contributed to the growth of automobile engines. This made Germany one of the two world leaders (along with the United States) of the second Industrial Revolution, which centered on chemistry, electricity, and petroleum-fueled vehicles.

In 1898 the British chemist William Crookes estimated that the human population was growing faster than agricultural resources. He predicted that as a result, the planet would be plunged into famine and ravaged by a world war before 1930 unless we found a way to create unlimited fertilizer, which would make it possible to plant more densely and more repetitively. Up until now the first harvest typically depleted the natural fertility of the soil, which meant that either land had to be left fallow to rest in between crops or that different systems (agroecology, among others) had to be used, such as crop rotation or the addition of natural fertilizers.

Fritz Haber was a citizen of a German state that had few colonies and was surrounded by potentially hostile neighbors, making it vulnerable to blockades. Most of its nitrate fertilizers were imported from Chile, which was overexploiting its guano reserves after Peru had exhausted theirs. Moved by his patriotic spirit, Haber took up the challenge. He created a process to synthesize nitrogen to create fertilizers! The idea involved taking the nitrogen that represents around four fifths of our atmosphere and transforming it into ammonia. Ammonium nitrate is water soluble and, as Haber discovered, can provide a significant source of plants' favorite food: nitrogen. Thanks to these synthetic high-nitrogen fertilizers, it was possible to resow the same land intensively every year, which radically increased food resources and saved humanity from the threat of famine.

Monkey could now produce his own nitrate fertilizers from the air. Eventually we did it so much that it would come to be a key sign of human presence in nature. The more nitrates there are in your water, the more likely it is that crops are grown nearby. Nitrates are also extremely widespread in the biosphere, along with plastic. They are found in all ocean waters, and the concentration of nitrates (sometimes linked with animal populations) fuels the proliferation of invasive algae.

It might have fed humanity, but the invention of synthetic nitrates did not prevent the world wars. In fact, it encouraged them. Although ammonium nitrate is not in itself explosive, when combined with fuel (among other possibilities) it can create powerful explosive devices, and Monkey the chemist set out to dabble in their creation. This was the initial intention of the patriot Fritz Haber. Without nitrates, there could be no gunpowder, no dynamite. That would mean no war for Germany and a potentially deadly weakness compared with rival states.

So in 1909 Haber managed to produce ammonia (NH_3) from atmospheric nitrogen and hydrogen. In 1913, in collaboration with the chemist Carl Bosch (1874–1940), part of BASF (which is still the number-one global chemical corporation with a €62 billion annual profit in 2018, roughly US$69 billion), he patented the process for synthesizing ammonia. Germany now had its own nitrates, and it could make as many bombs as it wanted.

Then the First World War broke out. Haber imagined that he could hasten victory by developing deadly gases that would seep through enemy lines and bring an end to the butchery. International treaties forbade this, but loopholes were found. German legal scholars emphasized that these regulatory texts only mention shells as a means for distributing prohibited poisons. They also rightly accused the French military of using tear gas in violation of international law. Haber was promoted to the rank of captain in 1915 and supervised the first use of poison gas in combat in Ypres, Belgium. The product was stored in canisters, and when the wind was blowing the right way, the vents were opened and 150 tons of chlorine gas, heavier than air, unfurled over the Allied trenches.

Along a front 7 kilometers (about 4.5 miles) wide, 5,200 soldiers were "liquidated," both figuratively and literally (the substance liquefies internal organs, particularly the lungs). And all for nothing, because the German military did not advance to make the most of the breach that had been opened, and it quickly closed again. For the chemists, chlorine gas was not efficient enough; in the years that followed they also enriched the arsenal with phosgene and yperite, most commonly known as mustard gas. Fritz Haber's wife, also a brilliant chemist, took her own life out of protest against her husband's criminal activities. He went on to establish "Haber's rule," the mathematical formula that calculates the exact dose of gas necessary to kill a human.

By 1919 the war was over. Although a war criminal, Fritz Haber received the Nobel Prize for chemistry in recognition of his work for "humanity's well-being." He would not be tried by the victorious Allies because now they also occasionally used mustard gas to put down Indigenous uprisings in the colonies. Emphasizing the illegality of chemical weapons in view of international law would mean depriving themselves of a very effective tool.

Intensive agriculture developed in Germany, which was isolated and in debt after the Allies forced it to shoulder the full burden of the cost of the war. Haber invested in crop maximization. His team developed the first synthetic pesticide, Zyklon A, which wiped out all predatory insects. In 1933 the Nazis came to power. Haber initially refused to see the danger. Then, like most Jewish intellectuals or those of Jewish origin, he planned his escape. He died in 1934 in Basel on the way to refuge to England. Many of Haber's relatives would be among the six million Jews, homosexuals, Gypsies, political opponents, and disabled people massacred between 1941 and 1945. A large proportion of the victims would be gassed using Zyklon B, made from hydrogen cyanide, which was initially designed as a pesticide by one of Haber's collaborators.

Deathlands

The Holocaust was part of a social construction that combined different ideas. First, there was the notion of "racial hygienics," which compared the German nation to a biological body attacked by diseases. This meant that the targets of violence could be equated to germs that had to be wiped out. This imagery resonated with eugenics and the notion of the hierarchical construction of races. Finally, there was the geopolitical idea of *Lebensraum*, or vital space, which comes from an obsession linked to the geographical situation of Germany. Surrounded by enemies, it lacked arable land to feed its population.[8] Hitler did not believe that intensive agriculture and new fertilizers could provide for the needs of the German people. He needed more land. His policy would be based on annexation, negotiation, and threats to extend control over land resources stretching beyond the northern and eastern borders of Germany.

Russia, on the other hand, became Communist after a long civil war that lasted until the early 1920s. The implementation of the new regime was accompanied by large-scale land appropriation policies, organized famine, and ethnic repression. Between 1932 and 1933, for example, hunger killed between four and five million people in Soviet Ukraine. Another three million succumbed under the Nazi occupation in the 1940s. Organized massacres were then reaching unprecedented levels: systematic hunting of Jews organized by the German army with the help of local police and militia; decimation of the Polish elites under the Nazi regime with the complicity of Stalin; massive bombing of civilian populations by the Germans and then by the Allies, in particular by the United States. Dresden and Tokyo became martyrs to this bombing and were wiped off the map under a deluge of fire from the heavens. Then Hiroshima and Nagasaki were ultimately obliterated by two atomic explosions in August 1945.

The Second World War brought humanity into a new era. It was by far the most murderous conflict in history both in terms of its global reach and the unlimited brutality displayed by all sides. It was the first war in which the life lost from disease was less than the number of those killed by weapons. Monkey learned to kill with a degree of efficiency far greater than nature. In order to progress in the Pacific region occupied by the Japanese, the American army also waged chemical warfare on the mosquitoes to save their troops from tropical diseases.

The Manhattan Project, which allowed the United States to develop the atomic bomb, represents a turning point in the history of science. It was unprecedented in terms of its funding (it had an unlimited budget), its resources (the number-one industrial power, the greatest concentration of great minds), and above all its organization (thousands of researchers working on separate problems without ever being made aware of the overall objective of their task). Together they produced a puzzle made up of numerous pieces that, once assembled, enabled the most destructive power imaginable to be harnessed. The Manhattan Project was the first complex research and development program in history. It was a triumph of technocracy. Henceforth, science embraced so many kinds of knowledge that no one person could claim to know all the different fields in detail. Epistemological considerations also became more complex.

In the postwar period, the fight against living things continued all over the planet. Although it was now theoretically possible to feed all humanity, famine persisted. What was once a consequence of the vagaries of nature, particularly of the climate, became a weapon. The major famines of the twentieth century were political. Mao's "great leap forward" to bring China up to speed with development in the West was an example of this. The countryside was left starving because all the agricultural production, including seeds, had been confiscated to feed the soldiers and the factory workers who had to produce large quantities of goods. An old peasant remembers,

> I had gone to see the production team head, Yao Dengju, and when I went into the production team office, I smelled the fragrance of cooked meat. He said, "Have some meat." I asked, "What kind of meat?" He answered, "Meat from a dead pig." I opened the pot and took out a piece. It was tender in my mouth. I said, "This isn't pork." He said that it was the flesh of a dead person that someone else had cut away, that it can been cut from a buried corpse, and that he had taken a piece and cooked it![9]

Like this man, many people were reduced to cannibalism to survive. China was plunged into a negationist madness. For fear of displeasing their superiors, party representatives at all levels exaggerated the harvests to be two, three, or four times greater than predictions. Province by province, the reports described amazingly bountiful harvests while the bowls in workers' canteens were empty and the state granaries remained full. The media reported miraculous productivity while people ate grass. Stealing an ear of wheat was punishable with immediate lynching. Between 1958 and 1962, famine wiped out thirty-six million people in China, mostly in the countryside. For official Chinese historiography, these people were the victims of drought.

Under Stalin, the Soviet Union would experience a similar fate. The all-powerful and paranoid leader led his country into an abyss. The starting point was a theory developed by the agronomist Trofim Denissovitch Lyssenko (1898–1976). Arguing that plants could be shaped by their environments and contrary to the dominant theories that emphasized hereditary genetics, Lyssenko certified that it was possible to make grain crops more productive by freezing the seeds before they were planted. From the 1930s, he convinced Stalin that his ideas were valid. Progressively he became the only agronomist to whom the "father of nations" would listen. Those who contradicted him ended up in the gulag. The teachings of so-called bourgeois genetics, on which the growth of postwar agricultural production in Western Europe and the United States were based, were prohibited from 1948, and Soviet research in biology came to a halt. Lyssenko's theory continued to be imposed on both farms and minds despite repeated failures. He issued disastrous instructions for sowing grain crops, planting millions of trees to restore Russian forests, and reproducing potato tubers. But he did it with such self-assurance that he managed to convince people that everything was going according to plan. His fall from grace did not come until 1964. In Putin's Russia, he is today considered a national hero, a martyr for science who was the first to have understood than notion of epigenetics (although Lyssenko in fact denied the existence of genes).

During the twentieth century, many nations waged wars against the environment in order to have an impact on population. This did not always go as planned. In the 2000s, for example, the US Secret Service supported the war that Colombia was fighting against the FARC (Fuerzas Armadas Revolucionarias de Colombia [Revolutionary Armed Forces of Colombia]) guerillas. The latter got most of their revenue from the "taxes" paid by coca farmers, or *cocaleros*, who prospered in the mountains controlled by the guerrilla movement. One of the primary "solutions" was to resort to systematic bombing with the herbicide glyphosate from high altitude so that the airplanes would not be vulnerable to shooting from the ground. It was important to ensure that the product was effective, so the standard doses were multiplied by five or six, but because they were released so high up, these pesticides spread out on the wind and became absurdly ineffective. They affected subsistence crops and banana plantations as much or even more than the coca fields. The forests lost their leaves and then their trees. The rivers no longer had any fish. The farmers and Indigenous communities were forced into exile. Those who resisted the best were the drug cartels: they paid geneticists to make the coca plants resistant to glyphosate, violating Monsanto's patents on its famous corn, which alone was able to resist the pesticides that wiped out other plants.[10] This investment cost them between €120 million (around US$150 million) and €160 million (around US$200 million; both figures based on the 2004 exchange rate), a drop in the ocean for these drug cartels, whose wealth is comparable to that of many countries.

Over the course of the last two hundred years, humans have extended their control over nature to unprecedented levels. Previously, this was driven by survival and the need to resist the unpredictability of the climate. But the ambition of total domination over nature now went alongside the goal of controlling other humans for the supposed benefits of a collective presented as being the only humanity worth preserving. The twentieth century was marked by gigantic messianic transformative projects. Stalinism, Nazism, and Maoism all set out to create a "new man." Liberal hygienist doctrines, particularly active in Anglo-Saxon and Scandinavian countries, sought to control poverty, mental weakness, and "vice" (a category that included various things, from homosexuality to belonging to certain ethnic or phenotypically defined groups) by any means possible. All these projects aimed to improve humanity and the social body as a whole. The revelation of Nazi atrocities meant that these ideologies progressively fell out of favor in the West after the war.

But the war against nature lasted much longer. It was not until the 1960s that the first global movements criticizing the impact of industry on nature emerged. The publication of Rachel Carson's *Silent Spring* in 1962 denouncing the ravages of DDT on biodiversity was emblematic in this respect. This new public awareness led to a protest movement that ultimately—ten years later—resulted in the prohibition of this product in the United States and then in most developed countries. But the chemical companies got organized. They found new resources, used lobbying against the environmentalist movements, and managed to progressively sideline them from debates in American politics.

The listlessness of these environmentalist groups goes some way to explaining why the threat of irreversible climate change, which appeared on the international agenda in the 1990s, did not lead to any significant change in behavior. Discussions around the need to reduce greenhouse gas emissions failed in the face of the two primary emitters' lack of enthusiasm and conflicting interests. On the one hand, until 2015, China intended to make the United States and other long-industrialized countries pay for the past emissions on which they had built their prosperity. On the other, the United States would have liked to see contributions to the funding of the fight against warming be more indexed on the volume of emissions.

Humanity has become addicted to industrial agriculture and its vast volumes of toxic fertilizers that their promoters argue are essential if we want to feed the world. This will cost us dearly. The chemistry of the land has been altered all over the world. The natural fertility of the soil has often been replaced by artificial productivity that is now dependent on increasing inputs. Results had to be optimized, plants had to be genetically selected for productivity. Today we use far fewer varieties of plants than the number of species that humans have domesticated over the last twelve thousand. Over the course of the twentieth century we have dilapidated more than half the genetic heritage accumulated by billions of farmers before us. They selected plants that resisted early cold snaps, others that could survive drought, and still others that survived pests and predators.

The results of this addiction are becoming increasingly clear on a global scale. According to the Intergovernmental Science-Policy

Platform on Biodiversity and Ecosystem Services report of 2019,[11] pesticides are one of the primary factors in the sixth extinction. The irremediable loss of biodiversity they provoke could lead to a sterilized world in which ecosystems are no longer able to perform their vital functions for the survival of animal species. If this scenario is confirmed, our descendants will not have enough resources to ensure their survival, and even the availability of oxygen and drinkable water could be affected by the end of the century.[12]

Today, a few major corporations now seek to control the living world as far as agriculture is concerned. Their strategy is based only on monocultures. For this they need bigger and more powerful machines to plow fields as far as the eye can see. They rely on economies of scale. This depends on more and more pesticides, herbicides, and fungicides: a relentless chemical war against parasites that mutate and resist the use of ever more complex toxins. Seeds are patented and genetically altered through increasingly sophisticated engineering. Farmers are prevented from saving seeds. At a global level, the inequalities between small and large farmers, as between rich countries and poor countries, have left peasants and farmers more vulnerable. In sub-Saharan Africa it is sometimes cheaper to buy a chicken imported from subsidized industrial farms in France than to buy one raised locally. Without Fritz Haber's inventions, the rural exodus that occurred in the twentieth century probably would have been less widespread, we would be less dependent on oil to feed ourselves, and farmers would be less beholden to large agrochemical corporations.

This might mean humanity would still be prey to widespread famines. But this assertion, often wielded by defenders of industrial agriculture, is now challenged by agronomic research that systematically shows how agroecology and organic farming are just as productive as industrial agriculture and infinitely more sustainable.

Dust and Misery

From the second half of the nineteenth century, Europe poured its demographic surplus into North America. During this period, forty-five million Europeans chose exile (of whom thirty-four million went to the United States). Immigrants were often poor, having paid the equivalent of several years' income for the ticket across the Atlantic. Frequently from rural families, they were prime candidates for participation in agricultural innovation, perfecting techniques that were already producing monocultures in Europe. But the scale of the North American continent was different; there were immense plains, powerful winds, and a serious risk of erosion.

Up until then, those plains had been covered with prairies, wildflowers, and woods. Converting them to intensive pasture and crop production led to progressive erosion the full extent of which would only be realized in the 1930s. On October 24, 1929, the New York Stock Exchange crashed, destabilizing societies around the world. The economic crisis of the Great Depression was echoed by a major ecological crisis, which became apparent in the transformation of vast swathes of previously fertile land into barren "dust bowls" where only tornadoes thrived.

Kansas, Colorado, Texas, Oklahoma, and New Mexico; a series of annual droughts struck the center and the south of the country. But these regions were already weakened by decades of inappropriate agricultural policy, and since the end of the First World War, the mechanization of agriculture had meant the land dedicated to growing crops had increased substantially. Between 1931 and 1939, 20 million hectares (about 50 million acres) of farmland were reduced to pebbles, dunes, and steppes. In Oklahoma alone, an estimated 440 million tons of fertile land became little more than dust. The journalist Timothy Egan put it like this: "Americans had become a force of awful geology, changing the face of the Earth more than the combined forces of volcanoes, earthquakes, tidal waves, tornadoes, and all the excavations of mankind since the beginning of history."[13] Subsequent studies have emphasized the fact that the monoculture of wheat and intensive pasture not only contributed to the development of dust storms in the twentieth century but also altered the climate at a continental level, considerably amplifying droughts. Palliser's Triangle in Canada experienced something similar. On these once fertile sandy soils, farming is now only possible with the help of major chemical inputs.

This process continues in the early twenty-first century. California is massively affected by it. The artificial lakes that provided the state's water supply are dry. Its forests go up in flames one after the other. Drastic emergency water restrictions have had to be implemented. Intensive agriculture, which is widespread, has been asked to reduce its water usage. Instead, it uses precious groundwater. This story has been repeated, with only slight variations, in Russia, China, and Africa.

The twentieth century was a time of paradox. Even as humanity expanded, our potential resources were destroyed. Arable land was left barren by unsuitable farm practices, by urbanization, and sometimes by industrial waste. And yet the development of farmland was still more substantial than its loss. But the clearing of new land is often accompanied by the destruction of many wetland or forest ecosystems. And finance has a good nose for a bargain. Given that arable land will reach its limits within a few decades, land grabbing, which consists in acquiring large expanses of farmland or forest in developing countries, has become a potentially highly profitable market. Freshwater resources are also similarly preempted. Certain Asian or Middle Eastern states and many private organizations, such as North American pension funds, are now speculating on the food security of the future.

Let us take a step back and translate this into macroeconomic terms. The economist Angus Maddison has produced an estimation of how global GDP has evolved since 1500.[14] Notwithstanding the reservations we might have about this kind of calculation, Maddison writes that the global economy was about 120 times larger in the late twentieth century than in 1500. But it followed an exponential curve that really accelerated only at the very end, most of this growth in fact occurring in the 1870s. Having grown by a factor of 3.5 over the course of the nineteenth century, in the twentieth it was multiplied by 20. This is an extraordinary achievement for humanity as a species, as John R. McNeill reminds us: as a global average, in 1990 we had "nine times more income per capita than our ancestors had in 1500, and four times as much as our forebears had in 1900."[15] The dark side of this, however, is that over the course of the twentieth century, a socioeconomic model based on competition and the acquisition of wealth was imposed both on individuals and on communities. This reckless social trend reached its peak in the neoliberal ideology that has, since the 1980s, established growth as the ultimate public policy goal to the systematic detriment of the environment and also, too often, of well-being.

The Silence of the Buffalo

Philip K. Dick (1928–1982) is a science-fiction author renowned for the often prophetic nature of his work. His book *Do Androids Dream of Electric Sheep?* (1968), adapted for the screen by Ridley Scott as *Blade Runner* (1982) is about robot "replicants" made to look exactly like humans. They are equipped with nonempathetic intelligence, which means tests can identify rebel androids who decide to go into hiding rather than be forcibly retired after a few years. Despite being programmed not to, some machines have developed emotions and therefore try to prevent their imminent extinction.

In this future, radioactive pollution has sterilized the earth. It began with the owls. Rick Deckard, the bounty hunter protagonist, remembers as a child seeing the ground being littered with owls, all dead, and no one knowing why. (In France today the same thing sometimes happens to swallows.) Now an adult and a recognized hunter specialized in the "neutralization" of synthetic humanoids, particularly rebel androids, Deckard has been contacted by one of the companies that control the world. His contact offers him an exceptional reward for a top secret mission: the last living owl.

The deal reveals a world of mirrors: there is no last living owl, even this one is artificial. The young woman who meets Deckard is herself an unwitting android. Ridley Scott pursues this plunge into mirages in the director's cut of *Blade Runner*, suggesting in the very last image that Deckard himself could be an android, programmed to remember a world in which the owls all died one day long ago.

When he imagined the sudden demise of the owls, Philip K. Dick had to have had in mind the fate of the passenger pigeon. In the nineteenth century, this species native to North America was the most widespread bird species in the world with populations several billion strong. Their flocks sometimes included a billion pigeons and even blocked out the sun. These birds had a clear impact on biodiversity through their consumption of seeds, millions of tons of droppings, and the deterioration of trees bent under the weight of their colonies. Farmers considered the species a threat to their crops, and from the 1860s it became fashionable to hunt them. Each seasonal migration was heralded by uninterrupted fire from shooters convinced they were saving the harvest. Of course, shooting into a dense cloud of birds was a sure thing. But in order to reproduce, these birds apparently need colonies of hundreds of thousands of pigeons. When their numbers were reduced, the colonies dispersed and declined. A few decades of shooting was all it took to exterminate the species. The last surviving passenger pigeon died in a cage in 1914.

Another animal that symbolizes the waves of extinctions brought about by humanity is the North American bison. The demographic trend of this massive creature can be summarized quite simply: before 1800, there were around sixty million bison; in 1885, there were only 325 left alive. Legend has it that hunters like William F. Cody, known as Buffalo Bill (1846–1917), were responsible for the slaughter of the herds. But this is only partially true. The United States' colonial wars against the American Indians of the Great Plains in the nineteenth century used famine as a weapon. Killing the bison during their migration meant starvation for the tribes, whose economy was dependent on hunting. Men like Buffalo Bill did make the systematic killing of bison into a profession for a time. But the three fundamental elements of their near extinction were the railways, agriculture, and livestock.

The new railways that crossed the United States from east to west were built on the trails that generations of bison had carved into the plains and mountains since time immemorial. So the passengers were well placed to observe the exclusive attraction of the Wild West railways. All aboard and bring your guns! The train stopped at the first sign of bison, the passages alighted or simply opened their windows. And fired at will. The bodies were left to rot, and the train continued on its journey. Parts of the carcasses were used, however. The tongue, a sought-after delicacy, was often removed. The hide, used in important crafts and trades that were exported to Europe, was also frequently collected. The bones were left to bleach in the sun. Meanwhile, herds of domestic cattle invaded the bisons' plains and pushed back the survivors or assimilated them through hybridization. The animals' bones were used to construct colossal pyramids or were ground into powder. This is how industrial agriculture began. Fertilizers paved the way for intensive monocultures and became necessary to feed the population to the point where it was no longer possible to imagine doing without. So when large-boned animals such as bison, and then whales, had all been massacred and the guano was all used up, even on the Patagonian beaches, then Fritz Haber and his synthetic nitrogen indeed became the savior of humanity.

In the mid-twentieth century, the Third Agricultural Revolution, known as the "Green Revolution," produced a boom in cultural production through the application of industrial methods for growing food. This was based on a constant exchange between the technologies for optimizing war and their application to the management of the biosphere. Techniques from industrial warfare would also be applied to livestock. Take the example of the first concentration camps. As Cuban people were fighting for their independence (1895–1898), the Spanish army locked them up with their livestock in camps surrounded by barbed wire and assumed anyone on the outside was a rebel. The idea was also adopted by the English in the second Boer War in South Africa (1899–1902) and then by the Germans when they planned the genocide of the Herero and Nama in Namibia (1904–1905). The twentieth century would also be the century of camps and displaced peoples. Populations who would have once been able to flee a conflict or a drought by migrating to other countries would soon be held en masse as borders became increasingly impenetrable and the management of refugees more and more problematic.

Today's industrial animal farming has sometimes been described in similar terms in light of the horrendous overcrowding to which poultry and livestock are subjected. The war against the living involves the total control of animal bodies, the concealment of images of their slaughter, and the negation of the torture we inflict on animals in our frenetic search for economic efficiency.

Let us go back to the American bison. Here is another number. Having narrowly avoided extinction, there are more than half a million of them left today. But are they in fact really bison? There are only twelve thousand purebloods left in the wild, kept in nature reserves. Three quarters of them have been struck by tuberculosis or brucellosis. A handful survive in zoos. The rest, the vast majority of what is now known as American bison, have been crossed with cows and have become livestock. The resulting animal has meat that—like bison meat—is low in cholesterol. But it grows fast, like a cow. These hybrids are packed into muddy feedlots, force-fed fatty genetically modified soy cakes imported from Brazil, and plied with antibiotics that make them gain weight and resist the diseases that would otherwise inevitably develop in this overcrowded, stinking hell. These creatures are known as "beefalos" and are reduced to their elementary status of beef steaks.

In response to human pressure, hybridization has tended to become the norm for wild animals, too, particularly those whose ecological niches are reduced to the point where their survival is under threat. In North America, for example, two new species have emerged in recent years. The "coywolf," a "natural" hybrid between the coyote and the gray wolf, with bits of dog thrown in, is the logical consequence of a dwindling wolf population. No longer able to find mates, the surviving wolves turn to their cousins to ensure reproduction. Similarly, there are also increasing reports of "grolars" or "pizzlies"—a hybrid between a grizzly and a polar bear. This animal is probably the result of climate change, which is pushing polar bears toward the south and their cousins the grizzlies.

If we asked our grandparents in the country, they would say they remember vast swarms of beetles flying over the fields, and crowds of frogs that would come out of the ground after rain. Amphibians are probably the group the most at risk in the sixth extinction, which began thousands of years ago and is reaching a spectacular peak with the Anthropocene. Amphibians are threatened by a microscopic fungus involuntarily spread by humans, which sometimes kills them even before biologists have been able to observe them in the wild. They are also affected by endocrine disruptors (see chap. 17). These creatures, which appeared 360 million years ago, breathe in part through their skin without a filter, so their hormone glands are particularly vulnerable to products likely to disrupt their reproductive abilities.

Today the countryside is silent. And this silence has conquered the planet insidiously. The soundscape ecologist Bernie Krause has spent forty years of his life recording the noises of the natural world.[16] He argues that in nature, the sounds of the elements (wind, water, etc.) combine with animal noises in a grand orchestra. Yet over the last century, biology as a discipline has sought to categorize the animal kingdom and classify each species independently. We are extremely creative when it comes to capturing the solitary warble of an endangered songbird. We will catalog the song, along with a photograph of the singer, under a Latin name and imagine we know everything about it. But that would be forgetting that the audible expression of a species can only be understood in the context of its biotope. A bird changes its song according to its interactions with the soundscape in which it lives.

Just as fiction sometimes gives us a better understanding of reality, reflecting it in all its cruelty, sometimes sound, rather than vision, best reveals the full extent of the attacks against biodiversity. Comparing the soundscapes of an American forest recorded a year apart, Krause shows the terrible impact of (even moderate) tree felling on biodiversity. Nearly all the insects, and the birds who feed on them, have been reduced to silence. Similar experiments documenting coral reefs a year apart also show the stunning collapse of their biodiversity. Studying this soundtrack of nature reveals environmental changes that would otherwise go unnoticed. All over the world, anthropization has made its mark. At the global level, half of all natural soundscapes have disappeared in the space of fifty years. And the background noise of civilization—the engines of cars, trucks, airplanes, and boats—can now be heard in 90% of biotopes.

We have forgotten the richness of sound. We have reached the point where we do not remember that some peoples, such as certain Indigenous tribes in Australia, were once able to navigate their environment using sound maps, analyzing the noises of insects to identify infinitely small variations in the physical environment.

How to Resolve an Environmental Crisis

Another chemist, the American Thomas Midgley (1889–1944) played an important role in the twentieth century's global alteration of the atmosphere. He discovered that it was possible to use a gas, Freon, to keep things cold. He also had the idea of adding lead to cars, which improved both combustion and the engine's durability. But both of these inventions came at a price.

Along with the car, television, and washing machine, the refrigerator became one of the four pillars of the consumerist dream of the postwar Western world. This vision of technological well-being was shared by all self-respecting families. But Freon gas, which gave these machines the power to keep food cold, is a chlorofluorocarbon, a CFC. This family of gases have remarkable chemical stability except when they rise into the stratosphere. Under the impact of ultraviolet rays, they decompose and release chemical agents that destroy the ozone layer. The ozone layer is, of course, what protects the biosphere from the sun's ultraviolet rays that would otherwise affect the genetic stability of living organisms, leading to skin cancers in humans, for example.

Scientists identified this problem as early as 1974. Environmentalist movements were already well developed in the United States, and they set out to raise awareness, first among the American public and then in Europe. The concern reached politicians, who suggested introducing incentives to replace this gas. Initially reticent, even denying the reality of the phenomena that had been observed, industrials eventually accepted the urgent need to find a less harmful substitute for CFCs. In 1985 the United Nations Environment Programme launched the Vienna Convention to protect the ozone layer. International cooperation led to a very significant decrease in the production of Freon gas, observable from 1988.

There were two important factors in this happy ending. First, chemical industries did not lose out in the replacement of CFCs. They rapidly developed and produced efficient substitutes, and the obsolescence of Freon-based machines was in fact a boon for sales. Concerned consumers got rid of their old appliances, which led to a significant increase in the production of new refrigerators and air conditioners. The second factor was that only a small number of countries were concerned, those whose populations could afford refrigerators in the 1970s and that had a significant system of industrial production: North America, Western Europe, Japan, and the USSR.

Since the 2000s, the ozone layer has slowly recovered. But we will have to wait till the end of the twenty-first century for it to be entirely fixed. Which is the blink of an eye in geological time. Our other global impacts will take infinitely longer to repair.

And yet societies have sometimes, rarely, been able to escape from the pollution produced by human activities. Remember John Evelyn, who wrote to King Charles II in 1661 about the cloud of smog that covered London (back in chap. 13)? Had he lived longer, he would have seen the situation get even worse. After 1780, industrial steam engines consumed endless and increasing masses of coal. Between 1840 and 1900, London's famous smog slowly killed at least 1.4 million Londoners by causing bronchitis and tuberculosis. And then came the reign of the automobile. After the First World War, lead was added to petrol. But it was not until after the Second World War that cars really took London by storm. In 1952, a particularly noxious yellow smog blanketed London, mostly due to the burning of poor-quality coal. Known as the Great Smog of London, it reportedly killed four thousand people in four days. The government declared it a flu epidemic before admitting that air pollution was responsible. Another eight thousand victims would die over the next month. In 1956 a clean air law was passed prohibiting the burning of coal in certain parts of the capital.

Slowly but surely, through regulation and restrictions, cities in developed countries managed to get air pollution under control. Tokyo prohibited diesel engines in the 1980s. Today Paris and other European capitals have embarked on a similar struggle to slowly depollute the atmosphere. Such supercities bring together ever-larger numbers of humans and therefore become pollution hotspots. The wealthiest cities are often able to take relatively efficient action. But the poorest, such as Delhi in India or Harbin in China, are simply overwhelmed by the extent of these environmental burdens. In 2012 alone, air pollution cut short the lives of seven million people in the world.

Having studied numerous environmental crises over the course of the twentieth century, John R. McNeill identified three golden rules, three conditions that have allowed certain societies to backpedal successfully in the face of major ecological damage.[17] In general, they had to have (1) a democratic political system allowing citizens to express themselves, (2) the possibility for politicians to demand change from polluting industries and implement restrictive regulations under pressure from citizens, and (3) a desire among companies to react positively to consumer demands.

But is humanity able to effectively fight against a process that is as extensive and complex as current global warming? First, it is important to be aware that this phenomenon will not be linear. The Earth-system* seems to have many feedback mechanisms that allow us to sporadically inverse, either globally or locally, the acceleration of the climate machine. Traces of these kinds of events can be seen in the past. There was the terrible cold wave of the Younger Dryas (chap. 3) 12,800 years ago, which can be analyzed as a thermal reaction resulting from an accelerated warming period. Large-scale melting of Canadian icecaps led to the interruption of warm currents in the Atlantic, plunging the Northern Hemisphere into cold.

But let us not be too hasty to wish for this kind of cooling. There are no direct testimonies from this period of course, but we can imagine the confusion of our ancestors in a world that saw annual average temperatures plummet by 7°C (12.6°F) in the space of a generation. The earth is an extremely complex machine, and our models do not provide reliable prognostics. The only thing we are certain of is that warming is unavoidable and that it will produce threshold effects. A threshold is a limit that we would do well to avoid crossing! In this respect, natural systems resemble human societies. Although they can manage change reasonably well, a sudden accumulation can disrupt their balance, and then things begin to break down. When fractures appear between members of societies and reach a point where a critical mass is passed, riots, revolutions, and other troubles begin. When the seawater in a tropical ocean gets warmer, this remains invisible until the temperature crosses the threshold of 26°C (78.8°F), at which point the thermal energy is converted into a cyclone. The very nature of thresholds makes them undetectable. No one can predict when a society will rise up despite those who claim that some particularly perspicacious expert "saw the crisis coming." Good luck to anyone who claims they can predict the social consequences of the warming that is underway when its physical evolutions are so difficult to foresee.

The international agreement signed at the COP21 (Conference of the Parties 21/2015 UN Climate Change Conference) meeting in Paris set out to limit warming to 2°C (3.6°F) for this century. Current predictions, based on the existing growth of greenhouse gas emissions and notwithstanding the uncertainty of political decisions to come, foreshadow an increase in annual average temperatures of between 3°C (5.4°F) and 4°C (7.2°F) or even 5°C (9°F) or 6°C (10.8°F), compared to the first measures in 1880. If we look at similar events in the past, it is highly probable that this warming will take place in a context of destruction, crisis, and collapse.

Monkey is playing while his house goes up in flames. But a large part of his brain continues to deny it.

16

THE BLIND FLOCK

The internet, child of the Cold War, has allowed us to create dense planetwide connections in real time. Monkey is now tangled in the web he wove. Could he become its slave?

The man opens his eyes. He is naked, half conscious, floating in space. There are suspended bodies all around him as far as the eye can see. Connected. He drifts back into sleep. Later he will remember this nightmare and realize that his life is staged, orchestrated by one or several Machiavellian demigods.

This is a familiar image because it is the starting point for several well-known science-fiction narratives. It is the opening scene of Philip José Farmer's *Riverworld* series and a central theme in the *Matrix* films. It also echoes the affirmations of the eccentric businessman Elon Musk, who claims to believe that we live in a synthetic world created by a giant computer. Like a character from science fiction, although quite real, Musk flatters himself he can design an "open source" artificial intelligence coded and controlled by humanity. He envisages humanity's adaptation to life on Mars. And he sells fully electric cars that will eventually drive themselves. All these innovations are liable to revolutionize our relationships with machines and with energy.

The vision of an unknowable world in which our destinies are controlled by obscure powers is now familiar to us all. It is the leitmotif of conspiracy theories, the scourge of our hyperconnected era. But what if an invisible force did guide our destinies—whether in the form of secret societies, aliens, or artificial intelligence? Can we keep the potential dangers of such myths at bay armed with reason alone? It will not be easy. Reality is now seen through the prism of the Sixth Revolution, the Digital or Media Revolution. It is an electronic hall of mirrors.

Our cognitive ability means we can imagine, reason, and remember. Initially we outsourced memory in the form of writing to free up space in our minds for imagination. We no longer needed to memorize 240 pages worth of information, we only had to remember where the book was kept in order to access the information it contained. Digital technology has exponentially expanded the possibilities of externalizing memory. A single keyword typed into a search engine provides access to limitless resources. But this technology goes further still. It plunges us into a virtual reality. We have entered a whole new world of stimuli that affect our cognitive functions.

Let us explore this Digital Revolution, which is an artificial evolution profoundly altering how we think.

The Triumph of the Media

Monkey has taken over the planet. He has shaken it up from top to bottom and changed all the rules. He has unlimited power over animals; he gets to decide who lives and who dies. There would be no more elephants if humanity had not decided to create nature reserves to fight against poachers and vaccination programs to ward off animal diseases. Monkey has meddled in the chemistry of the air, the water, and the earth. These vital elements have been transformed both on a molecular and a global level. Endocrine disrupters, plastics, aerosols; numerous chemical products have seeped into our environment and subtly, silently, modified our bodies and our minds (chap. 17).[1] Monkey has imposed his law on the world and made a deal with the Devil to do so. The digital (or media) revolution affects our perception of time and gives our consciousness access to new dimensions. In this sense the term *media* refers broadly to all the forms of communication we now have at our disposal.

At the dawn of universal religions, at the time of Confucius, Pythagoras, Isaiah, and Buddha, it was impossible to tell what long-term impact the discourses and practices of those prophets would have. History could have taken any number of different paths. Twenty-five centuries later we can look back and see that this Moral Revolution (see chap. 5) was associated with increased connections between many civilizations and societies. Universal religions joined together with universalist empires to spread a new idea of humanity as a cohesive group liable to act as one under a single ideology.

Back then humanity was divided; different more or less compatible universalist ideologies developed simultaneously and remained in competition with each other. Each religion argued that its version of salvation was the right one, which made the others necessarily wrong. None could reign supreme, because the law of fragmentation has always limited universalist ambitions. When a religion was imposed on too many people, there were always some who dissented, who challenged the whole. Protestants, for example, caused a breach within Christianity, setting up competition between doctrines as to who would chalk up the most conversions.

Today, ideological success is played out in the media. The rise of Christianity in the Roman world was partly linked to the success of codices, early bound books that meant the various versions of the Gospel in circulation could be standardized. The spread of Protestantism was based on the printing press, which made it possible to increase the dissemination of dissident ideas, outrun censorship, and put holy writings in the hands of as many people as possible. Martin Luther apparently liked to say that with a bible in hand, every Christian is the pope.

The Digital Revolution is likely to have a similar impact on humanity as the Moral Revolution did. But it will do so instantaneously because its two main consequences are to establish universal media and to abolish time. The so-called GAFAM companies—Google (created in 1998), Apple (1977), Facebook (2004), Amazon (1994), and Microsoft (1975)—are scarcely more than a few decades old. Yet they pride themselves on audiences comparable to those of the great political and religious ideologies and have annual revenues higher than the GDP of many countries. Moreover, these companies have global objectives and ideas about the future of humanity as well as the financial means to shape the seventh revolution: the Evolutive Revolution (see chap. 17), in which humanity is liable to either control its own evolution or lose control of it. This may be the stage for the final battle in this long war that has seen Monkey progressively bend nature to his all-powerful will.

Digital technology has altered both how we think and how we live.[2] On any kind of public transport you care to observe, you will see people staring fixedly at screens, tapping or swiping the glass, sometimes flaunting their conversation with a far-off speaker. In 1962 the science-fiction writer Fritz Leiber described a city cut off from the world after a nuclear war in his novella *The Creature from Cleveland Depths*.[3] In this story an inventor creates a machine that sits on people's shoulders and reminds them of their daily schedule to free their minds from trivial tasks. It does not take long for these machines to end up grafted to humans, stuck to their ears, dictating what they should do, and then what they should think, until their interactions come to constitute an intelligent neural network that enslaves the town. In 1962 this was an improbably pessimistic vision of the future. Today it seems more like a premonition of gated communities in which GAFAM-type networks reign supreme.

Today our environment is saturated with communication. The fortunate of the world, the globalized middle class, participate in this "democratized" media. Everyone with sufficient economic and educational resources can produce information on social networks, radically extending the range of audiences and exchanges. But, more profoundly, it means that there is nowhere on Earth you can escape this digital ecosystem. Even camping in the wilderness you can probably still have access to a network that can stream your images on the internet.

In less than thirty years, unbeknownst to himself, Monkey has been subject to an *exaptation*. This process is the opposite to adaptation. Adapting means learning to manage a new environment in order to better exploit it, or in any event to survive. *Exapting* means being modified by an environment that shapes you regardless of the strategies you implement to try and control it. Our far-off ancestors adapted to the African savanna by becoming bipedal, the first stage in a long evolution that would allow Monkey to become king of creation. But now we are conforming to a new technical environment deliberately shaped by certain human communities, specifically start-ups and privately funded research and development teams. We conform to it not because this environment is advantageous to us but because it is constructed according to free-market liberal norms. Needs and wants develop, are shaped, and become obsessions as soon as there are the technical means to satisfy them. These new artificial organs create new trends and new restrictions. The ubiquity of these communication technologies is such that they have come to shape human behavior. The need to "share" things with family and friends is so strong that we recreate virtual communities in social networks made up of countless "friends" or "followers" on Facebook or elsewhere. So tight is the grip of the screen that ensconced in the swiping and tapping, we no longer even acknowledge those physically around us.

These transformations affect what has been called the *noosphere*. This term, coined by Pierre Teilhard de Chardin (1881–1955), refers to all the thoughts, opinions, and concepts shared by humanity.[4] This noosphere was once abstract, fragmented, and diffuse, but now it has a concrete form. It exists in the exponential labyrinths of cables and endless interactions of electromagnetic waves that are constantly transmitting volumes of information so vast they absolutely defy quantification. Each of our actions in this universe creates a digital trail. If you post a photo online, it will be recorded on a server, perhaps sent to email addresses, and probably scanned by intelligent (or robotic) agents that will extract information that will in turn be stored on various hard drives to be analyzed later for global marketing, state surveillance, or the development of artificial intelligence.

Most of us are blissfully ignorant of the quasi-physiological processes that take place behind our screens, which are in fact simply the nerve endings of gigantic technical entities. Researchers working for Google are planning to implement safeguards on the artificial intelligence they are creating in case it gets out of control. Facebook, on the other hand, has a comfortable head start on the development of artificial intelligence because these programs are already studying the everyday lives of more than a billion humans on the firm's social network. This artificial intelligence compiles enormous quantities of behavioral data that allows it to emulate our emotions, our appearance, even *eventually* our very nature. We will soon be chatting online with synthetic beings indistinguishable from humans.

The Great Acceleration

Time is speeding up. Or, more precisely, our perception of time is shifting from linearity to immediacy. Let us call this the Great Acceleration, as a number of philosophers have done.[5] Time used to be continuous; it is now becoming fragmented. The media is dividing up our time according to more and more apparently urgent sequences. Everyone has felt that frenetic impulse to check their phone instead of concentrating on the task at hand. The boundaries of the private sphere are becoming increasingly blurred. Western middle classes are more and more subject to a range of connected ills: work is done outside working hours, because there is no "going home" in the digital world; fatigue is endemic, induced by the fragmentation of activities; and then there is burnout, which results from a person's energy being entirely consumed by their work, like wood in a fire.

We can see an emerging shift in our relationship with information. The world has become an object to look at rather than to see. Computers are now universal tools. They are the optimum of efficiency in receiving, treating, storing, and transmitting information. In this respect the system mimics human behavior. But it can also emulate all the rest: animal behavior, the evolution of institutions, or even atomic reactions. We are entering a new phase in our connections with externalized memories. Computers, which appeared such a short time ago, are already creating radical changes in our thought structures as well as in the functioning of our societies.

The very first human language communications were gestural and oral, a consequence of the Cognitive Revolution (see chap. 2). They were played out on the human body. The face, hands, voice, and gaze transmitted information to the brains of other humans, which were in turn progressively shaped by these interactions.

Indeed, the frontal lobes of our brains are perfectly adapted to this oral communication. Then writing appeared (chap. 4), the first complex external memory technique. Around 5,500 years ago, writing allowed us to externalize information onto clay, rock, vellum, papyrus, or paper. It was now possible to standardize law, "to engrave it into marble," to collect resources efficiently, to keep accounts, conduct inventories, collect taxes, structure increasingly complex commercial exchanges, manage more and more sophisticated territorial entities: cities, kingdoms, and empires. The state was born of writing.

The second phase of externalizing memory was the invention of printing, in China around 1,200 years ago, and around six hundred years ago the West. This mechanically led to the distribution of various tools for thought, communication, and economics: checks, banks, double entry bookkeeping, and joint-stock companies, among others. Capitalism was born of printing. So was experimental science, which was able to profit from the knowledge of previous thinkers. So was the crisis in religious certainty that swept the West with the Reformation. And so was the beginning of democracy and its progressive expansion, encouraged by the Republic of Letters. From the eighteenth century onward, the printing press made books—which were previously expensive and difficult to reproduce—affordable and (eventually) accessible to most. Books were the vectors of the rapid distribution of knowledge from Enlightenment Europe of the eighteenth century to the bureaucratic advances in the way European nation-states processed information in the nineteenth and twentieth centuries.[6]

Cybernetics, a science born during the Cold War for which everything is communication, allows us to draw a significant conclusion from this evolution: any upheaval in the relationship between the message and its form revolutionizes civilization. And that is what is happening today. We are living through a social revolution. Whatever we do for a living, most of us now type on a computerized interface, a keyboard or screen. But this revolution is also cultural: we speak a language that changes more quickly because of machine interaction. It is enriched, pushing us to develop jargon, to abbreviate, and to rely on predictive text. There is also a monetary revolution that has led some to advocate the end of physical currencies in favor of dematerialized ones, while others suggest we should replace them with local currencies. Other changes are more unpredictable. We experience things more intensely through images, which are more immediately accessible than the written word. But what might be the effects of this? Those who preach the most dynamic versions of global religions, whether evangelical Protestants or conservative Muslims, have massively occupied the internet, sometimes with devastating consequences. The clearest example of this is the terrorist group Islamic State (IS) which proved able to capture the imaginations of a wide range of people, often but not exclusively young, and manipulate them even to the point of suicide and slaughter. Unfortunately, this example may be a preview of what this third phase of externalized memory might look like once it is instrumentalized for religious propaganda.

This technology has the potential to become deeply connected to us. Unlike writing or printing, which enhanced memory with a physical support such as paper, digital memory is nonmaterial. This means that there is a potential for hybridization of emotion and artifact. Philip K. Dick explored this idea in another science-fiction story titled *The Little Black Box* (1964), in which he described an entity named Wilbur Mercer.[7] Consumers could enter into contact with Mercer's spirit by placing their hands in an "empathy box" and watching him endlessly climb a mountain of suffering under the onslaught of stones and insults raining down from an anonymous crowd. By sharing this martyr's pain through the black box, users felt relieved to the point where they forgot their earthly troubles. This vision of hell clearly evokes the passion of Christ presented in the Gospels, stories that were once chanted in the catacombs to unite the first Christian communities. There is no doubt that if it existed, such a device, offering both total immersion and redemption, would be extremely attractive.

The Media Revolution began with the design for the first computer, which was produced by the British codebreaking services under the direction of Alan Turing (1912–1954). The computer was a child of the war, invented to decode the encrypted ciphers of the German army. Similarly, the internet was initially derived from the

American military communications network Arpanet and was officially born in 1990 with the progressive connection of computers all over the world through shared protocols for transferring information.

The internet was therefore a belated product of the Cold War, and it remains crucial to military interests today. Its vital infrastructures are controlled by the United States. It is also important to bear in mind that the various organs of the internet may be threatened by cyberwarfare, which several countries are currently preparing for, including China, which now provides the majority of the electronic elements essential for the network. Our digital media revolution is therefore anything but permanent; on the contrary, its technological foundation is extremely fragile. The failure of digital infrastructures alone would be enough to provoke a degree of collapse.

The internet as we know it is only the beginning of the digital revolution. Up until now it has been limited to transcribing the lives of the wealthy in many countries into data and networks. But this revolution is changing us much more than we can change it. It is a technical metamorphosis that is not designed to adapt to humanity but is rather optimized to privilege a few of its actors. Those of us who wish to immerse ourselves in this revolution are subject to a process of exaptation.

Alongside this, the collection, sharing, treatment, and networking of private information in so-called Big Data all lead to increased control of individuals. Of course, the internet has created certain realms of freedom for those who master some of its protean potential. But the complexity of the network is such that it seems reasonable to fear that eventually the only actors able to influence the conception of the system will be major corporations and large states.

In the 1970s it was customary to predict that the digital revolution—which no one could yet really imagine—would mark the beginning of a new postindustrial era. This future was to be marked by the decline of industry, a reduction in pollutants, and a boom in the service sector. Although there has indeed been a transition toward services in developed countries, which is today reaching

India and China, industry has by no means been neglected. Never has humanity produced so many things. And, as always, the arrival of new economic sectors has not wiped out the old ones. Digital production has not replaced industrial production, it has simply added to it. Pollution has only grown. Industrial waste, fueled by planned obsolescence, which deliberately reduces the life expectancy of commercial goods,[8] has escalated dramatically. If they had been designed differently, digital networks could have evolved toward decarbonized economies organized around the distribution of clean energy, but they are now one of the primary producers of greenhouse gases. The emphasis on social networks, in which anyone can share large digital documents such as videos and photos,[9] has meant providers (GAFAM companies but also intelligence services, which, like the National Security Agency or private companies like NSO Group or Huawei, are accused of spying on all of us) have had to establish colossal databases to store and analyze this data. But accumulating and treating potentially unlimited volumes of exponentially increasing information ultimately involves astronomical cost, particularly in metals, electrical components, and above all energy. The internet is anything but a benign technology.

Although it was the United States, with help from the military and intelligence agencies, that developed the infrastructure of the global information society, major private corporations now control it. This is, of course, in keeping with neoliberal ideology that promotes minimal state intervention.

The Cost of What Is Free

Communications innovations may mean that we can work more efficiently, make society more democratic, or enable a greater distribution of knowledge. But that is not their primary objective. Facebook may be an excellent source of general or professional information, but the algorithm that manages it is programmed to sort through the flow of information you see. Its goal is to make you addicted to its services. It will therefore prioritize your interactions with those whom it considers—after a close analysis of your behavior—to be close to you. It thus aims to amplify our tendency for communitarianism, our propensity to prefer the company of those who are like us. Today's essential technical tools, in terms of programs and interfaces, do not serve the common good and are not designed to do so. The corporations that operate on the net aim only to constitute monopolies on the vital flows of information. Together they paint a vision of the world that is a sugar-sweetened caricature of our own, a rose-colored version intended to hypnotize us, like Narcissus, obsessed with our own reflection.

Alongside this, the internet also provides a global space for communications accessible to all. It was built, much to the displeasure of certain authoritarian states, around a potentially unlimited number of connections (any computer can serve as an interface and a relay). The democratic nature of the network that today connects more than half of humanity is also related to its history.

As we have seen, the internet was derived from Arpanet, the military network developed during the Cold War. However, a certain number of academics also worked on the project and managed to deviate it somewhat from its initial purpose for military communications. Influenced by the ideas of the 1970s counterculture, the scientists discreetly imposed their objective: the web had to belong to everyone, everywhere, and for everything. In its early phases, from the year 1990, the internet was designed to be accessible to the greatest number of people possible. The contributors shared a belief in the general interest and free access for all. Although profit-seeking companies have since commandeered whole swathes of the network, they do not know how to undo the initial choices that were made concerning universal free access. Or perhaps more accurately, it is no longer in their interests to try, having made this work to their advantage.

The internet is like a human brain, torn between contradictory principles. On the one hand, it allows more and more people to access enormous amounts of information, to communicate with each other in specific forums, and to express themselves publicly. The participative web—with its new media, blogs, forums, alternative sources of information (Snowden, WikiLeaks, Panama Papers, etc.), collaborative content (Wikipedia), and community networks—generates new political dynamics and unprecedented forms of emancipation.

On the flipside, however, the internet allows political and economic powers to conduct "soft" surveillance of citizens. We now live in a society of control. Each foray on the internet leaves a trace that is captured, stored, and processed in data packages retracing the activities of each user, constantly identifying his or her tastes, opinions, movements, and fantasies. Between these two opposing poles, between emancipatory freedoms and omnipresent surveillance, there is a tension that accelerates the expansion of the network.

The collection and sale of users' personal data has become the price to pay for "free access" to the internet. Economic actors like Google use this data to attract immense financial resources in the form of advertising. State actors such as the United States see this as a way of consolidating their geopolitical dominance in an unstable world. Mass surveillance is preventive surveillance. And it might occasionally serve the general interest. For example, in epidemiology, the appearance of keywords in emails (sore throat, fever) could allow Big Data to rapidly identify the beginnings of an epidemic. But this potential benefit supposes that individuals accept that their correspondence is read by machines. Indeed, this argument seems to be more a pretext to collect medical information (a fantastically profitable market for major corporations) than a genuine concern for your health.

However, data collection today is almost exclusively conducted for commercial and security purposes. There are invisible actors that monitor every move of the world's four to five billion internet users. These actors aim to either influence our purchases or anticipate certain actions that may be against the interests of those monitoring. We pay for our new freedoms by tacitly accepting to be either a consumer under influence or a potential terrorist under surveillance. The dictatorships of the twentieth century would have loved to have such a powerful, imperceptible tool for control. And it seems increasingly likely that Chinese nationalism, to give just one example, may transform the part of the web that is managed from its national territory into a net it can close to cut its population off from the rest of the world in a new kind of totalitarian nightmare.

The Three Theories of the Multitude

The French sociologist Andre Vitalis identified three analytic frameworks for the development of this Digital Revolution:[10] the theories of new empire, of cognitive capitalism, and of the post-media era. These three analytic theses are complementary. They are all based on the theory of the multitude, which postulates that communication networks play a decisive role in global politics.

The concept of the multitude is inscribed within a post-Marxist perspective based on the idea of a group of individuals and forces organized around communication-based and information-based forms of production. It is reminiscent of the old revolutionary proletariat that was structured around productivism in mines and factories. The internet is seen as the agent in this change, the environment that gives rise to a worldwide social shift, federating individuals into a multitude organized into cooperative decentralized networks in which local initiatives, democracy, and universal access to common goods constitute a credible alternative to the perspective of the unequivocal triumph of capitalism.

The theory of new empire is defended by Antonio Negri and Michael Hardt, who are leading figures in the alter-globalization movement.[11] These thinkers define *empire* as being "everywhere and nowhere," a power superstructure that is diffuse, decentralized, and deterritorialized. Globalized capitalism is the reference that drives the world. It has a decisive influence in a global space where nation-states have given up their sovereignty by accepting to submit to the interests of the market in keeping with neoliberal ideology.

The theory of cognitive capitalism analyses the advent of the third industrial revolution based on digital services, which are reliant on the omnipresence of electricity networks.[12] From this perspective, since 1975 cognitive capitalism has replaced industrial capitalism, which itself replaced merchant and slave-based capitalism. Yesterday's society was structured around the factory and the mine. Today's is built around screens, new spaces of nomadic production. The network society, which is made possible by information technology, has provoked an upheaval in social values. The actors in this network economy are presented as emphasizing creativity, autonomy, and collaborative work between peers, challenging hierarchical structures of traditional businesses as well as factory-based production practices such as repetitive, fragmented tasks. This shift, although insidious, has not failed to radically alter all aspects of the social space.

The theory of the postmedia era aims to emphasize the anthropological consequences of changes in the structure of the spread of information. Since the invention of writing, information has circulated vertically, from the upper echelons of society to the lower. Mass media belonged to the elites, those with capital (most newspapers are owned by private interests), or those in political power (the state controlled the television and radio channels). But since the 1980s a radical change has occurred: the horizontal circulation of information. The postmedia era is one in which the internet is seen as space in which every individual is potentially a media, a receiver and transmitter of information. This new sphere of communications has been superimposed over the old, or even hybridized with it. Seeking to broaden their audience, traditional media have now developed spaces for interaction with their users.

The development of writing did not alter the human brain, but it did mean parts of this organ adapted to this new function. It is far too early to know how the Digital Revolution may influence our neuronal circuits and our mental processes in the long term. That is assuming that the internet, which is a fragile artifact dependent on vulnerable technical and energy infrastructures, endures.

The Disruption of Networks

A computer does not merely reproduce reality. It reinvents it through images, sounds, and textual description. The virtual universe is a fiction that can exist autonomously outside of our minds. It has become standard practice to invent imaginary worlds in which we can meet and interact with others. We do this among our peers, of course, those who are like us. All media exists to produce consensus. Even when an online game plunges players into a violent war, those players share references with their adversaries. The totalitarian regimes of the twentieth century understood this and made it a priority to control radio and cinema to impose their ideas onto the people. Television long remained under the tutelage of the state because it was seen as having similar powers (the phrase "as seen on TV" even became a way of certifying the veracity of something). Today social networks can be fertile ground for mobilization in support of major causes, whether overthrowing a dictator or electing a populist. But they above all insidiously create closed spaces for "sharing" among members of a given group who have in fact been selected by algorithms. Limiting ourselves to what they show us of the world would mean erasing all perspective of otherness.

The digital environment has overflowed from our screens. It surrounds us, modifies our perception, and weaves us into a web of imperceptible signals. Our telephones are geotracked, our movements are filmed, our behavior is monitored. At the end of this road the nightmare of disruption lies in wait.[13] This term fundamentally refers to the fact that certain new economic actors disrupt traditional modes of production, just as Uber and Airbnb have done in breaking the monopolies of taxi drivers and hotels. By extension all behavior emanating from individuals with new media-fueled power is potentially disruptive. The archetypal example is that of the Uber driver and passenger, who both benefit from new economic and social opportunities; one is assumed to be making money, the other saving money, on the service. Their agreement, via new technical platforms, creates what is commonly called a new economy, which is therefore disruptive to the old.

But far from working for us, this disruption tends to be imposed on us with all the strength of a fait accompli. It has become difficult to avoid, because it is based on a global information network that is now literally able to remotely control us. Robots are everywhere, and their decisions have a direct impact on us. Take the example of high-frequency trading. Finance has an impact on every part of our lives; everything has become a subject of financial speculation, and over 85% of transactions are now conducted by robots at a frequency that far exceeds humans' capacity for control. Along the submarine cables that now form the neural network of the world, financial instructions crisscross the planet at a speed of 200 million meters (about 656 million feet) per second, or two thirds the speed of light. By extension, disruption describes the human feeling of a vertiginous loss of control. This feeling was familiar to the Americans who lost their houses to debt in 2008 in a financial snowball that shook the world economy.

As the world shifts toward the new dimensions of accelerated time and the digital revolution begins to affect our "disrupted" psyche, the seventh revolution is also underway. This Evolutive Revolution is torn between the twin opposing tendencies of artificially planned evolution and imposed evolution. Digital and evolutive, these two revolutions progress together, augmenting and intensifying their potential effects.

17

TOMORROW'S WORLD

Monkey must urgently gain an understanding of the Anthropocene if we are to survive. Will tomorrow make us miserable mutants or "amortal" gods?

Jack is an immensely famous journalist, a talk-show king. In the presidential race, this blasé TV star is constantly in the spotlight. His program is based on an interesting concept: a viewer calls in live to complain about some injustice suffered, and Jack then takes the microphone to call a powerful figure, live on air by video link, to justify and, if possible, repair said injustice. Jack plays the defender of the innocent.

But this particular evening, Jack's life will be turned upside down. Unbeknownst to him, he will be the first man to achieve biological immortality. But he will pay for it with his own humanity.

Jack and the Fourth Estate

Bug Jack Barron is a novel written by the science-fiction author Norman Spinrad, initially serialized in the British magazine *New Worlds* in the 1960s. Jack seems like a kingmaker, but he is really just a puppet of the system. Having been raised with communist ideals, he later adopts the most cynical form of liberalism. His talk show allows him to accuse whomever he wants. But he knows there are lines that cannot be crossed. Politicians and business leaders fear him but also appreciate what his show represents: a farce in which power appears to be challenged but where the Don Quixote defender knows how to pull his punches. Jack has set himself a line he will not cross. His elite guests must leave feeling chastened, shaken up by this pugnacious staging of controversial questions, which is punctuated by top-dollar advertising slots. But the world order must remain intact. This electronic pseudodemocracy must continue to elect the programmed president under the distant but omniscient control of magnates who control the true power of the planet: multinational corporations.

On this particular evening, Rufus W. Johnson phones in. He claims that the Foundation for Human Immortality has refused to accept his assets as payment for his future cryonic freezing. Rufus is black. He is accusing the foundation of racism. He claims that even if an African American has the half a million dollars required to have his (or her) body frozen for future resurrection, the organization makes it impossible for them to obtain a contract.

Jack is indignant. He tries to contact Benedict Howards, the foundation's president. Not available? John Yarborough then, the PR director, whose defense is swept away by Jack in an instant. Then it is the Governor of Mississippi's turn, Lukas Greene. He is also black, and one of the few African American politicians to be in the public eye. He argues that the foundation is not only racist but is also violating several federal laws by trying to have Congress pass an amendment that would legalize its monopoly on cryonic treatment and thus humanity's access to immortality. Greene is fighting for public, universal access to the "freezer system."

Jack calls a witness for the defense, the Senator Theodore Hennering, a partisan of Benedict Howards and his goal of monopoly over cryonics. The politician has trouble convincing anyone that the foundation is not racist. It does not seem to be going well for the company.

The next day, Jack is approached by Howards, the all-powerful head of the Foundation for Human Immortality. He offers him a pact with the devil: immortality in exchange for his support. His company is in fact already able to extend human existence indefinitely.

After a series of plot twists, Jack receives the immortality treatment only to discover that its ability to stop aging relies on harvesting glands from kidnapped children who receive a fatal dose of radiation. Innocents must therefore be sacrificed to achieve this "amortality." Unlike genuine immortality, preventing all death, "amortality" is the condition of a body that does not deteriorate over time but which can still die from injury or accident.

Becoming a god means accepting sacrifices.

This futuristic fable, written half a century ago, predicted the rise of mass media, high-profile African American politicians, spectacle democracy, videoconferencing, the transfer of compressed data files over a worldwide digital network, organ trafficking, and the possibility of cryonic freezing in the hope that technology will, eventually, be able to prolong life indefinitely. All these things are now reality. Only amortality has not yet been realized. Or has it?

The Temptation of Amortality

The seventh revolution will be evolutive. Humanity has acquired the tools to control evolution, and it has simultaneously altered its environment to the point where its own genetics are affected. We will begin by exploring the possibilities of controlled evolution available to Monkey before looking at uncontrolled evolutions that are already modifying our bodies and our minds.

Monkey now has a number of tools to shape evolution that open up a range of possibilities. Various kinds of prosthetics, psychotropic substances, and human-machine interfaces have already been lab tested and even commercialized, and all are liable to radically alter our biology. These nascent technologies aim to develop "augmented" humanity, to "upgrade" us as one might software. Let us use the acronym NBIC—nanotechnologies, biotechnologies, information technology, and cognitive science—to navigate this world of "improvement." This is the acronym used in work on transhumanism, an ideology that aims to transcend us all.

The promotors of transhumanism tell us that the NBICs are the tool kit for the voluntary transformation, or "improvement," of humanity. If we take them at their word, these four sciences provide a space in which their effects converge to push back the natural frontiers that still limit our possibilities. NBIC-augmented humanity is the ultimate extension of Monkey's capacity for mastery. It involves the following:

1. Mastery of the infinitesimally small. This is the N for nanotechnology, which reorganizes the mechanics of matter at the atomic and molecular level.
2. Mastery of reproduction and genomes. This is the B for biotechnology, with its genetically modified organisms that confer new properties onto the living world and stem cells with the capacity for cellular division at the foundation of all life and liable to sweep away any limits to repairing and augmenting the body.
3. Mastery of machines and communication. This is the I for information technology, which can be associated with the N for nano, to create quantum information science freed from the physical laws that chain our minds to the terrestrial world.
4. Mastery of cognition. This is C for the cognitive science of the brain, its structure and conformity, its cartography even, and control of consciousness, the most intimate layer of our being.

Monkey dreams of becoming an absolute master of matter, computation, thought, and the living world. He sees himself as an all-powerful, immortal, and omniscient god, self-referential and unlimited.

Even without these technologies we have become quasi gods. In two hundred years, we have gained powers the Ancient Greeks associated only with those on Mount Olympus thanks to our mastery over energy, the living world, and our environment. We have abolished distance because we can now fly around the world in a day (if we have the funds to do so). Thanks to our smartphones we can now see and hear scenes taking place on the other side of the world through live streaming applications. We now control our reproduction. We can fight diseases that used to be mass killers. We eat our fill and live longer and in better health than most of our ancestors. Some of us have even been into space. There are optimists who claim that this is just the beginning. And pessimists who, looking at the same scene, warn that we may be experiencing the last moments of humanity.

Let us take a closer look.

Nanotechnologies operate at the level of the infinitely small, where the laws of physics are different, where matter can acquire new properties, and where it is possible to imagine making nanorobots able to duplicate themselves. These minuscule, invisible machines may be able to do the impossible, serve as agents to repair the human body or hunt down cancers. Does that sound like science fiction? Mechanisms a billionth of a meter long (a nanometer) were already being studied in the 1980s, and prototypes were developed in the early 2000s. Today their building blocks, the invisible particles known as nanomaterials, are present in several hundred everyday products such as clothes, plumbing joints, telephone batteries, and so forth. They have entered our world insidiously. Their many promising physical properties make them highly sought after by industrial developers. But they are also decried by toxicologists who fear that these substances will turn out to be poisons liable to provoke genetic mutations and cancers—or even to replicate uncontrollably.

In biotechnology the major breakthrough came in 2003 with the first sequencing of the human genome. Since then the cost of genetic technology has declined considerably, and precision has improved, which has meant a substantial increase in advances in this field. Confirming the convergence of NBIC technologies, progress in the speed with which computers process information (they double their capacities every eighteen months) has led to an acceleration in this area. The border between the living and the digital worlds is perhaps being erased. Synthetic biology, for example, simulates genetic evolution using computer programs that perfectly replicate the attributes of primary living organisms, blurring the boundaries between the physical and nonphysical worlds. One day perhaps they will enable us to program bacteria as we would a computer program. In collaboration with nanotechnologies, biotechnology could produce living organisms that constitute organic material for industrial production, like bacteria that could be transformed into plastic or fuel. In the meantime, genetically modified (GM) organisms have been incorporated into many food plants but have also discreetly contributed to a wide range of medical products. For the moment, Europe seems to have resisted this wave of technology to a certain extent. But it feeds its meat animals GM soy cakes imported from Brazil. Other techniques, such as gene drives (the spread of a GM organism in the natural environment in order to deliberately alter the DNA of a specific species or even eradicate it) have already been implemented, for example, targeting disease-spreading mosquitoes.

Computer science has also benefited from improvements in the exploration of the infinitely small, and particularly the possibility (still embryonic today) to develop quantum computers that can use the properties of the atom to increase their power, perhaps even to create artificial intelligence?

Of the four NBIC disciplines, cognitive science has attracted the most investment. The studies of the human brain have benefited from progress in medical imagery. The convergence with progress in computer science has also made it possible to improve many areas related to machines, such as artificial intelligence, electronic simulation of emotions or voices, facial recognition, or the programming of robots with increasingly complex behavior. There are various projects underway aiming to digitally emulate the functioning of the human brain, from the Human Brain Project in Europe to the Brain Initiative in the United States.

In conclusion, these NBIC technologies are rich in both potential promises and potential environmental risks. The risks are accentuated by the fact that these technologies have developed over such a phenomenally short time that it is impossible to have an overview of what their impact on our everyday lives might mean in the long term. But the real stakes here go beyond the typical dialectic between potential advantages and perceived dangers. When taken together, these technologies aim to bring about the artificialization of the world and the disappearance of the boundaries between the living and the artificial—in other words, the birth of human-machines. They target the very nature of humanity in order to modify it.[1] It is this breach that paves the way for what is known as technological singularity.

Uncharted Singularity

NBIC technologies are just tools that humanity can use for any given purpose. But their convergence is a choice, made by an ideological movement known as transhumanism, that aims to transcend humanity, to "improve" it even to the point of transformation.

In 1959 the physicist Richard P. Feynman provocatively claimed that humanity would one day be able to store the entire Encyclopedia Britannica on the head of a pin. Today this is entirely banal, but back then it was met with hilarity and ridicule. That was a time when computers required colossal budgets, weighed as much as a semitrailer, consumed as much energy as one hundred trucks, and were less powerful than today's cheapest smartphone. It was prehistory.

In 1986 the engineer Kim Eric Drexler popularized the notion of nanotechnology,[2] which he saw as being a remedy to the erosion of mineral, agricultural, and societal resources anticipated in the Limits of Growth report (1972). An avid consumer of science fiction, Drexler claimed that science would soon be able to assemble atoms to save humanity from disease. He predicted the creation of molecular machines circulating within the body to repair diseased organs, nanomedicines made more efficient by their ability to place themselves precisely where the organism requires them, and nanobots able to recode DNA to fight against genetic diseases and cancer. For Drexler these inventions would eventually slow aging and then stop it and prolong life indefinitely. In the real world, since 1981 the scanning tunneling microscope has made it possible to observe particles at the atomic level. This tool has given humanity the ability to manipulate atoms. Drexler's vision gives us some ideas of its potential applications.

In the decades that followed, the US federal government supported the development of this technology. In the 2000s the National Science Foundation promoted the convergence of NBICs. The research and development budget of this organization is matched only by that of the military, space research, and the nuclear industry. A new world was unveiled, woven from promises of omnipotence, a world in which human desire could rebuild matter, atom by atom. This ability made it possible to create materials in symbiosis with the living world, to program genes able to repair themselves, to hybridize neurons with electronic connections, to subtly intervene in all aspects of matter, the organization of information, and the living world.

For their promoters, NBIC technologies are the guarantee of the survival of the human race. In a context where the global population will continue to grow over the course of the twenty-first century while agricultural resources will necessarily reach a production ceiling and where mineral and fossil resources will become more difficult to access, NBICs could make it possible to optimize the use of available assets so efficiently that they would become inexhaustible. This opens up unlimited possibilities with the ultimate objective being human beings with nanobots in their blood (to repair and maintain the body and constantly preserve it from degeneration) and whose consciousness is saved on external hard drives.

This new science aims to abolish both humanist and religious thought. Religion forced humans to be good in the interest of coexistence; humanism sought to improve minds through education; transhumanism aims to create artificial minds governed by their technological mastery of emotions. Religion founded moral communities obliged to respect transcendental rules. Humanism pushed societies toward constant improvement through social and political reform. Transhumanism considers that the application of technology will allow humanity to reach new levels of accomplishment, making society itself obsolete.

Does a god need others? Will transhumanism mean sacrificing solidarity, which is common to all the most developed species and which made Monkey the master of the world thanks to his ability to cooperate?

In the twentieth century, science has become so complex that even its deontology is complicated. As a result it is likely that what is technologically possible to do will eventually be done. Medical tourism will make it possible for people living in countries with ethical restrictions on particular practices to taste the joys of transgression and improvement elsewhere in countries with less restrictive laws. Alongside this, scientists are conditioned to think only in terms of knowledge, international competition, and publication in peer-reviewed journals. Ethical arguments may periodically be used to push back scientific programs—such as human cloning—that several international agreements have found to be not in our species' interests. This has not prevented scientific teams all over the world progressing in this area in order to create stockpiles of stem cells for medical purposes.

How do ethics weigh up against profit in a world dominated by neoliberal ideology? Humanity has been offered an attractive bargain. Temptation whispers in our ears: to the three basic obsessions we share with the other members of the animal kingdom—eat, rest, and reproduce—could we add a fourth desire specific to transhumanism: the desire for everlasting life? The apple that got Adam and Eve expelled from Paradise has taken on a sheen of artificial eternity.

One of the most visible figures of the transhumanist movement is Ray Kurzweil. This futurist, currently director of engineering at Google, argues that the prospect of augmented humanity—for example humans able to breathe underwater—is just around the corner. Human immortality is only a little farther down the road. What he calls the technological singularity, the moment when artificial intelligence surpasses the human brain in terms of cognitive power, is close to being attained as computer programs learn to simulate emotions and teach themselves by compiling resources and information, building their cumulative control of information through "deep learning."

What will happen when computers are smarter than us? Beyond that point, transhumanists say it is impossible to know what the future will bring, but they maintain that whatever happens, we will be able to keep control of our creations. For other authors,[3] artificial intelligence will destroy us, Terminator style.

Clearly we cannot know the future. But there are three major uncertainties here.

1. NBIC promises are far from having been achieved. Research will most likely lead to major discoveries but also probably to physical and technological dead ends. Such promises can only be fulfilled if the economic, social, political, and ecological conditions allow for it. The collapse of global finance, for example, could knock the feet out from underneath this research as surely as political decisions could throw such a Promethean project on the pyre.
2. From cryonics and people paying to have their bodies preserved until a hypothetical science has found a way to resuscitate them,

to the use of stem cells and high-tech prosthetics, all technology for improving or prolonging existence will always come at a price. Eventually, it is likely that the gap between those who have the means to artificially improve themselves and those who do not will become a chasm. On the one hand, there will be those who remain human, and on the other, perhaps those who have reached another dimension. Tomorrow, being ultra-rich may open the doors to humanity's ultimate dream: being superhuman.

3. The destruction of the environment may also hamper our mission to become gods and force a little modesty on us. Far from evolving voluntarily, we are now becoming transformed by our actions to the point where two classes of humanity might emerge. Perhaps a minority might hope to control its own evolution, but the vast majority remains exposed to undesired and unintentional genetic modification.

Monkey is also the master of fertility. Techniques for freezing sperm and oocytes, artificial insemination, and embryo transfers are already well established. This medically assisted procreation means we can fight infertility in both men and women, whether caused by age, physiology, medical treatment, or exposure to pollutants. This means that reproduction is becoming less and less biological and more and more societal. And research inevitably leads to options the law does not account for, including the ability to bring children into the world from the gametes of a deceased donor; to create clones for harvesting organs or cells for medical purposes; to choose the sex, physical, or behavioral characteristics of a child; and even to induce differentiation in germ cells in order to produce life from donors of the same sex.

Simple tests make it possible to detect the presence of genes that carry a predisposition for monogenic (linked to a single gene) diseases. This is the beginning of preventive genetic medicine, a market that several corporations are eyeing greedily. Beyond the predictable generalization of genetic testing, there are potential perspectives for treatment to correct, preventively or curatively, the DNA of a patient who may otherwise develop a pathology.

Once that has been accepted (who would deny victims of genetic diseases the use of these therapies?), the question becomes where intervention should stop. Improved athletes—who have had their DNA modified to enrich their blood with oxygen, for example—are perhaps already a reality.

Monkey is mutating. We only need to turn on the sports channel to see this. Athletes today are not what they once were. Robert Garrett, the gold medalist for shotput in the 1896 Olympic Games, weighed 81 kilograms (179 pounds) and was 1.80 meters (5 feet 11 inches) tall. The athlete who broke his world record in 2012 was Tomasz Majewski, a mountain of a man, 1.97 meters (6 feet 5.5 inches) tall and weighing in at a colossal 142 kilograms (313 pounds).[4] A little more than a century ago, Robert Garrett could also consider competing in the 400-meter race, which would be unthinkable for Tomasz Majewski, whose body mass does not correspond to that sought after by contemporary runners. To run, you need long muscles and a sleek body. To throw or lift, you need thick muscles on a strong bone structure, a massive body. In the early days of the Olympics, athletes collected medals in combinations of various sports. The Australian Teddy Flack won medals in both athletics and tennis in 1896. The same year, Danish sportsman Viggo Jensen excelled both in shooting and weightlifting. Today no one would think it possible for Roger Federer to win gold in the 400-meter freestyle.

This is what happens when you apply Darwin's survival of the fittest to the ruthless world of international competitive sport. Especially if you add the cultural elements of human evolution, which are controlled by our will and increasing physiological understanding. Athletes are psychologically conditioned not to doubt their abilities to push the limits; they are fed on scientific theories and nutritional calculations, literally exapted (not adapted) to achieve the best possible performance and sometimes (secretly) chemically augmented. They become optimized for specific tasks. Their bodies reflect society's efforts to hold up a strange mirror, that of the cult of performance imprinted into flesh itself. The musculature of Arnold Schwarzenegger is one of the icons of this mirage.[5]

Today, the bodies of Olympians or bodybuilders are prototypes for a possible future of augmented humans represented, among others, by transhumanism. Another aspect of this ideology envisages the simple replacement of the human species by machines, or cyborgs, or synthetic beings with both biological and nonbiological components. These two options—augmented humanity or humanized machines—are both possible paths for an evolution that is still partly controlled by human thought. We are living through the preliminary phases of this evolution today. The third option, which is also underway, is that of the unwitting mutant.

Strictly speaking, we are all mutants. The genetic heritage of every living thing has been altered by its environment. But the Anthropocene has changed more than just the planet. Monkey has poured so many different products into the environment that our bodies change faster than did those of our ancestors. The humanity of tomorrow, whose development will play out in a persistent fog of artificial pollutants, seems destined to be more feminine, less fertile, heavier and taller, perhaps less intelligent, and more subject to early puberty, autoimmune diseases, allergies, and cancers. In a way, we are all like today's Olympic athletes. Like them, we are significantly different from our ancestors of only a century ago, who were not confronted with an environment in which endocrine disrupters, solvents, plastics, and pesticides were omnipresent.

Unwitting Mutants

On Wednesday, January 18, 2017, at 11:24 a.m., I connected to the Chemical Abstract Service website.[6] Their registry provided a list of the 126,183,668 distinct chemical substances in the world, organic and nonorganic, including metal alloys, complex chemical compounds, mineral admixtures, polymers, salts, and solvents. When I left their website five minutes later, seventy-nine new entries had been added to the database.

A small number of these molecules will end up being commercially used and probably distributed around the world. And their effects may have major consequences, contributing to the biological changes humanity is currently experiencing (obesity, cancer, autoimmune disorders, as well as early puberty, feminization, higher rates of genital malformations and reproductive disorders, low sperm counts, increased rates of allergies, etc.). These are all due to a combination of factors: our lifestyle (more sedentary, less physically active), our diets (higher in fat, sugar, salt and chemical additives), and our exposure to invisible pathogens. Where our ancestors feared "miasma," a foul air that was thought to spread epidemics, we (rightly) fear airborne particles from diesel engines, phthalates from plastic packaging, and a host of other substances that our eyes cannot detect. We face a tidal wave of potentially harmful microscopic substances.

One of the first to raise the alarm on this subject was Theodora Colborn (1927–2014). She had an unconventional career, initially training as a pharmacist before moving to Colorado with her family to raise sheep. It was not until she was a grandmother that she returned to university, obtaining a PhD in zoology in 1985. She was commissioned to oversee a collective publication on the fauna of the Great Lakes region. In preparing this report she noticed that the top predators—birds, fish, mammals, and reptiles—contained high levels of man-made substances in their biological tissue and that they transferred these chemicals to their young. She also noted that these chemicals altered the development of certain organs; many male fish developed female characteristics, for example.

In 1991 Colborn organized a conference with twenty-one experts from fifteen disciplines who had made similar observations. Together they demonstrated that certain products provoke transgenerational alterations on the functioning of certain glands. This research would give rise to the term *endocrine disrupters* to describe substances that have the specific effect of interfering with hormones. Above all they showed that these substances do not obey the old adage, attributed to Paracelsus, that "the dose makes the poison."

Indeed, even exposure to an infinitely small quantity of these products can have devastating effects depending on the moment that the exposure occurs. For example, during cell division, and particularly embryogenesis, these substances have decisive effects in terms of fertility, sex differentiation, and immune-system function. Hormones regulate living beings much as a pilot flies a plane.

The functioning of a tadpole's sex organs, the moment a butterfly emerges from its chrysalis, the female pigeon's ovulation response to the display of the male, an antelope's fear when a lion attacks, or a human's blood pressure and blood sugar levels: all these are dependent on the hormones secreted by the endocrine system.

Colborn defended her research fiercely and spent the rest of her life fighting to raise awareness. She obtained the support of Al Gore, who was then the vice president of the United States, but he did nothing to fight the spread of these products and contented himself with appealing to industry to conduct new studies. Like Rachel Carson, the author of *Silent Spring*, Colborn would be defamed by lobbies who did everything to discredit her research. But her warning nevertheless led to worldwide investigation resulting in many countries banning bisphenol A from baby products, Canada labeling it a toxic substance in 2010, and France banning it in all packaging in January 2015, a decision that Europe will likely adopt in the near future. But for every step forward there are several steps back.

The time it takes to ban just one endocrine disruptor from a handful of symbolic products is a clear example of this. Bisphenol A has been mass produced since the 1960s. Today it is omnipresent in our environment. In the United States it is found in the urine of nineteen out of twenty individuals, a level that is more or less shared by most of the living organisms on the planet. In fact, its global production is increasing: more than a quarter of a million tons were produced in 2015. It is found in PVC pipes, the resin linings of cans (except the ones sold in France), and in the paper used in receipts or banknotes. We know that infinitely small doses of this product influence the behavior of mice for at least four generations. Bisphenol A is an xenoestrogen, it mimics the sex hormone estrogen.

This substance enters our bodies orally, when we eat, but also through contact with the skin and through the respiratory system as well. It affects several parameters of biological development. The higher the dose found in the urine of an American individual, the more likely it is that that individual will be obese. In clinical studies, the more of the substance a female rat absorbs, the fatter its young are. It is one of the suspects in the global obesity "epidemic" that has been observed in recent decades, but it is far from the only one. Dioxins from the burning of household waste, atrazine used in pesticides, polychlorinated biphenyl (PCBs) formerly used in insulation and lubricants are all also on the list of potential suspects.

Like many of these substances, bisphenol A contributes to the development of certain kinds of cancer, increases the probability of developing cardiovascular disease, and is very probably a risk factor in type two diabetes and liver dysfunction. Moreover, its progressive banning in certain products has led industry to replace it with bisphenol B or bisphenol S, which are suspected to be just as toxic.

Let us go back to the Great Lakes. In Canada, on the shores of the St. Clair River, the small Aamjiwnaang First Nation community is paying the price. Between 1994 and 1998, the proportion of male births in this community was only 82 for every 100 female live births where global averages are "normally" around 104 to 106 male births. The situation worsened between 1999 and 2003, with only 53 male babies born for 100 females. Like the salmon that they eat, the Aamjiwnaang people are giving birth to fewer males. They blame the pipelines that run over their lands and the proximity to the chemical plants in the area. Ontario's industry, concentrated in this area, has poisoned the air, the soil, and the water. Analysis has revealed dioxins, pesticides, lead, mercury, and many other toxic substances, some of which are present in concentrations several times higher than the limits set by the Canadian Ministry for the Environment.

This kind of situation has occurred—and is still occurring—in many places around the globe. In Japan in the 1950s, in a town called Minamata, cats began to die from atrocious convulsions. Their owners, too, were left agonizing in the hundreds because of poisoning from the outlet of a neighboring factory that poured mercury into the waters of the harbor, poisoning the fish. The fishermen had to endure forty years of suffering, neurological dysfunction, and brain damage before they were recognized as victims of "Minamata disease" for having eaten the fish that were "plucked" from the seas.

What does the global level tell us about what is happening to Monkey? We know that he is getting bigger and fatter. But that is not all. His ability to reproduce is also affected. Numerous studies demonstrate that young girls are reaching puberty more and more early. In the United States the average age of first menstruation has gone from age seventeen in the mid-nineteenth century to age fourteen in the mid-twentieth century. Today it hovers around age twelve. The onset of early puberty measured not by menstruation but by the appearance of breasts as early as age seven is more frequent today than it was in the past, affecting 10% of Caucasian girls and 23% of African American girls in the United States. Young boys are also affected by this early puberty, although it is more difficult to measure. Socioeconomic factors also probably have an influence on exposure to endocrine disrupters (mothers working in factories who are potentially exposed to pollutants, lower-quality processed food, industrial neighborhoods, etc.).

Boys have also suffered in another respect. For the last fifty years, studies in developed countries have shown a decrease in the level of live sperm contained in 1 milliliter (.03 fluid ounces) of semen between 1.5% and 2% per year! In other words, an average man today has half as many live sperm per ejaculation as his grandfather did in 1960. Once again, the suspected causes are endocrine disrupters, phthalates, bisphenol A, and pesticides. And of course lifestyle effects such as smoking. Eventually, this decrease in sperm may threaten the very survival of our species.

Another phenomenon is also in play here. Just like the fish in the Great Lakes, our baby boys have been subject to a degree of feminization. This biological term refers to the development of certain physiological traits ordinarily associated with females of the species. The anogenital (AG) distance is often used to measure this, the distance between the genitals and the anus being smaller in women. In men, a smaller than average distance is associated with lower levels of both semen volume and sperm count as well as a smaller penis size and underdeveloped testes. And the average AG distance seems to be decreasing. Certain studies have demonstrated a significant correlation between this decrease and the exposure of boys to phthalates in their mothers' blood while in the womb. The greater the phthalate rate (in the blood), the more the newborn appears feminized or even subject to genital malformations.[7]

Even our brain is being affected, with its 1,400 grams (3.1 pounds) of complex neural pathways. Average IQ, which is an indicator of certain aspects of intelligence, increased steadily over the twentieth century in industrialized countries. This is known as the "Flynn effect" after the political scientist and statistician James R. Flynn. This significant increase, which varies from three to seven points per decade depending on the country and the period in question, is correlated with an ever more simulating intellectual environment: longer formal education, gender equality, more parental involvement in education, and increased access to media and information. The Netherlands, for example, recorded a twenty-one-point increase in the IQ of individuals called up for military service between 1952 and 1982.

And yet since the turn of the twenty-first century, the Flynn effect has been inversed. In France the average IQ dropped four points between 1999 and 2009. This dip can be seen in all industrialized countries. It seems unlikely that it is due to a general "dumbing down" connected to the media or the internet. Instead, researchers are increasingly blaming pollutants, and particularly endocrine disrupters. We know that chemical pollution is both omnipresent and able to pass through the placenta, affect the development of the fetus's brain, and interfere with thyroid hormones. These hormones guide the expression of genes and have an important influence on the development of the endocrine system.

Could it be that our own brains are beginning to flounder just as Monkey is on the point of developing artificial intelligence?

There are two ways we can fight against these involuntary changes.

1. We can embrace the technology race. We can pursue Elon Musk's proposition, develop artificial intelligence, implant a chip into the jugular vein, let nanoparticles with a direct neural connection to the chip flow through our bloodstream, and transform humans into cyborgs. This would create a symbiotic

being, half-living–half-machine, whose cybernetic intelligence would make up for any biological deficiency.
2. We can restrict toxins. This is both simpler, healthier, and easier to imagine happening. But it runs counter to the interests of big business and industry, who do not hesitate to lie to preserve their profits. Dieselgate, the emissions scandal in which Volkswagen (and then diesel manufacturers around the world) was shown to have deliberately cheated on pollution tests for their engines, speaks volumes. Thanks to neoliberalism, the private sector is responsible for conducting tests validating the market safety of all the substances that this entrepreneurial system devises, whether in pharmacology, the chemical industry, or agriculture. Everyone focuses on their own tasks, obeys orders to keep their jobs, and silently participates in the spread of thousands of potentially harmful molecules around the world liable to combine into cocktails that increase their impact. Whistleblowers are few and far between, and the gauntlet they must run (at the very minimum legal charges and personal attacks) is sufficient to deter even the most determined.

We are changing our environment, and it is changing us in return. If we cut down a forest, its pathogens will seek new environments to survive. From time to time, we catch a virus and our body has to adapt. This is one of the iron laws of evolution. By modifying biotopes at an unprecedented level and at cataclysmic speed, Monkey has opened Pandora's box. We have created 126 million new molecules of which thousands are already spread all over the planet, and they are liable to come together to form new unpredictable combinations. It is in their nature to fundamentally affect the reproductive ability of all living organisms as well as our immune systems and even our emotions. Microparticles of plastic are now mixed into the sand on Pacific Islands where there is no human life at all. Pollutant particles are now trapped in Antarctic ice. DDT has been found in polar bear fat.

What should we make of this? Is it a nightmare? Monkey should be screaming in his sleep.

Let us return to the so-called real world, the world of moderation. For many experts involved in these processes, the problems of the Aamjiwnaang community, along with the failure of the Fukushima nuclear reactors, are exceptions that should not make us change our behavior. No need for alarm. We will find solutions. Our mastery of artificial insemination will make up for lower sperm counts. Storing excess carbon (which we do not yet have the faintest idea how to do effectively) will prevent global warming. And our descendants will be able to find "solutions" for nuclear waste ... probably, if we leave them enough energy to do so.

If there is an animal that symbolizes our future in this "best of all possible worlds" it is the cow.[8] Over the years the proud aurochs of the past have become placid cud-chewing bovines. And then industrial agriculture emerged and reshaped them from A to Z. Artificial insemination initially allowed them to select the best breeders to the point where a species like the Holstein (the famous black-and-white dairy cow) now suffers from inbreeding because a single bull has effectively served to inseminate hundreds of thousands of cows. All without ever actually seeing one. To avoid the loss of any precious genetic material, the bull is made to ejaculate on a castrated male so that all the semen can be collected. And cows are given extra estrogen to redress their naturally low rates of ovulation so that they get pregnant faster.

In the 2000s the mapping of cattle DNA accelerated this process. Algorithms decide which bull's sperm will produce a calf of a given temperament. It is possible to incorporate variables like ease of birth, the calf's sex, lack of horns, or muscle development for butchering. Automatic milking at atrocious frequencies and the concentration of animals in a single barn require preventive antibiotics to avoid disease and incidentally to facilitate weight gain (even if it means making antibiotics less effective and increasing human obesity problems). The beasts are fattened up using corn or soy cakes, which the cows have trouble digesting but which increase their weight gain. Their lives end on an abattoir production line, industrialized to maximize profits and block out suffering. On top of all this, both the air and the water become heavily polluted. The system generates an abundance of tasteless meat and more or less digestible low-cost milk.

Descartes imagined the animal-machine. Our society created it and genetically modified it. At first glance, Rosita is an ordinary Argentinian cow. But she is also a cow with no father and six mothers. She is a chimera, a clone, originally a stem cell extracted from a cow, given the genetic information of a second cow, inseminated to begin her development in the uterus of a third cow, then transferred to complete her growth in a fourth cow. She was also given genes from two human females so that her milk would contain enzymes previously found only in human mother's milk. So, is she the first bioreactor? This is what some biologists call the cows of the future, who will be potentially genetically modified to make certain medicines such as insulin.

From the very beginning, the domestication of animals was based on the control of reproduction, which allowed us to select for certain characteristics in the young. The bull had to be less strong and less aggressive than his aurochs ancestors, and the cow had to produce more milk. It is logical that today we use our newfound powers to extend this modeling of the living world. Moreover, we are now able to produce changes so quickly that it is even possible to lengthen cows' teats so that they are better suited to automatic milking machines!

We have come a long way from our first steak tartare 3.3 million years ago.

Widespread changes in diet resulting from the productivist model developed in the West now mean that excess calories constitute the planet's largest food problem. While eight hundred million people suffer from food insecurity, more than two billion are overweight, and among them seven hundred million are considered obese. Once collective imagination associated wealth with corpulence, but now it is the opposite: the rich have the means, both nutritional and recreational, to be able to stay svelte and healthy. Those who lack those means often have to content themselves with poor-quality food (fatty, sugary, highly processed food rich in additives and endocrine disrupters).

China, a Metaphor for an Accelerated Future

From deathly poverty to tomorrow's number-one world power, China's trajectory is a summary of our era since 1979. Let us take a look at the international framework in which the "Chinese miracle" took place. Over the course of the twentieth century, up until the 1990s, economic inequalities between countries continued to grow, extending the Great Divergence that began in the early nineteenth century (chap. 12). Everything changed when the USSR collapsed. A "Great Convergence" inversed the secular rhythms of the economy. The world's poor countries "woke up," and the shift occurred. Now the poster child for unfettered liberalism, China is becoming liberalized under the attentive control of a communist state anxious to adopt the best capitalism has to offer (economic growth) without what it considers unnecessary (democracy).

Toward the end of the twentieth century, China morphed into an investor's dream, a state that invested unlimited amounts into countless infrastructures: thermal power stations and hydroelectric dams to achieve flows of inexhaustible energy, factory cities exploiting uprooted rural populations, colossal harbors pouring constant flows of merchandise into standardized shipping containers. It took a decade for the communist utopia to become the promised land of free trade.

Progressively accompanied by the rest of Asia, Chinese growth made up the lost ground that had previously separated those in rich countries from those in poor countries. This occurred alongside a significant increase in internal inequalities. In 1980 all Chinese people were poor, with the possible exception of the party elites. In 2017, two hundred million to four hundred million were members of the middle class. Their income was equivalent to that of European middle classes, which meant their buying power was much larger. But the majority of the 1.3 billion people in China remain poor. The Chinese economy has become one of the most unequal in the world. The Gini coefficient, which is used as a statistical indicator of inequality in a given society, hovers around 0.2

for egalitarian societies such as Denmark, 0.3 for France, and 0.5 for China, with its highest value of 0.6 for Brazil. Yet there is redistribution in Asia. If you are not poor yourself, you pay domestic workers to contribute to collective wealth. This is the implicit rule in the Asian social contract.

In Russia, India, and elsewhere this is a common scenario: societies where people used to be equal in their poverty have become richer but profoundly divided in the distribution of their gains. There are two main consequences here. The first is that even as a Grand Convergence reduces the gaps between national economies expressed in per capita GDP, societies are also torn by more and more pronounced internal inequalities.[9] This creates fertile ground for unrest. The rise in global violence, religiosities, and populisms can be correlated with this increase in the disparity of wealth distribution. The paradox is that humanity has never produced as much wealth as it does today, yet that wealth has never been shared by so few. Just 1% of humanity controls half global wealth.

The second consequence of the Grand Convergence is the changing middle class. In 1980, there were five hundred million people in the global middle class, primarily from Western Europe, North America, and Japan. Communist welfare states—the USSR, China, even Poland, Syria, and Cuba—ensured that a large proportion of the world's population had free access to health care and a relatively reliable food supply. The wave of neoliberalism brought them chaos in the 1990s. But most of the affected societies found their way out between 2000 and 2015. In China, India, Brazil, Russia, and a few other countries, a new dynamic middle class emerged that contributed to the growth of these economies. In 2009, the OECD (the Organisation for Economic Co-operation and Development) estimated that 1.8 billion people were middle class,[10] and that of them, more than 1 billion lived in developing economies.

Today these middle classes represent around 40% of humanity. And they control 45% of global wealth. This means that more than two thirds of humanity must make do with just 5% of wealth. The OECD predicts that by 2025 one human in two will have access to middle-class status. When this happens, China will probably be the largest economic power in the world. Global economic wealth, which left Asia in the eighteenth century, will have come home. But pollution will make it there first.

There is an orange-brown cloud that covers much of Asia, which hikers can contemplate from the Himalayan peaks. It is known as the Asian Brown Cloud. Its shadow stretches from India to China, home to two thirds of humanity. The toxic cloud has left its black signature on glaciers as well as on lungs. It is considered responsible for at least four million premature deaths in Asia every year, and it is partly produced by local populations, who in their poverty burn whatever they can get their hands on for cooking and heating, everything from wood to rubbish. But it is also the result of the fact that for a thirty-year period China was (once again) the workshop of the world. When you buy a smartphone that is made in China, you subcontract your ecological footprint to Asia, paid for by the burning of coal required for its electricity. But now China is producing for its internal market, a gigantic Moloch in the shape of prosperity for hundreds of millions of new consumers. In recent years the proportion of pollution resulting from the production of goods for export has dropped below 20%. Henceforth, it is Chinese consumers who buy "made in China," and in so doing they are limiting the life expectancy of their children through air pollution, dust resulting from massive erosion, millions of tons of soot, and soil drenched with heavy metals.

Like the rest of the world, China gets most of its energy from coal, burned to produce electricity. Its appetite appears infinite, to the point where it continues to build two thermal power stations every week, which up until recently was probably justified. But evidence from late 2016 suggested that these new power plants were not needed for providing energy. The regime continues to create them because it needs to provide work for its inhabitants and to use its resources to guarantee the social contract. Yet the regime has shown that it is concerned about ecological issues. Most protests in China are linked to the environment, but they are driven by NIMBY—"not in my backyard." This attitude has long been dominant in the West, based on the idea that pollution is an acceptable cost for the additional prosperity it will bring as long as it does not affect me personally.

China has also paradoxically become the world leader in green energy. It is becoming clear that a planned, state-controlled economy is able to outrun its competitors. In the early 2000s, Germany had a decent head start on solar technology. But China was able to steal a number of its patents and use them to create its own internal market. Then it invested and constructed. All this was relatively straightforward: the state controls banks and industry, all consolidated under the authority of the party, as well as formidable financial reserves for times of recession. There is no equivalent to social security in China, and citizens save an average 40% of their income, a world record.

The one-child policy, decreed by Mao, has possibly contributed to a "masculinization" of the Chinese population because tradition considers that sons take care of their parents while daughters take care of their parents-in-law.[11] From the moment the state decided that parents were allowed only one child, the number of girls declined: killed or abandoned at birth, sometimes not declared, often aborted. From the 1980s, mobile ultrasound units roamed the countryside. For a modest sum it became possible to know the sex of one's fetus before birth. Although it was illegal to abort for gender reasons, this could be overcome. The doctor noted there was a "problem," money changed hands, and the female fetus was smudged out.

Asia is missing around two hundred million women. The Chinese one-child policy is far from being the only cause. Similar things happen in Pakistan, Afghanistan, and India. The Hindu tradition favors boys, who are seen as being responsible for lighting (and paying for) their parent's funeral pyre, and as in China, for taking care of them in their old age. Raising a daughter, on the other hand, is akin to "watering your neighbor's garden" according to an Indian proverb, because a daughter will join her husband's family and provide him a considerable dowry. The upwardly mobile Indian middle classes are those who practice selective abortions the most often. First, because they can pay, and second, because the dowries they have to pay for a daughter's marriage reach astronomical levels to the point where families are ruined. This has many consequences, including the traffic of women around Asia for forced marriage or prostitution, the mafia seeking out their prey in poor peripheral countries like Nepal, the Philippines, or Vietnam. It also means there are cohorts of single men who cannot find wives and who might eventually be conscripted into armies, possibly leading to an increase in belligerence in Asia. Finally, it means that whole societies continue to regard women as property even as women become more educated and fight to have their rights recognized. India, for example, like many other developing countries, is now trying to forge a path between the heritage of rape culture and the defense of fundamental rights.

The expansion of the world's middle class has also come at the cost of widespread pollution. Of course we must celebrate the fact that so many people are now free from poverty. But belonging to the middle classes also means keeping up with the Jones's, showing off one's ability to consume. If everyone aspires to the consumerist model dominant in America, humanity's depredation of the planet's resources is simply unsustainable.

In recent years, the rise of populist leaders around the world, from Donald Trump in the United States to Narendra Modi in India to Jair Bolsonaro in Brazil, has further delayed the decisions that humanity must necessarily, inevitably, make to protect what little is left of our ecological heritage. These new leaders dismantle environmental and social protection programs under the pretext of stimulating economic growth and in so doing only accelerate the coming disaster. Although predictable, it is impossible to know exactly what forms this will take when certain thresholds related to the climate, the biotopes, or both are crossed. Prediction is always a difficult, speculative art, but many of the scenarios we can extrapolate from the current tendencies have the merit of raising the iconoclastic question of whether China, today's number-one global polluter, might tomorrow become the symbol of environmentally responsible policy.

Well before President Trump's election could have been seriously envisaged, Naomi Oreskes and Erik M. Conway used anticipatory fiction to explore the political consequences of global warming.[12] The scenario plays out in the year 2393. Western societies have collapsed, and Chinese archaeologists are studying them and reconstructing the sequences that led to their downfall. In Oreskes and Conway's vision, China's advantage was its ability to impose its decisions on its people through its authoritarian regime, while liberal mercantile societies proved unable to anticipate risk, mobilize citizens, and change the course they had set for the cliff's edge. All their technology, their science, and their media could not save them.

This should by no means be read as an apology for any given political regime. Through this pedagogical work of science fiction, the authors imply that endless debates on climate change result in delaying urgent decisions because our democratic societies are gangrened by industrial lobbies. This raises the specter of a green totalitarianism; the damage inflicted by warming in the near future may oblige humanity to submit to an environmental dictatorship, including all nations, in order to save civilization from collapse. The only way of preventing this outcome would be the immediate application of decisions to inverse the processes involved in extreme global warming. Imagine if our democracies continue to fail to impose concrete decisions on humanity to effectively limit greenhouse gas emissions, the use of natural resources, and the destruction of ecosystems. Then dictatorships may be the only credible solution to ensure our survival.

Capital's Blind Spot: The Climate

The year 2016 was the hottest ever recorded since measurement began back in 1880. We are currently living through an increase of 1.2°C (2.2°F), and yet, against all the odds (and most of the increasingly alarming scientific evidence), international bodies still hope to limit warming to +2°C (+3.6°F) by the end of the century, a target they hope to reach without taking measures that are overly restrictive for states.

The ultimate indicator is the level of carbon dioxide (CO_2) in the atmosphere. As we have seen, this greenhouse gas is produced by combustion. Its increase is the most important anthropogenic factor in global warming, and its variations are closely correlated with fluctuations in the climate. During the ice ages that occurred in the last two million years, the amount of carbon dioxide in the atmosphere remained below two hundred parts per million (ppm). The warming that has occurred over the last 10,000 years brought it to 260 ppm. This then began to slowly increase after the Industrial Revolution. In 1958 Charles David Keeling raised the alarm based on the increase of this measure, which had then reached 300 ppm. In 1990, just as international negotiations to keep warming within acceptable limits were taking place, this carbon density reached 350 ppm. In May 2013, we crossed the symbolic threshold of 400 ppm. To find equivalent concentrations of carbon in the atmosphere in the past, we have to go back three million to five million years, to when *Australopithecus* roamed the African savanna. Back then, global temperatures were 3°C (5.4°F) to 4°C higher, the poles were 10°C (18°F) warmer, and the sea levels were perhaps 20 to 30 meters (66 to about 100 feet) higher. This constitutes an equilibrium that our descendants will reach in two or three centuries—if they are lucky. But for this to eventuate, we their ancestors would have to succeed in halving or even quartering our carbon emissions between 2020 and 2030, and for the moment these emissions continue to rise relentlessly. Unless . . . we miraculously manage to find a way to store the carbon we have released, and continue to release, into the atmosphere.

Between 1990 and 2013, human-origin carbon dioxide emissions increased 60%. As a result, polar ice has melted faster than predicted, and the acidification of oceans is also more significant than anticipated by the average scenarios put forward by experts. What we know is that at the rate things are going, by the end of the century the world will be between 4°C (7.2°F) and 6°C (10.8°F) hotter, sea levels will be 1 to 5 meters higher (3.3 to 16.5 feet) or even more, and the impacts of natural disasters will be entirely unpredictable.

Climate scientists are panicking. In a record consensus in scientific debate, 99% of them agree that the planet is warming beyond a reasonable level and that humanity is responsible. And yet due to a suicidal process of negation, we are not taking action against this threat except through the waltz of international meetings and conferences and the bumbling moves of politicians.

One step forward: China has committed to decarbonizing its economy in the coming decades. Two steps back: it continues to build thermal power stations, and Donald Trump has promised to prioritize American industry, to the detriment of what he sees as the "fake" problem of greenhouse gas emissions.

Dominant economic models are all based on the idea that change is linear. But they do not take into account the thresholds that have already been crossed, the record-breaking violent winds, the ever more extreme forest fires, the extinctions, and the loss of biodiversity linked to the massive fragmentation of ecosystems. Our economy is at war with life on Earth. The driving force of competition burns only one fuel: conflict.

Is it possible to stop this machine? Many authors have tried to imagine what this might look like. One of the most convincing attempts is Naomi Klein's *This Changes Everything*. Let us take a broad look at her proposed solutions: (1) break all the rules laid down by new liberalism, (2) limit the power of corporations and financial markets, (3) (re)construct local democracies and economies, (4) learn to take only what we need rather than what we want, and (5) grant fundamental rights to all living things, in stages if necessary.

In an earlier book, Naomi Klein described the functioning of neoliberal economy as a war machine.[13] She emphasized that since the 1970s, the business world has systematically taken advantage of crises (like environmental damage, wars, national disasters, or stock market crashes) to impose measures intended to enrich a small minority. Such measures include deregulation, cuts to social expenses, massive privatization, state policies to limit civil liberties and social action, and even systematic violations of human rights justified by "exceptional" situations.

Developing countries shoulder the full burden of this. Although they are not responsible for the degradation of the climate, they are the most affected by crisis policies and are the most vulnerable to the effects of warming. Often situated in tropical zones, precisely where the climate disruptions are the most substantial, the poverty of these populations makes them all the more vulnerable. A fragile political system is less able to face up to natural disasters. A planet structured around humanity and solidarity would urgently transfer resources to help the most vulnerable populations anticipate future cataclysms. We just focus on closing borders.

Global warming is already a huge challenge to our capacity for empathy. Should we help refugees? Is it not in Europe's interest to invest in helping Africa prepare for environmental risks if only to prevent the instability that we know will eventually threaten the geopolitical equilibrium? At the very least, limits should be put on speculative land grabbing of arable farmland, the predatory behavior of certain elites should be punished, women's emancipation should be supported everywhere, and human rights should be uncompromisingly defended. That would be the strict minimum that would allow us to envisage a viable world for our children.

For want of action, we are building a world in which rich countries will have to barricade themselves behind more and more inhuman walls of technology and intervene with increasing violence to protect their resources and their interests. Monkey will continue to do what he does best: oppress his fellows, monitor them, trample their rights to ensure his own prosperity, and continue to find the right excuses to justify this and delegate violence so that everyone can have a clear moral conscience. The current global economic system is designed to maximize the profits of corporations and marginalize the interests of the state, which is cordially asked to support this annexation. Every human cog in this machine—you, me, —accepts his or her role in this collective undertaking in exchange for a more or less tiny fraction of the profit that is generated. These gains are obtained through an extremely complex mechanism that is ultimately devouring the earth. Our historically unsurpassed prosperity is based on the massive and untenable overexploitation of natural resources.

There are initiatives, often at the local level, that could help reduce our impact. A range of interesting solutions have been tested around the globe.[14] Historically, we can see that fundamental change in environmental practices, particularly when a society feels threatened, can occur very rapidly. For example, from 1942, when national resources were mobilized fighting Japan and Germany, the United States government called on citizens to support the war effort by growing food at home. Two thirds of the civilian population got involved, and together they managed to produce half of all the vegetables consumed during the war by creating twenty million new "victory" gardens.

Since the end of the Cold War, two contradictory political processes have structured the collective evolution of the world. On the one hand, there is the farce of the climate agreements, marked by repeated failures, while in direct opposition there is the triumph of liberalization, with the North American Free Trade Agreement (signed in 1994) and the World Trade Organization (created in 1995). Although global wealth has reached unprecedented levels, every free-trade agreement signed is an invitation to pollute more, to produce more, and to increase trade deregulation.

Monkey has entrusted the planet's future to fancies driven by profit.

Let us rewind to a symbolic scene. A sweltering 37°C (98.6°F) day in Washington, June 23, 1988. The air conditioning was broken in the room where the climatologist James Hansen declared, before a United States Senate committee, that he was "99%" sure that rising global temperatures were due to human activities. The sense of urgency was similar to that which led to the prohibition of the insecticide DDT after Rachel Carson published her book *Silent Spring* in 1962. The Intergovernmental Panel on Climate Change (IPCC) was created in November 1988. And then . . . nothing really happened. Major international gatherings, such as the Rio Summit in 1992, merely reiterated the existence of the problem. Nearly thirty years later, the state of the climate is disastrous.

The year 1989 brought the end of state communism in Europe, and the burgeoning global awareness of environmental issues could have led to a different outcome. But it was not to be. Neoliberalism enabled the spread of Western consumerism to the entire planet just at the very moment we were beginning to realize that this behavior was unsustainable. Monkey has become schizophrenic. He wanted to save the planet and sign international agreements on climate change in which state commitments were voluntary and nonadherence went unpunished. This is of course not the case for trade agreements, such as those of the World Trade Organization, which come with heavy sanctions for infringements. In other words, environmental measures were systematically subordinated, or even sacrificed, to mercantile interests because the two are generally antagonistic. So what if international trade is a source of greenhouse gases? Putting up barriers would mean breaking international law and incurring penalties. That is why, in a frenetic search for profit, a T-shirt goes around the world twice on average before being worn. The cotton is bought at the lowest possible price, woven where the workforce is the least expensive, sent on container ships into a lawless world where ship workers are exploited, profits are filtered into tax havens, and pollutants and waste are directly discharged into the sea. All this occurs at ridiculously low cost because of the obscure functioning of the economy and the implicit blessing of states who have given up any substantial attempts at regulation in this area. Environmental protection is sacrificed on the altar of the unrelenting promotion of free trade and the search for profit.

Monkey covers his eyes, his ears, and his mouth. He wants to not hear, not see, and not condemn a world that has gone mad, where the international violence we thought was gone for good has resurfaced, where air pollution cuts short the life of more and more people, and where food is increasingly abundant but is paid for in changes to our bodies.

And Monkey dreams of immortality. In his heart of hearts, he knows he will never be so close to achieving his dream. Is he Icarus or Prometheus? Will he plunge from the sky or transform into a god?

Only our children will find out.

CONCLUSION

Humanity is waging war on our planet. This is the poisoned apple of our history, the price to pay for our unprecedented evolutionary success. We have to understand this. And end the war. Urgently.

We can cultivate the land without drenching it in pesticides and synthetic fertilizers. We can live alongside animals without locking them up in barbed-wire pens like concentration camps and stripping them of rights. We can live comfortably without the mindless consumption of latest-generation smartphones that will be obsolete in two years, made of precious metals and alloys so complex they are nonrecyclable.[1]

Even with a population of twelve billion people, we can live on Earth without destroying and breaking up ecosystems—by far the primary cause of today's global mass extinction. We urgently need to establish economics and ecology based on empathy and resist the dominant ideology that promotes a perpetual war of each against all and of all humanity against the planet.

We have to accept that capitalism is one myth among others. Like them, it has recurring themes that have become the defining elements of our artificial dream. It is based on the belief that economic growth heals societies of all ills and guarantees full employment, that free trade optimizes private interests and allows everyone to live a better life. It promotes the idea that states are less competent than the private sector, that public services must be slashed and privatized, and that unfettered competition enables both workers and companies to reach their full potential. It sees natural resources as free, even marketable. And according to this myth, solutions will be found—even if we have to reshape the earth through geoengineering—and we will save biodiversity with redemptive speculation.[2]

But this myth is destroying the earth because it is based on a deadly but unexamined truth. Over thousands of years, humanity has become shaped for war. It is in keeping with our evolution that economics today is seen as a study of a total war—with entrepreneurial strategies, captains of industry, mercenary raiders and traders, hostile takeovers, conquests of new markets, social negotiations designed to dismantle systems of wealth redistribution, and formerly rich regions pillaged and left destitute.

If our history has shown that Monkey can be a killing machine, a hyperpredator, that is only one of his roles. Like Janus he has a second face; he can be full of compassion and empathy. We have taken this sentiment to its extreme, and it has made us the kings of creation. It has fostered a group spirit that led to the creation of culture. It allows us to perform miracles using cooperation to the point where we can break the natural laws to which the rest of the living world is subject.

Compassionate killers. What monsters we have seen in this blockbuster production: Alexander the Great, Qin Shi Huang, Genghis Khan, Napoleon. These great men (and they were all men) are inscribed in the pantheon of our memory proportionally to the number of deaths provoked by their excesses. For every Gandhi, how many Hitlers, Stalins, Maos, or Pol Pots must fill our history books?

Let's fast forward through the film again.

We became human by testing guerrilla warfare through the Physiological Revolution that shaped our bipedal bodies and large brains. Tools changed us, gave us meat to chew. We first controlled energy through fire. That meant we could burn forests, destroy large animals, and colonize and modify all the biotopes on Earth thanks to the Cognitive Revolution. This made us the kings of the world. All that remained then was to control the rest, systematically exploit it, tame it.

Between the Agricultural Revolution, which began around twelve thousand years ago, and the Moral Revolution around 2,500 years ago, we had to deal with the whims of nature, holding fast to our little plots of land, sometimes fighting with our neighbors when survival looked grim. Yet our numbers grew. From around 500 BCE, universalist empires and religions helped us to form strong connections with each other. Societies became more coherent and had a greater impact on the environment. Yet Monkey's war against nature was still very local. We could scratch away at the forests, but every so often a natural disaster or epidemic would strike, sometimes decimating a continent, and the forest would take back its own.

Global connections began in 1500 when Spanish and Portuguese ships created links around the world. Tiny European states forged themselves into great powers through intracontinental conflicts. Their aggression allowed them to dominate the affluent seas of Asia. The microbial massacre that depopulated the Americas gave them the resources from two continents. Fueled by the slave trade and the demand for sugar, slavery became a preindustrial system. Between 1500 and 1800 Europe sought to control the entire world without having the means to achieve land victories against the great Ottoman, Persian, Mughal, and Chinese empires or against the African states.

So it contented itself with ruling the waves. And getting rich. And training at home. The progressively liberal economy and the mechanisms of state debt allowed the European powers to wage tireless war, mobilizing ever more troops. The French Revolution, which gave rise to military conscription, was the culmination of this excess of violence. After 1789, war was definitively "democratized"; it became the duty of every man.

The Industrial Revolution was boosted by the battalions of British peasants who flowed into the cities. It began with an economic war in India that drained the country of its infrastructure for production. It then continued with industrial and technical progress in weapons and pharmacology that enabled Europe and the United States to take control of the world. The first phase of the Industrial Revolution was based on organization and energy. Coal gave humanity its strength, work became a production line, and the planet imperceptibly began to get warmer.

The second phase of the Industrial Revolution extended energy transformations into the world of chemistry. Plastics, petrol, and pesticides revolutionized agriculture, evolving alongside unprecedented demographic growth. In two hundred years the human population went from one billion to more than seven billion people! This growth was so large that the socioeconomic logic on which it was based became unquestionable. The West dominated the world through a model based on war that provided for its own comfort and food security. Chemistry enabled the West to be more efficient killers both of other humans (chemical weapons) and so-called pests (insecticides). Chemicals were sold as essential to feed the world, and this falsehood has deeply rooted itself into our psyches. The Third Agricultural Revolution—known as the "Green Revolution," which took place in the mid-twentieth century—launched a war against the living.

Nuclear power is also technology for warfare that was adapted to the production of civilian energy. It relies on matter being bombarded with particles, an operation conducted within industrial superstructures that are so complex that their functioning and overall consequences are beyond the bounds of comprehension for an individual human. Genetic engineering, information technology, exploration of cognitive sciences, the space race—all of these are extensions of this rationale of war and complexity. But there is an exception to this logic in the countercultural ideology among the early developers of the internet, designed as a universal network based on free and open collaboration.

From the 1980s neoliberal ideology drowned out alternative thought. State communism, which degenerated into monstrosities under Mao and Stalin, was cast aside. Monkey was now only concerned with achieving a better life for himself. Developing countries, from China to India, set out to replicate the living standards of the West. The middle class became the new standard for humanity: well fed, often overfed, and slowly poisoned. We dream of miracles, caught up in the everyday technological magic of the digital revolution (long-distance communication, media omniscience, etc.), even as we share immersive videos of natural disasters and humanitarian crises, watched in the indecent comfort of one's own home.

CONCLUSION 383

Looking for the Way Out

War remains the world's dominant paradigm. We are conditioned to think in terms of aggression and defense, and as a result we are exhorted to prepare for a permanent state of belligerence against the environment even as the effects of global warming hit us harder and harder.

Monkey must wake up. Urgently. It is still possible to change the way we see the world, but this means changing the way we see ourselves and our nature. We must no longer dream of immortality, or amortality, but rather foster the full potential of our altruism, extending it to all humanity, and indeed all living creatures. But every passing day brings us closer to the cliff's edge.

In a science-fiction novel titled *Deathworld*, Harry Harrison imagined a planet on which the entire ecosystem fought back against the humans in the mining colony set up to exploit the planet's resources. Humans adapted. Each survivor had to become a prodigy of muscles, reflexes, and hypervigilance. The human community absorbed colossal budgets for weapons, obsessed with the idea of maintaining the colony whatever the cost, even in a world where every blade of grass became a poisoned razor and tree roots could thrust up from the earth to strangle their victims. An outsider arrived in this world and it nearly killed him. And then as he moved away from the colony, he realized that this planet was in fact a paradise in which all forms of life lived together in harmonious coexistence. *Deathworld* was a super organism, a symbiotic system that secreted its own antibodies. It fought humanity as a human would fight cancer: by total war.

We are making the earth into a *Deathworld*.[3] We just have to realize this before our war takes us beyond a point of no return. And we have to find a solution to the problems we have created.

There are solutions. In *This Changes Everything*, Naomi Klein provides a nonexhaustive list:[4] consume less, and more locally; prioritize agriculture that respects the environment; tax luxury businesses to maintain essential service sectors (health, education);

develop public transport and renewable energy; phase out subsidies for fossil fuels and industrial agriculture; put an end to tax havens; fight inequality, particularly gender inequality; give up nuclear energy, which is dangerous and produces terrifying pollution that future generations will not be able to manage; protect and restore ecosystems; end planned obsolescence, criminalize it if necessary; tax financial transactions and regulate speculation, particularly by introducing rules to manage automatic trading; reduce working hours, which would mechanically lead to a drop in greenhouse emissions; ban countries who do not respect certain legal norms (gender equality, freedom of minorities, etc.) from international bodies; entrust toxicological testing of chemical products to independent state-controlled institutions not associated with industry; introduce international frameworks to subject research in genetics and cognitive sciences to rigorous ethical standards; and finally, to help Monkey act more quickly and effectively, explain how these measures are essential and promote them.

There are only a few decades left for the democratic implementation of these measures. After that, it is likely that humanity will have to act more radically if it is to survive.

It is impossible to launch such rescue measures while so many people remain unaware that it is neoliberal ideology that is driving the catastrophe. At the extreme opposite, a small number of people who promote this doctrine continue in the hope that they will be able to surf on the crest of the tsunami, transforming themselves into technological gods, and push this model to its most extreme limits. But its only limits are the limits of what the planet can take.

For just a little while longer, Monkey can dream of changing the world. We can dream of consolidating the improvements the West saw between 1945 and 1980 and that the rest of the world has begun to see since then: fewer and fewer people living in poverty, increased life expectancy, greater food security, less violence. But the inversion of all these measures since 2011 should be a warning to us. Our ship is taking on water. It might not necessarily sink, but we need to quickly take control of the helm before a minority pretending to act in the interests of all sends us to the bottom of the sea.

Tomorrow it will be too late to start bailing. First-class passengers have nothing to worry about; they think they still have time to finish their champagne before becoming gods. They always manage to find the last lifeboats anyway. They may become amortal perhaps, potentially at the cost of their humanity. But evolution does not care; its only concern is the survival of a well-adapted organism. Economy class passengers, however, should be very worried.

We urgently need to declare an end to this war against the planet if we want Monkey's saga to end well, if we want our children to see real owls, in real forests, in the future.

EPILOGUE TO THE ENGLISH EDITION
Two and a Half Years after the French Edition . . .

We have reached the last scene in Monkey's blockbuster. In two years, our ship has moved even closer to its wreckage. The alarms are deafening. There are now countless books full of scientific warnings published every year with extensive arguments explaining why humanity is on the path to suicide.[1]

Monkey invented the myth of the suicidal lemming for a 1958 Disney documentary film about large rodents who ran blindly to their collective deaths. To get these images, the crew set up a gyroscopic platform to fling a few thousand captive lemmings into the air. But, unlike Monkey, lemmings are not really suicidal animals. When their population gets too large, after an unusually fertile spring, it feeds the growth of predator populations, foxes and snowy owls. Once the lemming population has dropped, the predators look for food elsewhere. The Malthusian check, and not mass suicide, is nature's most effect tool for regulating populations.

Or it used to be. Until Monkey meddled with it and got around the laws of evolution with culture to the point where they are today being completely rewritten. But if Monkey thinks he is clever, he has blinded himself. He is the lemming now. There is no need for a crystal ball here. One look at the scientific literature is all it takes to realize that several critical ecological thresholds will be crossed over the coming century. Some of them are already happening. There are no longer any climate scientists who seriously believe global warming will be limited to +1.5°C (2.7°F) compared to the reference temperatures of the late nineteenth century. Greenhouse gas emissions that have already been released put us on track to cross the +1.5°C (+2.7°F) threshold in the 2030s. And if we do not reduce these emissions by at least half, if not three quarters, by 2030, then we will cross the +2°C (+3.6°F) threshold between 2050 and 2060. The climate models all converge on this.

And then what?

Then chaos. These models cannot simulate what will happen after +2°C (+3.6°F), because several tipping points are likely to be crossed, like the melting of the permafrost, the frozen ground that would release enormous quantities of methane, an extremely potent greenhouse gas. There is nothing to be gained from contemplating such scenarios. It would be like prematurely committing humanity to palliative care. We still have one last attempt, one last breath before we go under.

We have twelve years.[2] Twelve years not to save Earth from disaster (which has already happened) but to prove that being human means being deserving of this world and saving what can still be saved.

When the French edition of *Cataclysms* was first published in April 2017, the initial reaction from the public was to ask why it was "so pessimistic." But only a few months later, people had stopped asking that question. It had made way for other words, other unformulated fears. We were beginning to hear talk of the collapse of biodiversity. It was becoming clear that there were far fewer insects being squashed on windscreens or fluttering in gardens. For good reason. A German study found that in just twenty-seven years, more than 80% of the biomass of flying insects has disappeared in Europe.[3] And the measurements for this study were taken in nature reserves, so the figure is undoubtedly higher in reality.

Yet biodiversity is what allows us to survive!

Without our wetlands, old-growth trees in the rainforests, or plankton, Earth will be incapable of attenuating the extreme climate events that will soon ravage it. And before the end of the century, before today's children reach old age, the air we breathe, the water we drink, and the food we eat will all be affected without the complex ecosystems we rely on to produce and purify it.

When Philip K. Dick wrote *Do Androids Dream of Electric Sheep?*, this scenario was pure science fiction. But now it is just science. And science tells us, with astonishing concord, that Earth is at great risk of becoming a dead land.

Once environmental degradation took place too slowly for an ordinary human to notice it. This is known as ecological amnesia. No one alive today can remember what the oceans were like in the seventeenth century, when there were twice as many whales as there are today. This nostalgia becomes palpable when grandpa sighs, his eyes wet with tears, remembering the millions of scarab beetles that would swarm in autumn, while his grandchildren listen, round-eyed, wondering what strange kind of drone this could be. Any bird lover knows that in the last two decades certain once familiar species have all but gone. With my own eyes I have seen frogs, snakes, swallows, and crickets disappear from my land. My house and my garden run onto fields that are sprayed with pesticides, where hedges have been dug out and flattened to make way for ever-larger tractors spraying ever more herbicides, fungicides, insecticides. One might reread Rachel Carson and weep.

But no matter.

We will roll the credits of this blockbuster you have been watching. In order of their appearance, here are the actors who have shared the screen with Monkey.

The San people. Ever more excluded, some obstinately seek to return to the desert, while authorities want to keep them out. They are forbidden from digging wells where their ancestors— also driven off the best land—were free to roam. The last hunter-gatherers are dying out and with them the last memories of our past, these three million years during which humans lived nomadically in nature.

Elephant. She once ruled the world, and without her we would not be human, because we now know she paved the way for us, leaving Africa six million years ago and blazing the trail we would follow. Two hundred years ago, before colonization, there were twenty million of her kind left in Africa. In 1979, when the whole continent had achieved independence, there were only 1.3 million. Today, there are fewer than four hundred thousand left, and their population dwindles by 8% every year. Poaching is a major cause, of course, but so is the economic growth of Africa. All over our planet, the extension of roads, housing, and agriculture gnaws away at what is left of the wild.

Wheat. He is doing well. His bargain has paid off. Accepting domestication has meant he has evolved along with Monkey's demographic growth. But it also means he is condemned to die, and in the near future. Shhh! Don't tell him! Today wheat does well because humans have modified his genome and accepted putting ten calories of hydrocarbons into the soil (petrol for tractors, fertilizers, and other chemical products) to extract just one calorie of digestible wheat. But petrol will probably run out eventually, and before that its use will have to be rationed more carefully. Global warming (partially due to the industrial production of wheat) will make wheat less productive, less nutritious, and more vulnerable to drought, of which there will be more. Soon our industrial wheat will die, and we will cry famine unless we manage in the very near future to transition to more resilient forms of agriculture, such as agroforestry and the cultivation of less homogenous crops, the end of intensive livestock farming, and a return to pasture.

Civilization. She is a major actor in this story. Despite her longevity we have seen that she is mortal. The idea of civilizational collapse has caused a media storm recently and reawakened old fears. Probably because it teaches us something fundamental. We are trapped in the idea that we can live forever, we have blocked out death from our consciences, and we have glossed over the industrialized slaughter in the abattoirs and the agony of our elderly in institutions. The collapse of civilizations is that oversight that comes back to haunt us, reminding us that entropy lies in wait.

Money. Now the uncontested king of the world. What a path from his humble beginnings 2,500 years ago. Along with his cronies, State and Ideology, he has us wrapped around his finger. Civilization is in fact a combination of these "3 Ms": missionaries, militaries, and merchants. Money fuels trade and increases the concentration of wealth.

Religion, or Ideology more broadly. She makes the inequality created by Money acceptable. All people are equal before God, or the Market, or the State. The invention of the nation between the seventeenth and nineteenth centuries was the promise of equality for all in a world that would henceforth be dominated by the state

State. He provides protection, taxation, and redistribution. Empire made way for the nation-state, but perhaps only temporarily. The global wave of populism expresses a nostalgia for power that could bring older models back to light. Over the course of history, the 3 Ms have reconfigured themselves many times. Missionary-military-mercantile. Today various combinations of these ideal types coexist around the world according to three main patterns. In the West, the nation is subject to politics, which is in turn controlled by the economy. In Saudi Arabia, Wahhabism governs the state, which in turn governs the economy. And in China, an authoritarian state dominates the national community and capitalist entrepreneurs. Three models for a single hegemony. Now the boundaries of the playing field are clear. We have reached the limits of the planet.

Germs. These were the next costars to appear. Monkey vanquished them. But germs are tough little guys, obstinate. They are the most indestructible element in the living world and the most likely to evolve in radically new ways. They will outlive us. But we have forgotten what it is like to live in a world where a scratch could bring down a king, where septicemia was everywhere, where two in three children would die before their fifth birthday. Today smallpox, mumps, and flu no longer terrify us. But what about tomorrow? If resistance to antibiotics increases (which is possible, even probable, given industrial overuse). If, for economic reasons, the world produces 90% of its medicine in India and China and developed states are unable to insure their autonomy and basic medicines (which is already the case in most European countries). Then Germs may again become a major player. And the apocalyptic epidemics of the past will once again become our greatest fear.

The Worm. So humble, yet so essential for the fertility of the soil. In dirt which has been sprayed with chemicals on an industrial scale over the last three decades, worm populations have dropped to a tenth or twentieth of what they were. This is what scientific studies suggest at least, and my shovel seems to confirm it.

Mercury. It is still around. We are now able to study more and more precisely the evolution of its concentration in biotopes. It has doubled over the course of the twentieth century and continues to rise. The primary source is the burning of coal, which is still growing, particularly in China. Mercury is a bio accumulator, which means it remains in the food chain, so the higher up you are as a predator, the more mercury is concentrated in your system. It is also a neurotoxin and an endocrine disruptor. Tuna sashimi anyone? But there is hope. Unlike other pollutants, Mercury disappears relatively quickly if we stop releasing it. We just have to manage to do that.

Forest. Once the word for the world, now receding everywhere. Nothing new under the sun. Soy plantations eat away at Amazonian biotopes, and the last orangutans wander hopelessly among the remains of their forests clear-cut for palm plantations.

Smog. It is still here, too, although it receded in Europe before master Diesel and others revived it. It also lessened in China when the authorities decided to protect the major cities and save the social consensus. But it accumulates ever more densely over India and Africa. Air pollution leads to millions of deaths every year and a procession of incommensurable suffering. How can such human cost be calculated?

Volcanoes. They are still around too. We supplanted them as the active agents in geology, but they know their reign will return. Ours is merely temporary. Pray that they do not sneeze. That would make our problem seriously worse.

Computers. Now all-powerful. In China there are now four hundred million electronic eyes, brains that recognize the world. In Sweden there are people who insert microchips under their skin to control their environment. The cyborgs are already here.

The Future. The last player in Monkey's saga.

In June 2017, shortly after *Cataclysms* was first published, two research studies came out. The first brought together work by UN demographers suggesting that by 2100 there will be 4.5 billion people in Africa.[4] The other, by an international team of climate scientists, modeled what Africa would look like in 2100 according to a "business as usual" scenario—+3.7°C (+6.66°F)compared to late nineteenth-century reference temperatures. The computer crunches the figures and the diagnosis is clear: most of the continent will be uninhabitable, the humidity will be unsuitable for human survival for two thirds of the year.[5] One does not have to be a geopolitician or philosopher to understand what happens when 4.5 billion people are supposed to live somewhere it is impossible to survive. Interdisciplinary analysis will be required to anticipate and prepare for this future, but humanity persists in preferring atomized knowledge. Climatologists, demographers, geopoliticians, and philosophers work assiduously but independently, each in their own world.

We face two distinct scenarios. Choose which one you prefer.

From where we stand today, we may still have a possibility for action. But for the moment we are frozen, like the wise monkeys we are, hands over our eyes, over our ears, over our mouths.

The film has reached its final scene. But the ending is still yet to be written. Our boat is taking on water, and it is difficult to imagine us regaining dry land. Will we face a tidal wave? A wall of water rearing up in front of us? Or will the water just continue to slowly, inexorably, rise?

Unless there is a last-minute plot twist, Monkey's blockbuster will soon be over.

There is one final shot.

APPENDIX A: GLOSSARY

Glossary terms are identified by an asterisk the first time they appear in the text.

AFRO-EURASIA: The continental whole made up of Europe, Asia, and Africa. Unlike the term *Old World*, which is purely geographic and developed by European explorers in contrast to the *New World* of the Americas, the term *Afro-Eurasia* emphasizes the connections that unify this ensemble.

ANTHROPIZED: An adjective to describe an environment transformed by human activity.

ANTHROPOCENE: From the Greek meaning "the age of man." It describes the current geological period in which human actions have a greater impact on the environment than any other factor. Most researchers consider it as beginning with the Industrial Revolution, although some go back to the Agricultural Revolution, and others associate it with the control of atomic energy.

COMMON ERA: The notation system for time used here. Before the Common Era (BCE) and Common Era (CE) are secular terms numerically equivalent to the BC and AD system but which avoid reference to Christianity.

CULTURE: The range of behavior and knowledge taught to humans within a given society that shapes part of their behavior and interactions. Another variable part is shaped by genetics and the environment.

EARTH-SYSTEM: Earth-system sciences take a holistic perspective on the organization and interactions of Earth's "spheres"—atmosphere, biosphere, hydrosphere, etc.—as well as their different subsystems and the influence of human activity on the whole. It draws on a range of natural and social sciences.

EL NIÑO: A climate anomaly that disrupts the circulation of ocean currents in the Pacific and low and high zones of atmospheric pressure to the point where the Asian monsoons are considerably affected, as are, by extension, wet and dry seasons in many tropical and temperate zones.

EPIGENETICS: A discipline in biology that studies how genes are modified by the environmental context, acquiring characteristics that can be transmitted from one generation to another. By extension, this term also includes processes by which these genes change following interactions with the environment or with other genes.

GEOENGINEERING: The ensemble of techniques that seek large-scale artificial modifications of the earth's environment (atmosphere, climate, biotopes, marine environments). Certain researchers and politicians see this as a possible solution to the degradation of the environment in general and the climate in particular. Others express their alarm at the intrinsic risks in such procedures, whose complexity is well beyond our capacities for simulation.

HOLOCENE: The geological era that began 11,700 years ago and that saw humanity spread throughout the planet and begin to substantially alter the biosphere with agriculture and livestock.

HOMINID: A member of the great ape family: orangutan, gorillas, chimpanzees, and humans

HOMININAN: A member of the subtribe Hominina: humans and our closest extinct relatives, *Australopithecus*, but excluding chimpanzees.

HOMOGENOCENE: The homogenization of the planet's ecosystems involuntarily brought about by humans from the point that the world was unified in the sixteenth century.

the term *Middle East*, which is more geopolitical and includes all the countries between Egypt in the west and Iran in the east, Turkey (and Caucasia) in the north and Yemen in the south. The notion of the Near East is even more polysemic because it is sometimes understood as the Middle East, sometimes as the Levant.

MALTHUSIAN CHECK: From the philosopher Malthus, who proposed that there was a limit to population growth in which each technological innovation or agricultural crop that improved productivity would lead to demographic growth that would in turn produce food shortages and famine and a return to a state of equilibrium.

MEDIEVAL WARM PERIOD: A period of global warming that occurred between the ninth and thirteenth centuries.

MESOAMERICA: A geocultural area including the southern half of Mexico and part of Central America to northern Costa Rica, where a number of civilizations (Olmecs, Aztecs, Maya) flourished between 1300 BCE and 1518 CE, before the arrival of the Spanish. They were linked by shared cultural characteristic: maize agriculture, shared calendars and religious beliefs, and ritual architecture including stepped pyramids, among other things.

MONKEY: With a capital *M*, he is the metaphorical protagonist of this story, representing humanity as a whole.

PLEISTOCENE: The geological period marked by the glacial periods running from what is considered to be the beginning of *Homo*, around 2.6 million years ago, to the emergence of agriculture, around 11,700 years ago, and the beginning of the Holocene.

PYROCENE: From the Greek, "age of fire." A concept put forward by the historian Stephen J. Pyne, who postulates that humanity, over thousands of years, has shaped the natural environment by repeated controlled burning.

SIXTH EXTINCTION: The current episode of mass extinction that began several thousand years ago and that has considerably increased in pace in recent years. Where the five other previous phases of extinction over the past 550 million years were produced by natural causes, this sixth wave is due to human activity.

WORLD/S: The notion of worlds refers to the homogenous geographic level regrouping the largest possible number of people connected by their interactions. We can therefore talk of worlds in the plural (Chinese,

Aztec, etc.) before their interconnection with European expansion in the sixteenth century. From a long-term environmental perspective, the world has been constructed progressively in parallel with human expansion given that our species affects its milieu at the same as it discovers it.

WORLD-SYSTEM: A term popularized by the sociologist Immanual Wallerstein, it refers to a space that is economically autonomous, limited in size by the time required for its occupants to interact. A world-system is constructed as an integrated ensemble of political and cultural spaces structured in four parts: core, semiperiphery, periphery, and others.

APPENDIX B: CHRONOLOGY

TABLE 1   Evolution of the world's population between 10,000 BCE and 2050 CE

3 million years before the Common Era (BCE)	**Physiological Revolution**: Appearance of *Homo* in Africa followed one million years later by the expansion of *Homo erectus* throughout Eurasia.
Between 500,000 and 50,000 BCE	**Cognitive Revolution**: *Homo* learns to control fire, language, and art, a prelude to more and more efficient collaboration. *Homo* becomes *sapiens*.
Around 65,000 BCE	The arrival of humans in Australia, leading to the progressive but total extinction of the megafauna there around 40,000 BCE. Large-scale use of controlled burning for clearing land began as early as 36,000 BCE.
Between 30,000 and 16,000 BCE	Dogs are domesticated. Humans arrive in the Americas, leading progressively to a quasi-extinction of the megafauna.
Around 15,000 BCE	The end of the ice ages and the beginning of the current temperate period.
11,000 BCE	The first genetic alteration of plants by human selection.
10,800–9700 BCE	The Younger Dryas cold period.
9700 BCE	The beginning of the Holocene and the **Agricultural Revolution**: Progressive domestication of the living world (agricultural, farming livestock) all over the planet

8000–3000 BCE	The Sahara is covered in prairies.
Around 3500 BCE	The invention of writing in Mesopotamia; the emergence of city-states, taxes, and then empires; the development of metalwork and long-distance trade.
Around 2550 BCE	A man, later to be known as Otzi, dies in the Alps. His body is discovered by hikers some 4,500 years later.
Around 1200 BCE	The systemic collapse of Bronze Age civilizations in Greece and the Levant, including Assyria, the Hittite Empire, Elam, the Kassites, and the weakening of Egypt.
Around 500 BCE	**Moral Revolution**: The foundation of universalist ideological systems (Confucianism, Taoism, Buddhism, Hinduism, monotheisms, Ancient Greek philosophy), which developed alongside the rise of universal empires. The creation of paper money. The invention of blast furnaces and metal plowshares in China.
Around 320 BCE	Alexander the Great connects Greece, Egypt, and Persia. The Greeks acquire elephants.
Around 267 BCE	The reign of Ashoka begins in India.
221 BCE	Unification of China by Qin Shi Huang, first Emperor of the Qin Dynasty
27 BCE	Augustus proclaims the Roman Empire.
11 Common Era (CE)	A flood in Huang He province provokes a famine in China against a backdrop of political unrest.
165–180 CE	Epidemics sweep through the Roman and Chinese empires.
189–220 CE	Collapse of the Han Empire followed by the fragmentation of China.
376–476 CE	Fall of the Western Roman Empire.
602–628 CE	Byzantine-Sassanian Wars, which favored the growth of Islam leading to the Golden Age of Islam in the eighth century. The Arab Agricultural Revolution contributed to the spread of food plants and farming techniques.

950 CE	The Medieval Warm Period begins (ends 1250).
1048 CE	The dams protecting the Chinese Empire from the Huang He break. This is the beginning of the collapse of the Northern Song Dynasty, whose capital Kaifeng is taken by northern enemies in 1127.
1257 CE	Eruption of the Samalas volcano, which contributed to the beginning of the Little Ice Age around 1300.
Around 1315–1330 CE	The Little Ice Age brings a wave of famine that extends from Europe to China. The beginning of the black plague in Asia, which would claim one third of Eurasia's population.
1492	Christopher Columbus arrives in the Americas. The Columbian Exchange begins: epidemics brought by Europeans wipe out 90% of the Indigenous populations of the Americas. Flora from Afro-Eurasia invade the biotopes of the New World, and American food plants are exported to the Old World to feed global demographic growth.
1545–1800	The silver mined in Potosi radically alters the global economy. Mercury begins poisoning the Andes.
1600	Foundation of the British East India Company, which will eventually conquer India, then one of the two wealthiest countries on the planet. The beginning of capitalism.
1600–1715	Global crisis due to renewed intensity of the Little Ice Age. The world's average temperature drops by 1°C (1.8°F). Harvests suffer and threaten political power and stability everywhere.
1618–1648	Thirty Years War, ending with the Treaty of Westphalia that laid the foundation for modern notions of sovereignty and the nation-state.
1627	The last aurochs dies.
1789	The French Revolution breaks out in part because of famine resulting from poor harvests provoked by hot summers and cold winters.

1812	Typhus wipes out Napoleon's Great Army.
1815	Eruption of Tambora in Indonesia, resulting in the "year without a summer" (1816).
1820	End of the Little Ice Age. Early phases of the Industrial Revolution and the **Energy Revolution**. Humanity begins to burn coal and become a major geopolitical actor influencing the planet. Beginning of the Anthropocene.
1830–1890	Easter Island is depopulated, and famine kills millions in China and India.
1859	The "Carrington Event" solar storm.
1909	The German Chemist Fritz Haber discovers how to synthesize nitrogen, paving the way for industrial agriculture and the wide-scale production of explosives.
1918–1920	The Spanish Flu kills between sixty and one hundred million people.
1945	Nuclear energy is mastered.
1962	Rachel Carson publishes *Silent Spring*, and public opinion begins to become aware of environmental issues.
1972	Donatella and Dennis Meadows and their colleagues publish the Limits of Growth report on the simulation of economic and population growth in a world of finite resources.
1977	The smallpox virus is eradicated.
1985	An international agreement brings an end to the destruction of the ozone layer by CFC gases.
1990	The beginning of the internet and of the **Digital Revolution**.
2016	The hottest year on record, with an increase of 1.2°C (2.16°F) above the twentieth-century baseline. The hottest twenty years on record have all occurred in the last twenty-two years. Biodiversity is plummeting, forests are shrinking, air pollution is taking lives all over the world. The **Evolutive Revolution** is beginning. As our

	bodies and minds are modified by many of the chemical substances we consume, technology is preparing for the possibility of extending human existence.
2019	After two years of increasingly alarming publications, climate scientists and biologists are stunned. The former are reeling that nothing is being done about the greenhouse emissions responsible for global warming. It now seems unavoidable that we will pass the 2°C (3.6°F) limit in the 2060s at the latest, at which point the threshold effects cannot be modeled or predicted. The latter lament the collapse of the biotopes, which is occurring infinitely more quickly than predicted, even in the most pessimistic scenarios. The uniformization of biotopes by humans, exacerbated by global warming is on track to wipe out all complex life-forms within the coming century.

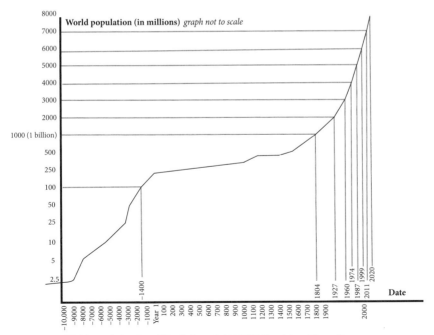

FIGURE 1 World population (in millions) since 10,000 BCE. Graph is not to scale.

For the period between 10,000 and 1000 BCE, the data comes from the History Database of the Global Environment (Hyde), the Dutch Ministry for the Environment, 2007. For a summary, see World Population Estimates," *Wikipedia* (http://en.wikipedia.org/wiki/world_population_estimates).

For the period between 1 and 1999 CE, I conducted a compilation of data in Angus Maddison, *Contours of the World Economy, 1–2030 AD: Essays in Macro-Economic History* (New York: Oxford University Press, 2007), 376, corrected by certain estimations drawn from John R. McNeill and William H. McNeill, *The Human Web: A Bird's-Eye View of World History* (New York: W. W. Norton, 2003); Jacques Bertin et al., *Atlas historique de l'humanité* (1997; Paris: La Martinière, 2004), and Jean-Noël Biraben, "L'évolution du nombre des hommes," *Population & sociétés*, no. 394 (October 2003) (https://www.ined.fr/fr/publications/editions/population-et-societes/l-evolution-du-nombre-des-hommes/).

Comments

The data for the period between 10,000 and 1000 BCE provides an idea of the density of the human population and its progression, fueled by the expansion of the Agricultural Revolution. The farther back in time we go, the more speculative this data is; the margin of error doubles when we go back several thousand years.

In the first century of the Common Era, Roman and Chinese census documents estimate the populations of these empires around sixty million people each.

In the sixteenth century, epidemics wiped out between forty and eighty million Indigenous people in the Americas. This demographic deficit was leveled out by strong population growth in China and India, which were each then home to 150 million people.

Around 1804 we crossed the symbolic threshold of one billion people in the world. The above estimates probably smoothed over certain epidemiological events such as the Justinian Plague in the sixth century CE, the black plague in the fourteenth century, and the consequences of the Columbian Exchange in the sixteenth century.

From the end of the nineteenth century, the world population began to grow exponentially. Although our calculations have become more precise, even the census results of the global population must be considered with a margin of error of 5%–10%. Some births are not registered, some governments underdeclare minority populations, and census errors remain frequent.

According to the average UN projections (United Nations, Department of Economic and Social Affairs, Population Division, *World Population Prospects: The 2015 Revision, Data Booklet, 2015*, https://population.un.org/wpp/Publications/Files/WPP2015_DataBooklet.pdf) over the coming decades, if the general trends remain unchanged (which is unlikely given the massive changes underway), the world population should peak between ten and twelve billion in the second half of the twenty-first century before slowly declining. It is worth noting that in recent years, these projections have been regularly increased.

NOTES

Introduction

1. To use the expression coined by zoologist Desmond Morris, *Naked Ape*.
2. Wu Cheng'en, *Journey to the West*.
3. These revolutions are also chrononyms, specific periods that, much like geographic areas (which are attributed capital letters by virtue of the fact that they are toponyms), have a specific location—temporal rather than spatial.
4. For a presentation of the methodological approaches, see Testot, *L'histoire globale*.
5. This was published by Sciences Humaines Éditions in November 2019 as a book titled *La nouvelle histoire du monde*. The most remarkable contributions in this area include Harari, *Sapiens*; Diamond, *Guns, Germs, and Steel*; Morris, *Why the West Rules*.
6. A new generation of French scholars emerged in the wake of the pioneering studies on the climate by Emmanuel Le Roy Ladurie, Grégory Quenet, Christophe Bonneuil, Jean-Baptiste Fressoz, François Jarrige, Thomas Le Roux, and Jean-François Mouhot, among others.
7. A scenario evoked in Klein, *This Changes Everything*.
8. For French perspectives on counterfactual history, see Deluermoz and Singaravelou, *Pour une histoire des possibles*. See also Besson and Synowiecki, *Écrire l'histoire avec des si*.

Chapter One

1. Heinrich, *Why We Run*, 8–9.
2. Monotremes, marsupials, and placentals are the different kinds of mam-

mals (animals who feed their young on mother's milk), the results of trial and error in mother nature's innovation, between 100 and 200 million years ago, when Earth was still dominated by reptiles. Monotremes are characterized by the fact they possess both a beak and the ability to lay eggs. Today, the only remaining species, echidnas and platypus, are found in Australia and New Guinea. Marsupials, whose presence in the Northern Hemisphere has been traced back 115 million years ago, have a very short gestation period and give birth to premature young who then develop in a pouch. With a few exceptions in South America, marsupials are now found only in New Guinea and Australia, where they have long been the dominant fauna with no competition. The third group, placental mammals, of which humans are a member, seems more recent than the other groups. It also needs more food to carry its young to full term.

3. See De Waal, *La politique du chimpanzé*, and Picq, *L'homme est-il un grand singe politique?*

4. Lieberman, *Story of the Human Body*.

5. Werdelin, and Lewis, "Temporal Change."

Chapter Two

1. His original text, in French, is available online. See Cuvier, "Mémoire sur les espèces d'éléphants vivantes et fossils par le citoyen Cuvier" (https://www.biodiversitylibrary.org/page/16303001#page/175/mode/1up). For a contextualization of his writings see Candegabe, *L'incroyable histoire de l'éléphant Hans*. The description of Longueuil's expedition is recounted by Elizabeth Kolbert in *Sixth Extinction*.

2. He was not the first to formulate the concept of extinction, which had already been developed by the Greek philosopher Empedocles and the Roman scholar Lucretius and then during the Renaissance by the humanist Bernard Palissy. On this, see the fascinating book by Julien Delord, *L'extinction d'espèce*.

3. Harari, *Sapiens*.

4. Pyne, *World Fire*.

5. These figures are based on the data presented by Valérie Chansigaud in *L'homme et la nature*. Other figures can be found; for an overview in English, see the Wikipedia page on quaternary extinction events (https://en.wikipedia.org/wiki/Quaternary_extinction_event). For a description of this megafauna, see the blog TwlightBeasts: https://twilightbeasts.wordpress.com.

6. Various indicators, particularly the results of archaeological digs at Pedra Furada in Brazil, suggest that the first humans arrived in the Americas at least thirty thousand years ago.
7. A genetic study has recently suggested that the San broke off from the rest of humanity around one hundred twenty thousand years ago. If this date is confirmed, it will invalidate the Toba hypothesis of a cognitive revolution after this date.
8. Roebroeks et al. "Use of Red Ochre by Early Neanderthals."
9. On this subject, see the excellent "L'homme de Neandertal et l'invention de la culture," special issue of *Dossier pour la science*.
10. According to the island rule laid out by the biologist J. Bristol Foster in 1964, animal populations isolated on islands see their size fluctuate according to available resources. Restricted to a limited territory, large mammals such as elephants become miniature. Conversely, rodents, turtles, or birds, which can more intensely exploit the environment, become gigantic as long as there are no predators sharing the new territory with them. Gigantic or miniature, all of these populations are vulnerable to the arrival of humans and their associates (rats, cats, dogs). This is epitomized by the fate of the dodo, which died out from its native Mauritius in the seventeenth century.
11. The oldest sewing needle has been dated to around fifty thousand years old. Whittled out of a bird bone, it was found in 2016 in the Denisova cave in Russia, where both Neanderthals and Denisova lived. *Sapiens* might not have invented the needle, but we certainly made good use of it.
12. Doughty, Wolf, and Malhi, "Legacy of the Pleistocene Megafauna Extinctions."
13. Malhia et al., "Megafauna and Ecosystem Function," 843.
14. Hayden, *Naissance de l'inégalité*. See also Guy, *Ce que l'art préhistorique dit de nos origines*.
15. Pleonexia is the condition of insatiable avarice or covetousness. See Dufour, *Pléonexie*.
16. See, for example, Martin, *Twilight of the Mammoths*.

Chapter Three

1. Harari, *Sapiens*.
2. To explain the Younger Dryas, some authors blame phenomena other than subduction in the Gulf Stream, citing volcanic eruptions, the consequences

of the extinction of American megafauna, or the crash of a large meteorite. All of these are phenomena liable to cause global cooling through a modification of the composition of the earth's atmosphere.

3. Braudel, *Civilisation matérielle, économie et capitalisme*.

4. The oldest demonstrated use of ceramics has been found at the archeological site of Dolní Věstonice in today's Czech Republic. Around twenty-five thousand years ago, clay figures representing a woman and animals were made and fired here and then intentionally broken, possibly for religious or symbolic reasons. This invention of pottery in Europe seems to have left no legacy; ceramics would not return to Europe until eight thousand years ago, imported form far-off Anatolia in the Agricultural Revolution's bag of tricks.

5. Xuehui Huang et al., "Map of Rice Genome Variation." Before this study, the debate was heated. Some researchers believed that Asian rice had been domesticated separately in the Ganges Valley and along the Yangtze river, others that it had developed exclusively around the latter. Over the course of millennia of domestication, it became hybridized with several wild rice species, leading to the creation of distinct subspecies. A second, independent domestication took place with another species in West Africa around 2,500 years ago.

6. Wittfogel, *Oriental Despotism*.

7. See, in particular, the wealth of detail provided in chapter 3, "La forêt: archéologie et environnement," of the book by Stéphanie Thiébault, *Archéologie environnementale de la France*.

8. Cauvin, *Naissance des divinités, naissance de l'agriculture*.

9. Schmidt, *Göbekli Tepe*.

10. Built ten thousand three hundred years ago, this tower stood outside the village and does not seem to have been intended for defense. It was an ostentatious building, demanding much more work and materials than an ordinary house, and it was possibly used for community meetings or ceremonies. For an analysis, see Aurenche, "La tour de Jéricho, encore et toujours."

11. Scott, *Art of Not being Governed*, and *Against the Grain*. Scott's thesis extends the ideas of the anthropologist Pierre Clastres, *La société contre l'état*, on Amazon societies. For an analysis of the Zomia thesis, see Delalande, "Zomia, là où l'état n'est pas."

12. See Theofanopoulou et al., "Self-domestication in *Homo sapiens*."

13. Testot, *Homo canis*.

14. Caesar, *De bello Gallico XXVIII*.

15. United Nations "World Population Prospects 2019: Highlights."
16. Since the publication of the first metastudy on terrestrial biomass in 2018 (Bar-On, Phillips, and Milo, "Biomass Distribution on Earth"), it is possible to have consolidated figures even though the margin of error remains substantial. Although this report expresses biomass in gigatons of carbon (GTC), I have used real weight because for complex biomass (mammals and birds), the differences in the proportion of carbon and water in the animals' bodies is much less significant, and the figures are more accessible.
17. An OECD study from autumn 2016 suggests that the generation born in Russia at the time of the Soviet collapse, between 1985 and 1995, measure on average 1 centimeter less at adulthood than the previous generation and the one that followed it. Individuals have therefore paid for the destruction of public health systems and the decline in available food during this decade.

Chapter Four

1. A mineral is described as native when it is found in a near pure state on or near the surface of the ground. It is easy to exploit in this state and does not require much founding to be usable.
2. Pure copper melts at 1,085°C (1,985°F), but the use of additives can lower its melting point. This is true for all metalwork industries, including iron.
3. For a defense of this thesis see Ruddiman, *Plows, Plagues, and Petroleum*.
4. See Jeanne, "Des chars aux éoliennes, irremplacables 'terres rares.'"
5. Cline, *1177 B.C.*
6. Cline, 139.
7. Wolfe, *Proust and the Squid*.
8. This is why the classic analysis by André Leroi-Gourhan remains relevant; see Leroi-Gourhan, *Les religions de la préhistoire*.
9. Alleton, *L'écriture chinoise*.
10. Magni, *Les Olmèques*.

Chapter Five

1. For an anglophone introduction to this epic, see John Brockington, *Comprehensive Guide to the Mahābhārata (and the Rāmāyaṇa)*. For a dramatic adaptation, see Jean-Claude Carrière, *The Mahābhārata*. See also the English translation in ten volumes by Bibek Debroy, *The Mahabharata*.

2. *The Bhagavad Gita*, translated by Winthrop Sargeant, 15, 11, 13,27.

3. In the original meaning of the Latin word *fanum*, "little temple," which is suggestive of the nature of fanaticism: a belief in an exclusive revelation that leads to actions that are only meaningful within the sphere of that belief. Fanatics are those who withdraw from humanity and shut themselves up in a dialectic space they see as sacred, indisputable, and transcendent.

4. Victoria, *Zen At War*

5. Jaspers, *Origin and Goal of History*.

6. On this subject see the works of primatologist Frans de Waal.

7. For an introduction to Chinese religions, see Javary, *Les trois sagesses chinoises*.

8. For a biography of this figure, see Thierry, *La ruine du Qin*.

9. For an understanding of how much our contemporary cinematographic, literary, and other cultural imaginaries are conditioned by these references, reading Alain Musset's *Le syndrome de Babylone* is a must.

10. The reader will find an exhaustive panorama of ancient thought on the relationship between nature and humanity in the canonical work by Clarence J. Glacken, *Traces on the Rhodian Shore*. On animals in antiquity, see the excellent collection of work by Jean-Louis Poirier, *Cave canem*, and Elisabeth de Fontenay, *Le silence des bêtes*.

11. This saga is narrated by Stephen Greenblatt, *The Swerve*.

12. For a near exhaustive account of these acts of destruction, see Bàez, *Universal History of the Destruction of Books*.

Chapter Six

1. To fully grasp this inexorable retreat of elephants from China, see the map in Elvin, *Retreat of the Elephants*, 10.

2. Pliny, *Natural History*, 8.21.

3. Graeber, *Debt: The First 5000 Years*.

4. Up until 107 BCE, when Marius established a professional army.

Chapter Seven

1. Khaldun, *The Muqaddimah*, 12. For a brilliant analysis and application of this theory to the history of the last two thousand years, see Gabriel Martinez-Gros, *Brève histoire des empires*.

2. On the decline of Roman comfort as an indicator of collapse, see Ward-Perkins, *Fall of Rome*.

3. Ward-Perkins, 88.

4. See Harper, *Fate of Rome*.

5. See, in particular, Veyne, *Quand notre monde est devenu chrétien (312–394)*; Brown, *Through the Eye of a Needle*.

6. Procopius, *History of the Wars*, 467, 471.

7. Keys, *Catastrophe*.

8. Dendrochronology is the statistical dating of tree rings, which allows us to trace the climates of the past.

9. For a description of this battle, see Farale, *Les batailles de la région de Talas*.

10. Watson, "Arab Agricultural Revolution."

11. See Fauvelle-Aymar, *Le rhinocéros d'or*.

12. Perhaps in this burning, the Bantu took over from the San, and possibly even from earlier peoples who may have burned these hunting grounds since time immemorial so that game would proliferate, much like the Aboriginal people in Australia (chap. 2). This is the hypothesis defended by Pyne in *World Fire*. He argues that much of the vegetation typical of South Africa has long since (perhaps since 1.6 million years ago) evolved to adapt to repeated anthropogenic burning.

13. Hartwell, "Revolution in the Iron and Coal Industries."

14. Zhang, "Traditional Chinese and the Environment." See also, for an overall perspective, Elvin, "Three Thousand Years of Unsustainable Growth."

15. Juvaini, *Genghis Khan*, 105. See also May, *Mongol Conquests*.

16. For a fascinating history of the spread of the postal system, see Gazagnadou, *La poste à relais en Eurasie*.

17. Lavigne et al., "Source of the Great A.D. 1257 Mystery Eruption Unveiled."

18. Chaunu, *L'expansion européenne*.

Chapter Eight

1. De Carvajal, *Discovery of the Amazon*.

2. Lake Texcoco was drained between the eighteenth and twentieth centuries to combat flooding in Mexico.

3. Liebmann, Joshua Farella, Christopher I. Roos, Adam Stack, Sarah Martini,

and Thomas M. Swetnam, "Native American Depopulation, Reforestation, and Fire Regimes in the Southwest United States, 1492–1900 CE," PNAS Direct Submission, January 25, 2016 (http://www.pnas.org/content/113/6/E696.full.pdf).

4. Lewis and Maslin, "Defining the Anthropocene."

5. Ferdinand Magellan experienced this directly in 1521. In this same year, when the Aztec confederation disintegrated, the explorer and his armored soldiers, equipped with muskets and steel swords, were killed in battle while fighting a chieftain in Mactan Island in the Philippines whose military technology was several hundred years behind his own.

6. The Mongol Yuan dynasty was replaced by the Chinese Ming dynasty in 1368.

7. On this subject, see Crouzet, *Christophe Colomb*.

8. Grmek, *Les maladies à l'aube de la civilisation occidentale*.

9. This history is the subject of a magnificent work by Gruzinski, *The Eagle and the Dragon*.

10. It is said that in his frustration, the Mongolian general catapulted the cadavers of plague victims into the city of Caffer to contaminate it, which would constitute the first known act of biological warfare. But there are no reliable sources to attest to this. However, we do know that since ancient times, wells were habitually poisoned with rotting meat to weaken enemy armies.

11. Bos et al. "Eighteenth Century *Yersinia pestis* Genomes."

12. McNeill, *Mosquito Empires*.

13. This history is presented by Stephan Talty in his book *The Illustrious Dead*.

14. See Spinney, *Pale Rider*.

15. Petrarch, adapted from George Deaux, *The Black Death 1347*, 92–94.

Chapter Nine

1. See Mann, *1493. Uncovering the New World Columbus Created*.

2. Quoted in Mann, 39.

3. Mann, 39.

4. Crosby, *Colombian Exchange*.

5. Crosby, *Ecological Imperialism*.

6. This is superbly described in Hämäläinen, *The Comanche Empire*.

7. In his book *The Other Slavery*, Andrés Reséndez calculates that European colonizers captured or bought a total of 2.5 to five million Indigenous slaves between the sixteenth and the nineteenth centuries from local tribes over the Americas as a whole. He estimates between 147,000 and 340,000 victims for North America, between six hundred thousand and 1.4 million for Mexico and Central America, with the remainder in South America.

8. Quoted in Mann, *1493. Uncovering the New World Columbus Created*, 181.

9. See Dockès, *Le sucre et les larmes*.

10. Mintz, *Sweetness and Power*, 25.

11. *Zanj* is the Persian word for "black," and the origin of the name of the town Zanzibar, the "Black Coast," hub of the slave trade in the Indian ocean.

12. In his book *The Birth of the Modern World (1780–1914)*, Christopher A. Bayly defines what he calls archaic globalization as the period between the sixteenth and the nineteenth centuries. It began with the Portuguese and Spanish empires followed by the Dutch empire in the seventeenth century and then the British empire in the eighteenth and nineteenth centuries. Bayley, *Birth of the Modern*, 44.

13. Sloterdijk, *L'heure du crime*.

Chapter Ten

1. Quoted in Robins, *Mercury, Mining, and Empire*, 225.

2. Coca is the plant from which cocaine is extracted; it acts as an appetite suppressant and painkiller. In the seventeenth century, coca consumption was so damaging to the Andes that it was punishable by excommunication. The church allowed one exception to this however; consumption was authorized in Potosi, where it was essential to the functioning of the mine.

3. Corsairs are distinct from pirates in that they work for a hostile state. These private shipowners thus had a license that legitimized their actions, and they paid taxes on what they confiscated.

4. Brook, *Mr Seldon's Map of China*.

5. This tradition, which is found in many countries, was the source of inspiration for the contemporary notion of "environmental commons." These systems often combine common property with collective usage rights and include all forms of organization, traditional or modern, through which

beneficiaries autonomously and often collectively manage resources and environments, whether these are pastures, forests, wetlands, moorlands, watercourses, fisheries, mineral springs, or fuel sources.

6. Brook, *Vermeer's Hat*.

7. This was technology that they mastered well before European powers did, as we can see in the voyages of Admiral Zheng He. In the first half of the fifteenth century, this Muslim Chinese eunuch, in the service of the Ming dynasty, conducted seven expeditions, some to East Africa and to Jeddah in Arabia. His flagships were more than one hundred meters long and with up to nine masts. They mastered technology such as onboard artillery, magnetic compasses, and watertight compartments for increased buoyancy and resistance against ramming or shipwreck. The Ming gave up their maritime expansion and these technological innovations around 1450, preferring to dedicate their resources to conquests on the continent.

8. This term refers to the maritime coasts from the China Sea to the Indian Ocean, a zone that provides access to two-thirds of the human population, the heart of world trade for the last two thousand years. See Gipouloux, *La méditerranée asiatique* and Beaujard, *Les mondes de l'océan Indien*.

9. Sallmann, *Le grand désenclavement du monde, 1200–1600*.

10. Parker, *Military Revolution*.

11. The funds needed for this expedition were raised using the model of trade operations between the Italian city-states, which allowed them to fund maritime expeditions. Clearly the boundary between trade and war was sometimes vague, particularly on the sixteenth-century seas.

12. On this subject, see Stanziani, *Bâtisseurs d'empires*, and Douglas E. Streusand, *Islamic Gunpowder Empires*.

13. Hanson, *Carnage and Culture*.

Chapter Eleven

1. Timothy Brook, *Troubled Empire*, 250.

2. British historians Eric Hobsbawm (1917–2012) and Hugh Trevor-Roper (1914–2003) emphasized in the 1950s that the period between 1640 and 1690 was exceptionally dense in violent political, economic, and social events. For Trevor-Roper, this phenomenon was connected to the rising power of centralized states challenged by local political communities. In spite of their diverging opinions, both of these historians chose to call this

moment the "general crisis." The term *global crisis*, first used by Geoffrey Parker, links these social, economic, and political dimensions to environmental factors. Following Timothy Brook, we have chosen to extend this period over the Maunder minimum between 1640 and 1715.

3. Geoffrey Parker chose to focus on the period between 1640 and 1690 in his book *Global Crisis*. This period was indeed rich in climate disasters linked to global cooling, but this periodization minimizes the impact of cold episodes in the first half of the seventeenth century.

4. Lewis and Maslin, "Defining the Anthropocene."

5. See Diamond, *Collapse*, for this thesis and McAnany and Yoffee, *Questioning Collapse* for its refutation.

6. Zhang et al., "Climate Change and War Frequency in Eastern China."

7. In English in the French original.

8. Maddison, *The World Economy: A Millennial Perspective*, and above all *The World Economy: Historical Statistics*.

9. An expression made famous by the title of Kenneth Pomeranz's book *The Great Divergence: China, Europe, and the Making of the Modern World Economy*.

Chapter Twelve

1. In *La forêt des 29*, Irene Frain provides a fictionalized biography of the life of Jambho, the founder of the Bishnoi community, and concludes with a description of the 1730 massacre. Franck Vogel has also conducted a documentary on the Bishnoi (*Rajasthan, l'âme d'un prophète*) as well as a self-published book (*Bishnoïs: Écologistes depuis le XVe siècle*). In English, see Jain, *Dharma and Ecology of Hindu Communities: Sustenance and Sustainability*.

2. As the British historian Edward P. Thompson demonstrates in his book *Whigs and Hunters*.

3. Food and Agriculture Organization of the United Nations, *The State of the World's Forests*, Rome, 2012, http://www.fao.org/3/i3010e/i3010e00.htm.

4. Hopkirk, *Great Game*.

5. McNeill, *Rise of the West*.

6. Jones, *European Miracle*.

7. Pomeranz, *Great Divergence*.

Chapter Thirteen

1. The original text is available online (https://archive.org/stream/fumifugium00eveluoft?ref=ol#mode/2up).

2. This remark from Evelyn is printed in the prefatory note of the text.

3. Adapted from the *Ballad of Gresham College*, stanza 23 (https://en.wikisource.org/wiki/Ballad_of_Gresham_College).

4. On the concept of preindustrialization (also known as protoindustrialization) see Jan de Vries, *Industrious Revolution*.

5. This was Max Weber's key hypothesis in *Protestant Ethic*.

6. The steam engine as it appeared in the United Kingdom seems to have been inspired by a Chinese invention dating back to the eleventh century. See Norel, "Techniques chinoises."

7. Some historians argue that this "British agricultural Revolution" proceeded and sustained the Industrial Revolution, providing a basis without which the latter could not have occurred. See Mark Overton, *Agricultural Revolution in England*.

8. An anecdote relayed in Charles C. Mann's book, *1493: Uncovering the New World Columbus Created*.

9. Quoted in Mann, 178.

10. Descartes, *Discourse on Method*, chap. 6.

11. Harari, *Homo Deus*.

12. It is important to distinguish authoritarian regimes from dictatorships. A dictatorship implies full power for a limited number of people, often just one, without control, limitation of powers, or scheduled duration of these powers. The authoritarian regime fills a grey zone between democracies and dictatorships. They tolerate some opposition and cede certain aspects of political power.

13. Mitchell, *Carbon Democracy*.

14. See, in particular, Mouhot, *Des esclaves énergétiques*.

15. See the enlightening article by Jean-Marc Jancovici, "How Much of a Slave Master Am I?"

16. Although Kwame Anthony Appiah argues that honor and not moral motivation specific to certain groups was the deciding factor in abolitionism. See Appiah, *Honor Code*.

17. Aline Helg demonstrates that slaves were also autonomous actors in this

history. Their strategies for resistance weighed heavily on the evolution of slavery as an institution. See Helg, *Plus jamais esclaves!*

Chapter Fourteen

1. D'Arcy Wood, *Tambora*. See also Klingaman and Klingaman, *Year without Summer*.
2. Lines 1–9, quoted in D'Arcy Wood, *Tambora*, 67.
3. Lavigne et al. "Source of the Great A.D.1257 Mystery Eruption Unveiled."
4. In his introduction, the historian Patrick Boucheron, evokes the eruption of Kuwae as the only event having a worldwide effect in the fifteenth century. See Boucheron, *Histoire du monde au XVe siècle*.
5. Winchester, *Krakatoa*.
6. For a biography, see Clark, *Sun Kings*.
7. Or even up to four to eight occurrences per thousand years according to a report by the insurance company Lloyds, which was commissioned to establish the risk of a similar event reoccurring. This text evokes various risks, from sewage overflows due to the failure of electric pumps to food riots, among others. It expresses considerable concern for the financial markets that would suffer from a prolonged electrical blackout. Lloyds and Atmospheric and Environmental Research, *Solar Storm Risk to the North American Electric Grid*.
8. For a good introduction to the island's history, see Orliac and Orliac, *Des dieux regardent les étoiles*. See also, Fischer, *Island at the End of the World*.
9. Diamond, *Collapse*.
10. Tainter, *Collapse of Complex Societies*.
11. Métraux, *Easter Island*, 38.
12. Davis, *Late Victorian Holocausts*, 37.
13. Davis, 245.
14. Leclerc, *The Epochs of Nature*, 98, 124.
15. For a description of this kind of event, see Coquery-Vidrovitch, *Le Rapport Brazza*. In English, see Hochschild, *King Leopold's Ghost*.
16. Grove, *Green Imperialism*.
17. Revelle and Suess, "Carbon Dioxide Exchange."
18. Oreskes and Conway, *Merchants of Doubt*.

Chapter Fifteen

1. On the use of oxymorons as a device for social control enabling regulation of political and environmental ideas, see Méheust, *La politique de l'oxymore*. He took this approach further in Méheust, *La nostalgie de l'occupation*.
2. McNeill, *Something New Under the Sun*.
3. Ehrlich and Ehrlich, *The Population Bomb*. This book raised the question of human overpopulation and predicted that we were on the way to a world population of seven billion by the year 2000 (which proved to be accurate). It also predicted a global famine in the 1980s and 1990s (which proved completely false). The authors have since updated their theories in Ehrlich and Ehrlich, *The Dominant Animal*.
4. Meadows et al., *Dynamics of Growth in a Finite World*.
5. "What has been will be again, what is done will be done again; there is nothing new under the sun." Ecclesiastes 1:9.
6. Readers thirsty for an exhaustive presentation can find references to numerous books in the appendixes, particularly those that explicitly mention the Anthropocene.
7. Fabrice Nicolino presents a biography of Fritz Haber at the beginning of his book *Un empoisonnement universel*.
8. On this subject see Snyder, *Bloodlands*, and *Black Earth*.
9. Jishen, Mosher, and Jian, *Tombstone*, 41.
10. Related in Vadrot, *Guerres et environnement*.
11. Intergovernmental Science-Policy Platform on Biodiversity and Ecosystem Services, "Nature's Dangerous Decline 'Unprecedented': Specied Extinction Rate 'Accelerating,'" 2019 (https://www.ipbes.net/news/Media-Release-Global-Assessment).
12. Barnosky et al., "Approaching a State Shift in Earth's Biosphere."
13. Egan, *Worst Hard Time*, 127.
14. Maddison, *The World Economy: A Millennial Perspective*.
15. McNeill, *Something New Under the Sun*, 5.
16. Krause, *Great Animal Orchestra*.
17. McNeill, *Something New Under the Sun*.

Chapter Sixteen

1. A polluted future anticipated by John Brunner in his novel *The Sheep Look Up*.
2. For a slightly unnerving and conservative reading of this, see Simone, *Presi nella rete*.
3. Leiber, *Creature from Cleveland Depths*. This short novel was also published as the first part of the collection *Night of the Wolf*. This collection of four short stories extends the author's reflection on future societies that have become inhumane through technology and environmental destruction resulting from a nuclear war.
4. Teilhard de Chardin, *Vision of the Past*.
5. See Rosa, *Alienation and Acceleration*. The term is also used by environmental historians to describe the acceleration of socioeconomic indicators that are associated with the Anthropocene. See also Steffen et al., "Trajectory of the Anthropocene."
6. For a description of the development of central nervous systems in nation-states, see Gardey, *Écrire, calculer, classer*.
7. Mercer reappears in Dick's *Do Androids Dream of Electric Sheep?*
8. Latouche, *Bon pour la casse*.
9. Expressed in octets—or kilo (Ko), giga (Go), tera (To), or peta (Po) octets—digital size refers to the quantity of data stored and transmitted for a given piece of information.
10. Vitalis, "La 'révolution numérique.'"
11. Hardt and Negri, *Empire*. This book is the first part of a trilogy that is completed by *Multitude* and *Commonwealth*.
12. Moulier-Boutang, *Cognitive Capitalism*.
13. Stiegler, *Age of Disruption*.

Chapter Seventeen

1. The philosopher Günther Anders launched a premonitory reflection on this "Promethean" issue. *Die Antiquiertheit des Menschen* (The outdatedness of human beings) is a two-volume work that has been published in several editions in German but has not been translated into English. It was translated into French as *L'obsolescence de l'homme*.
2. Drexler, *Engines of Creation*.

3. Truong, *Totalement inhumaine.*
4. Olds, "Survival of the Fittest."
5. See Momcilovic, *Prodiges d'Arnold Schwarzenegger.*
6. Chemical Abstract Service (https://www.cas.org/).
7. See among other publications Swann et al., "Decrease in Anogenital Distance."
8. On this subject, see the excellent documentary by Jean-Christophe Ribot, *Et l'homme créa la vache.*
9. See Bourguignon, *The Globalization of Inequalities.*
10. The middle class can be defined as people who can save a significant portion of their income for investment (education, property, etc.).
11. Recent studies have thrown doubt on what until 2018 appeared self-evident: the declining birth rate may have nothing to do with the one-child policy but rather be to do with increasing living standards. Thailand, for example, has experienced an identical demographic trajectory without any state intervention but with economic growth per capita identical to China. People tend to have fewer children as their wealth increases.
12. Oreskes and Conway, *Collapse of Western Civilization.*
13. Klein, *Shock Doctrine.*
14. For an overview, see the excellent panorama by Bénédicte Manier, *Un million de révolutions tranquilles.* For an introduction to this question, see Cyril Dion and Mélanie Laurent's film *Tomorrow.*

Conclusion

1. On this point, see Bihouix, *L'âge des low tech.*
2. See Bonneuil and Feydel, *Prédation.*
3. Lovelock, *The Revenge of Gaia.*
4. Klein, *This Changes Everything.*

Epilogue

1. For the six-month period between January and June 2019 alone, a large number of books were published in France on the theme of collapse. A nonexhaustive alphabetical list includes Amicel, *Que reste-t-il de l'avenir?*;

Barrau, *Le plus grand défi de l'histoire de l'humanité*; Bohler, *Le bug humain*; Bihouix, *Le bonheur était pour demain*; Gancille, *Ne plus se mentir*; Latouche, Jouventin, and Paquot, *Pour une écologie du vivant*; Semal, *Face à l'effondrement*; Vargas, *L'humanité en péril*.

2. According to IPCC special report, "Global Warming of 1.5°C."

3. Caspar A. Hallmann et al., "More Than 75 Percent Decline over 27 Years in Total Flying Insect Biomass in Protected Areas."

4. United Nations, "Key Findings and Advance Tables," on United Nations, Department of Economic and Social Affairs, Population Division, *World Population Prospect: The 2017 Revision*.

5. Abstract available in Camilo Mora et al., "Global Risk of Deadly Heat"; for North Africa, see also J. Lelieved et al., "Strongly Increasing Heat Extremes."

BIBLIOGRAPHY

Alleton, Viviane. *L'écriture chinoise: Le défi de la modernité*. Paris: Albin Michel, 2008.

Ameisen, Jean-Claude. *Sur les épaules de Darwin: Retrouver l'aube*. Paris: France Inter, Les liens qui libèrent, 2014.

Amicel, Gérard. *Que reste-t-il de l'avenir? Entre posthumanité et catastrophe*. Rennes: Apogée, 2019.

Anders, Gunther. *Die Antiquiertheit des Menschen*. Munich: C. H. Beck, 1956. Translated by Christophe David as *L'obsolescence de l'homme*, vol. 1, *Sur l'âme à l'époque de la deuxième révolution industrielle* (Paris: Ivrea; Editions de l'Encyclopédie des Nuisances, 2002), and vol. 2, *Sur la destruction de la vie à l'époque de la troisième révolution industrielle* (Paris: Fario, 2011).

Appiah, Kwame Anthony. *The Honor Code: How Moral Revolutions Happen*. New York: W. W. Norton, 2010.

Appleby, Joyce. *The Relentless Revolution: A History of Capitalism*. New York: W. W. Norton, 2010.

Aries, Paul. *Une histoire politique de l'alimentation: Du paléolithique à nos jours*. Paris: Max Milo, 2016.

Armstrong, Karen. *Compassion: Twelve Steps to a Compassionate Life*. London: Bodley Head, 2011.

Armstrong, Karen. *The Great Transformation: The World in the Time of Buddha, Socrates, Confucius and Jeremiah*. London: Atlantic Books, 2006.

Auberger, Janick, and Peter Keating. *Histoire humaine des animaux: De l'antiquité à nos jours*. Paris: Ellipses, 2009.

Aurenche, Olivier. "La tour de Jéricho, encore et toujours." *Syria: Archéologie, art et histoire* 83 (2006): 63–68. http://syria.revues.org/172.

Auzanneau, Matthieu. *Or noir: La grande histoire du pétrole*. Paris: La Découverte, 2015.

Bàez, Fernando. *A Universal History of the Destruction of Books: From Ancient Sumer to Modern Iraq*. Translated by Alfred McAdam. New York: Atlas, 2008.

Bairoch, Paul. *Victoires et déboire: Histoire économique et sociale du monde du XVIe siècle à nos jours*. 3 vols. Paris: Gallimard, 1997.

Barbier, Frédéric. *L'Europe de Gutenberg: Le livre et l'invention de la modernité occidentale (XIIIe-XVIe siècle)*. Paris: Belin, 2006.

Barnosky, Anthony, Elizabeth A. Hadly, Jordi Bascompte, Eric L. Berlow, James H. Brown, Mikael Fortelius, Wayne M. Gertz, et al. "Approaching a State Shift in Earth's Biosphere." *Nature* 486 (2012): 52–58.

Bar-On, Yinon M., Rob Phillips, and Ron Milo. "The Biomass Distribution on Earth." *Proceedings of the National Academy of Sciences* 115, no. 25 (June 19, 2018): 6506–11. https://www.pnas.org/content/115/25/6506.

Barrau, Aurélien. *Le plus grand défi de l'histoire de l'humanité: Face à la catastrophe écologique et sociale*. Neuilly-sur-Seine: Michel Lafon, 2019.

Baudart, Anne. *Naissance de la philosophie politique et religieuse*. Paris: Le Pommier, 2016.

Baudet, Jean C. *Curieuses histoires de la Pensée: Quand l'homme inventait les religions et la philosophie*. Waterloo: Editions Jourdan, 2011.

Bayly, Christopher A. *The Birth of the Modern World 1780-1917*. Oxford: Blackwell, 2004.

Beaujard, Philippe. *Les mondes de l'océan Indien*. 2 vols. Paris: Armand Colin, 2012.

Bellah, Robert N. *Religion in Human Evolution: From the Paleolithic to the Axial Age*. Cambridge, MA: Belknap Press of Harvard University Press, 2011.

Bentley, Jerry H. *Old World Encounters: Cross-Cultural Contacts and Exchanges in Pre-Modern Times*. Oxford: Oxford University Press, 1993.

Bentley, Jerry H. and David Christian, eds. *World Environmental History*. Great Barrington, MA: Berkshire Publishing Group, 2012.

Berthoz, Alain. *La vicariance: Le cerveau créateur de mondes*. Paris: Odile Jacob, 2013.

Bertrand, Romain. *L'histoire à parts égales: Récits d'une rencontre Orient-Occident*. Paris: Le Seuil, 2011.

Besson, Florian, and Jan Synowiecki, eds. *Écrire l'histoire avec des si*. Paris: Rue d'Ulm, 2015.

The Bhagavad Gita. 25th anniversary ed. Translated by Winthrop Sargeant and edited by Christopher Key Chappel. Albany: State University of New York Press, 2009.

Bihouix, Philippe. *L'âge des low tech: Vers une civilization techniquement soutenable*. Paris: Le Seuil, 2014.

Bihouix, Philippe. *Le bonheur était pour demain: Les rêveries d'un ingénieur solitaire*. Paris: Le Seuil, 2019.

Bilimoff, Michele. *Histoire des plantes qui ont changé le monde*. Paris: Albin Michel, 2011.

Birdzell, L. E., Jr., and Nathan Rosenberg. *How the West Grew Rich: The Economic Transformation of the Industrial World*. New York: Basic Books, 1986.

Bohler, Sébastien. *Le bug humain: Pourquoi notre cerveau nous pousse à détruire la planète et comment l'en empêcher*. Paris: Robert Laffont, 2019.

Boltanski, Luc, and Eve Chiapello. *Le nouvel esprit du capitalisme*, Paris: Gallimard, 2011.

Bonneuil, Christophe, and Sandrine Feydel. *Prédation: Nature, le nouvel Eldorado de la finance*. Paris: La Découverte, 2015.

Bonneuil, Christophe, and Jean-Baptiste Fressoz. *L'evénement anthropocène: La terre, l'histoire et nous*. Paris: Le Seuil, 2013.

Boqueho, Vincent. *Les civilisations à l'épreuve du climat*. Paris: Dunod, 2012.

Bos, Kirsten, Alexander Herbig, Jason Sahl, Nicholas Waglechner, Mathieu Fourment, Stephen A. Forrest, Jennifer Klunk, et al. "Eighteenth Century *Yersinia pestis* Genomes Revealed the Long-Term Persistence of an Historical Plague Focus," *Elife* 21 (2016). https://elifesciences.org/content/5/e12994v1.

Boucheron, Patrick, eds. *Histoire du monde au XVe siècle*. Paris: Fayard, 2009.

Boumediene, Samir. *La colonisation du savoir: Une histoire des plantes médicinales du "Nouveau Monde" (1492-1750)*. Vaulx-en-Velin: Les editions des mondes à faire, 2016.

Bourguignon, François. *The Globalization of Inequalities*. Translated by Thomas Scott-Railton. Princeton, NJ: Princeton University Press, 2015.

Bouvet, Jean-Francois. *Mutants: A quoi ressemblerons-nous demain?* Paris: Flammarion, 2014.

Braga, José, Claudine Cohen, Bruno Maureille, and Nicolas Teyssandier. *Origines de l'humanité: Les nouveaux scénarios*. Montreuil: La ville brûle, 2016.

Braudel, Fernand. *Civilisation matérielle, économie et capitalisme, XVe-XVIIIe siècle*. Vol. 1. *Les structures du quotidien: Le possible et l'impossible*. Paris: Armand Colin, 1979.

Brockington, John. *A Comprehensive Guide to the Mahābhārata (and the Rāmāyaṇa) and Scholarship on Them*: *The Sanskrit Epics*. Leiden: Brill. 1998.

Brook, Timothy. *Mr Selden's Map of China: The Spice Trade, a Lost Chart & the South China Sea*. London: Profile Books, 2013.

Brook, Timothy. *The Troubled Empire: China in the Yuan and Ming Dynasties*. Harvard, MA: Harvard University Press, 2010.

Brook, Timothy. *Vermeer's Hat: The Seventeenth Century and the Dawn of the Global World*. New York: Bloomsbury, 2008.

Brown, Peter. *Through the Eye of a Needle: Wealth, the Fall of Rome, and the Making of Christianity in the West 350-550 A.D*. Princeton, NJ: Princeton University Press, 2012.

Brunet, Michel. *D'Abel à Toumaï: Nomade, chercheur d'os*. Paris: Odile Jacob, 2006.

Brunner, John. *The Sheep Look Up*. New York: Harper and Row, 1972.

Burbank, Jane, and Frederick Cooper. *Empires in World History: Power and the Politics of Difference*. Princeton, NJ: Princeton University Press, 2010.

Burke, Edmund, and Kenneth Pomeranz, eds. *The Environment and World History*. Berkeley: University of California Press, 2009.

Caesar, C. J. *De bello Gallico and Other Commentaries*. Translated by W. A. MacDevitt. Project Gutenberg; Everyman's Library, 2004. First published 1915 by Dent, London. https://www.gutenberg.org/files/10657/10657.txt.

Campagne, Armel. *Le capitalocène: Aux racines historiques du dérèglement climatique*. Paris: Editions Divergences, 2017.

Candegabe, Philippe. *L'incroyable histoire de l'éléphant Hans des forêts du Sri Lanka au Muséum d'histoire naturelle*. Paris: Vendémiaire, 2016.

Carré, Jacques. *La prison des pauvres: L'expérience des workhouses en Angleterre*. Paris: Vendémiaire, 2016.

Carrière, Jean-Claude. *The Mahābhārata: A Play Based upon the Indian Classic Epic*. Translated by Peter Brook. New York: Harper and Row, 1987.

Cauvin, Jacques. *Naissance des divinités, naissance de l'agriculture: La révolution des symboles au Néolithique*. Paris: CNRS Editions, 2010.

Chalvet, Martine. *Une histoire de la forêt*. Paris: Le Seuil, 2011.

Chansigaud, Valérie. *L'homme et la nature: Une histoire mouvementée*. Paris: Delachaux et Niestlé, 2013.

Chaunu, Pierre. *L'expansion européenne: Du XIIIe au XVe siècle*. Paris: PUF, 1969.

Cheng'en, Wu. *Journey to the West*. Translated and edited by Anthony C. Yu. Chicago: University of Chicago Press, 1983.

Cipolla, Carlo M. *Guns, Sails & Empires: Technological Innovation & the Early Phases of European Expansion 1400–1700*. New York: Minerva Press, 1965.

Clark, Gregory. *A Farewell to Alms: A Brief Economic History of the World*. Princeton, NJ: Princeton University Press, 2007.

Clark, Stuart. *The Sun Kings: The Unexpected Tragedy of Richard Carrington and the Tale of How Modern Astronomy Began*. Princeton, NJ: Princeton University Press, 2007.

Clastres, Pierre. *La société contre l'état: Recherches d'anthropologie politique*. Paris: Editions de Minuit, 1974.

Claval, Paul. *L'aventure occidentale: Modernité et globalization*. Auxerre: Sciences Humaines Editions, 2016.

Cline, Eric H. *1177 B.C.: The Year Civilization Collapsed*. Princeton, NJ: Princeton University Press, 2014.

Collaert, Jean-Paul. *Céréales: La plus grande saga que le monde ait vécue*. Paris: Rue de l'échiquier, 2013.

Coppens, Yves. *Histoire de homme et changements climatiques*. Paris: Collège de France; Fayard, 2006.

Coppens, Yves. *Préludes: Autour de l'homme préhistorique*. Paris: Odile Jacob, 2014.

Coquery-Vidrovitch, Catherine. *Le rapport Brazza: Mission d'enquête du Congo; rapport et documents (1905–1907)*. Neuvy-en-Champagne: Le passager clandestine, 2014.

Coquery-Vidrovitch, Catherine. *Les routes de l'esclavage: Histoires des traites africaines, VI*^{e}*-XX*^{e} *siècle*. Paris: Albin Michel; Arte Editions, 2018.
Cosandey, David. *Le secret de l'Occident: Vers une théorie générale du progrès scientifique*. Paris: Flammarion, 2008.
Crosby, Alfred W., Jr. *Children of the Sun: A History of Humanity's Unappeasable Appetite for Energy*. New York: W. W. Norton, 2006.
Crosby, Alfred W., Jr. *The Colombian Exchange: Biological and Cultural Consequences of 1492*. London: Praeger, 2003.
Crosby, Alfred W., Jr. *Ecological Imperialism: The Biological Expansion of Europe, 900–1900*. New York: Cambridge University Press, 2004.
Crouzet, Denis. *Christophe Colomb: Héraut de l'apocalypse*. Paris: Payot, 2006.
Cuvier, Georges. *Memoirs on Fossil Elephants and on Reconstruction of the Genera Palaeotherium and Anoplotherium*. New York: Arno Press, 1980.
D'Arcy Wood, Gillen. *Tambora: The Eruption That Changed the World*. Princeton, NJ: Princeton University Press, 2014.
Davis, Diana K. *Resurrecting the Granary of Rome: Environmental History and French Colonial Expansion in North Africa*. Athens: Ohio University Press, 2007.
Davis, Mike. *Late Victorian Holocausts: El Nino, Famines, and the Making of the Third World*. London: Verso, 2001.
Deaton, Angus. *The Great Escape: Health, Wealth, and the Origins of Inequality*. Princeton, NJ: Princeton University Press, 2013.
Deaux, George. *The Black Death 1347*. New York: Weybright and Talley, 1969.
Debeir, Jean-Claude, Jean-Paul Deléage, and Daniel Hémery. *Une histoire de l'énergie: Les servitudes de la puissance*. Paris: Flammarion, 2013.
De Carvajal, Gaspar. *The Discovery of the Amazon*. Translated by Bertram Tamblyn Lee. Special Publication, no. 17. New York: American Geographical Society, 1934.
Delalande, Nicolas. "Zomia, là où l'état n'est pas." *La vie des idées* (March 20, 2013). http://laviedesidees.fr/IMG/pdf/20130320_zomia.pdf.
Delorde, Julien. *L'extinction d'espèce: Histoire d'un concept et enjeux éthiques*. Paris: Publications scientifiques du Muséum national d'histoire naturelle, 2010.
Deluermoz, Quentin, and Pierre Singaravelou. *Pour une histoire des possible: Analyses contrefactuelles et futurs non advenus*. Paris: Le Seuil, 2016.
Demoule, Jean-Paul. *La révolution néolithique dans le monde*. Paris: CNRS Éditions, 2010.
Demoule, Jean-Paul. *Les dix millénaires oubliés qui ont fait l'histoire: Quand on inventa l'agriculture, la guerre et les chefs*. Paris: Fayard, 2017.
Denhez, Frédéric. *Les pollutions invisibles: Ces poisons qui nous entourent*. Paris: Delachaux et Niestlé, 2005.
De Puy Torac, Pierre. *L'homme: Coauteur de l'évolution*. Versailles: Editions Quae, 2014.

Descartes, René. *Discourse on method and related writings.* Translated by Desmond M. Clarke. London: Penguin, 2003.

De Vries, Jan. *The Industrious Revolution: Consumer Behavior and the Household Economy (1650 to the Present).* Cambridge: Cambridge University Press, 2008.

De Waal, Frans. *La politique du chimpanzé.* Paris: Edition Du Rocher, 1992.

Diamond, Jared. *Collapse: How Societies Choose to Fail or Succeed.* New York: Viking, 2005.

Diamond, Jared. *Guns, Germs and Steel: The Fates of Human Societies.* New York: W.W. Norton, 1997.

Dick, Philip K. *Do Androids Dream of Electric Sheep?* New York: Ballantine Books, 1968.

Dion, Cyril, and Mélanie Laurent, dirs. *Tomorrow.* Paris: Mars Films, 2015. 118 min.

Dockès, Pierre. *Le sucre et les larmes: Bref essai d'histoire et de mondialisation.* Paris: Descartes, 2009.

Doueihi, Milad. *Pour un humanisme numérique.* Paris: Le Seuil, 2011.

Doughty, Christopher E., Adam Wolf, and Yadvinder Malhi. "The Legacy of the Pleistocene Megafauna Extinctions on Nutrient Availability in Amazonia." *Nature Geoscience,* August 11, 2013. http://www.yadvindermalhi.org/uploads/1/8/7/6/18767612/doughtyngeo1895.pdf.

Drexler, Kim Eric. *Engines of Creation: The Coming Era of Nanotechnology.* New York: Anchor Books, 1986.

Dufour, Dany-Robert. *Pléonexie, [dict.: "Vouloir posséder toujours plus."].* Lormont: Le bord de l'eau, 2015.

Dugain, Marc, and Christophe Labbe. *L'homme nu: La dictature invisible du numérique.* Paris: Plon; Robert Laffont, 2016.

Du Roy, Olivier. *La règle d'or: Histoire d'une maxime morale universelle.* 2 vols. Paris: Les éditions du Cerf, 2012.

Egan, Timothy. *The Worst Hard Time: The Untold Story of Those Who Survived the Great American Dust Bowl.* New York: Horton Mifflin, 2006.

Ehrlich, Paul R., and Anne H. Ehrlich, *The Dominant Animal: Human Evolution and the Environment.* Washington DC: Island Press, 2008.

Ehrlich, Paul R., and Anne H. Ehrlich. *The Population Bomb.* New York: Ballantine Books, 1970.

Eisenstein, Elizabeth L. *The Printing Revolution in Early Modern Europe.* Cambridge: Cambridge University Press, 1983.

Elvin, Mark. *The Retreat of the Elephants: An Environmental History of China.* New Haven, CT: Yale University Press, 2004.

Elvin, Mark. "Three Thousand Years of Unsustainable Growth: China's Environment from the Archaic Times to the Present." [Basis of the Annual Lecture of the Centre for Modern Chinese Studies at St Antony's College, Oxford, May 11, 1994.] *East Asian History,* no. 6 (December 1993): 7–46. http://www.eastasianhistory.org/sites/default/files/article-content/06/EAH06%20_02.pdf.

Evelyn, John. *Fumifugium: or The Inconveniencie of the Aer and Smoak of London dissipated. Together With some Remedies humbly Proposed by J.E. Esq. to His Sacred Majestie, to the Parliament now assembled.* London, 1661. https://archive.org/stream/fumifugium00eveluoft?ref=ol#mode/2up.

Fagan, Brian. *Floods, Famines, and Emperors: El Nino and the Fate of Civilizations.* New York: Basic Books, 1999.

Fagan, Brian. *The Little Ice Age: How Climate Made History, 1300–1850.* New York: Basic Books, 2000.

Farale, Dominique. *Les batailles de la région de Talas et l'expansion musulmane en Asie centrale: Islam et Chine; Un choc multiséculaire.* Paris: Economica, 2006.

Fauvelle-Aymar, François-Xavier. *Le rhinocéros d'or: Histoires du moyen âge africain.* Paris: Alma, 2014.

Fernandez-Armesto, Felipe. *Civilizations: Culture, Ambition, and the Transformation of Nature.* New York: Simon & Schuster, 2002.

Ferrao, José E. Mendes. *Le voyage des plantes & les grandes découvertes.* Paris: Chandeigne, 2015.

Fischer, Steven Roger. *Island at the End of the World: The Turbulent History of Easter Island.* London: Reaktion Books, 2005.

Fontenay, Elisabeth de. *Le silence des bêtes: La philosophie à l'épreuve de l'animalité.* Paris: Fayard, 1998.

Fraine, Irene. *La forêt des 29.* Neuilly-sur-Seine: Michel Lafon, 2001.

Frankopan, Peter. *The Silk Roads: A New History of the World.* London: Bloomsbury, 2015.

Fressoz, Jean-Baptiste. *L'apocalypse joyeuse: Une histoire du risque technologique.* Paris: Le Seuil, 2012.

Gancille, Jean-Marc. *Ne plus se mentir: Petit exercice de lucidité par temps d'effondrements écologique.* Paris: Rue de l'échiquier, 2019.

Gardey, Delphine. *Écrire calculer, classer: Comment une révolution de papier a transformé les sociétés contemporaines (1800–1940).* Paris: La Découverte, 2008.

Gazagnadou, Didier. *La poste à relais en Eurasie: La diffusion d'une technique d'information et de pouvoir; Chine-Iran-Syrie-Italie.* Paris: Editions Kimé, 2013.

Gernet, Jacques. *Le monde chinois.* Vol. 1, *De l'âge de bronze au moyen âge, 2100 avant J.-C-Xe siècle après J.-C.* Paris: Armand Colin, 1972.

Gipouloux, François. *La méditerranée asiatique: Villes portuaires et réseaux marchands en Chine, au Japon et en Asie du Sud-Est.* Paris: CNRS Éditions, 2009.

Glacken, Clarence J. *Traces on the Rhodian Shore: Nature and Culture in Western Thought from Ancient Times to the End of the Eighteenth Century.* Berkeley: University of California Press, 1967.

Goody, Jack. *The Domestication of the Savage Mind.* Cambridge: Cambridge University Press, 1977.

Graeber, David. *Debt: The First 5000 Years.* New York: Melville House, 2011.

Gras, Alain. *Le choix du feu: Aux origines de la crise climatique*. Paris: Fayard, 2007.

Greenblatt, Stephen. *The Swerve: How the World Became Modern*. New York: W.W. Norton, 2011.

Griffon, Michel, and Florent Griffon. *Pour un monde viable: Changement global et viabilité planétaire*. Paris: Odile Jacob, 2011.

Grmek, Mirko. *Les maladies à l'aube de la civilisation occidentale*. Paris: Payot, 1983.

Grove, Richard. *Ecology, Climate and Empire: Colonialism and Global Environmental History, 1400–1940*. Cambridge: White Horse Press, 1997.

Grove, Richard H. *Green Imperialism: Colonial Expansion, Tropical Island Edens and the Origins of Environmentalism, 1600–1860*. Cambridge, Cambridge University Press, 1993.

Grundmann, Emmanuelle. *Ces forêts qu'on assassine*. Paris: Calmann-Lévy, 2007.

Gruzinski, Serge. *The Eagle and the Dragon: Globalization and European Dreams of Conquest in China and America in the Sixteenth Century*. Polity: Cambridge, 2014.

Gualde, Norbert. *Comprendre les épidémies: La coévolution des microbes et des hommes*, Paris: Les Empêcheurs de penser en rond, 2006.

Guilaine, Jean. *Cain, Abel, Otzi: L'héritage néolithique*. Paris: Gallimard, 2011.

Guillaume, Jean. *Ils ont domestiqué plantes et animaux: Prélude à la civilisation*. Versailles: Editions Quae, 2011.

Guy, Emmanuel. *Ce que l'art préhistorique dit de nos origines*. Paris: Flammarion, 2017.

Hache, Emilie, ed. *De l'univers clos au monde infini*. Bellevaux: Editions Dehors, 2014.

Hallmann, Caspar A., Martin Sorg, Eelke Jongejans, Henk Siepel, Nick Hofland, Heinz Schwan, Werner Stenmans, et al. "More Than 75 Percent Decline over 27 Years in Total Flying Insect Biomass in Protected Areas." *PLoS One*, October 18, 2017. https://doi.org/10.1371/journal.pone.0185809.

Hämäläinen, Pekka. *The Comanche Empire*. New Haven, CT: Yale University Press, 2008.

Hanson, Victor Davis. *Carnage and Culture: Landmark Battles in the Rise of Western Power*. New York: Anchor, 2007.

Harari, Yuval Noah. *Homo Deus: A Brief History of Tomorrow*. London: Harvill Secker, 2016.

Harari, Yuval Noah. *Sapiens: A Brief History of Humankind*. London: Harvill Secker, 2012.

Hardt, Michael, and Antonio Negri. *Commonwealth*. Cambridge, MA: Belknap Press of Harvard University Press, 2009.

Hardt, Michael, and Antonio Negri. *Empire*. Cambridge, MA: Harvard University Press, 2000.

Hardt, Michael, and Antonio Negri. *Multitude: War and Democracy in the Age of Empire*. London: Hamish Hamilton, 2005.

Harper, Kyle. *The Fate of Rome: Climate, Disease, and the End of an Empire*. Princeton, NJ: Princeton University Press, 2017.

Harrison, Harry. *Deathworld*. London: Sphere, 1973.

Hartwell, Robert. "A Revolution in the Iron and Coal Industries in the Northern Sung 960–1126AD." *Journal of Asian Studies* 21, no. 2 (1962): 153–62.

Hayden, Brian. *Naissance de l'inégalité: L'invention de la hiérarchie durant la Préhistoire*. Paris: CNRS Éditions, 2013.

Heinrich, Bernd. *Why We Run: A Natural History*. New York: Harper, 2002.

Helg, Aline. *Plus jamais esclaves! De l'insoumission à la révolte, le grand récit d'une émancipation (1492–1838)*. Paris: La Découverte, 2016.

Herlehy, David. *The Black Death and the Transformation of the West*. Cambridge, MA: The President and Fellows of Harvard College, 1997.

Hobhouse, Henry. *Seeds of Change: Six Plants That Transformed Mankind*. London: Pan Macmillan, 2002.

Hochschild, Adam. *King Leopold's Ghost: A Story of Greed, Terror and Heroism in Colonial Africa*. London: Pan Books, 2012.

Hopkirk, Peter. *The Great Game: On Secret Service in High Asia*. London: John Murray, 2016.

Huang, Xuehui, Nori Kurata, Xinghua Wei, Zi-Xuan Wang, Ahong Wang, Qiang Zhao, Yan Zhao, et al. "A Map of Rice Genome Variation Reveals the Origin of Cultivated Rice." *Nature* 490 (2012): 497–501. https://doi.org/10.1038/nature11532.

Huchet, Jean-François. *La crise environnementale en Chine: Evolution et limites des politiques publiques*. Paris: Les Presses de Sciences Po, 2016.

Iliffe, John. *Africa: History of a Continent*. Cambridge: Cambridge University Press, 1995.

Jain, Pankaj. *Dharma and Ecology of Hindu Communities: Sustenance and Sustainability*. London: Routledge, 2016.

Jancovici, Jean-Marc. "How Much of a Slave Master Am I?" Last modified August 1, 2013. https://jancovici.com/en/energy-transition/energy-and-us/how-much-of-a-slave-master-am-i/.

Jarrige, Francois, and Thomas Le Roux. *La contamination du monde: Une histoire des pollutions à l'âge industriel*. Paris: Le Seuil, 2017.

Jarrosson, Bruno. *Humanisme et technique: L'humanisme entre économie, philosophie et science*. Paris: PUF, 1996.

Jaspers, Karl. *The Origin and Goal of History*. London: Routledge and Kegan Paul, 1953.

Javary, Cyrille J. D. *Les trois sagesses chinoises: Taoïsme, confucianisme, et bouddhisme*. Paris: Albin Michel, 2010.

Jeanne, Ludovic. "Des chars aux éoliennes, irremplacables 'terres rares.'" *Conversation*, September 7, 2016. https://theconversation.com/des-chars-aux-eoliennes-irremplaceables-terres-rares-64788.

Jishen, Yang, Stacy Mosher, and Jian Gao. *Tombstone: The Untold Story of Mao's Great Famine*. London: Penguin 2013.

Johnson, Chris. *Australia's Mammal Extinction: A 50,000 Year History*. Melbourne: Cambridge University Press, 2006.
Jones, Eric L. *The European Miracle: Environments, Economies and Geopolitics in the History of Europe and Asia*. Cambridge: Cambridge University Press, 1981.
Juvaini, 'Ala-ad-Din 'Ata-Malik. *Genghis Khan: The History of the World Conqueror*. Translated by J. A. Boyle. Seattle: University of Washington Press, 1997.
Kempf, Hervé. *Comment les riches détruisent la planète*. Paris: Le Seuil, 2007.
Kempf, Hervé. *Fin de l'Occident, naissance du monde*. Paris: Le Seuil, 2013.
Keys, David. *Catastrophe: An Investigation into the Origins of the Modern World*. New York: Ballantine Books, 1999.
Khaldun, Ibn. *The Muqaddimah: An Introduction to History*. Translated by Franz Rosenthal. Princeton, NJ: Princeton University Press, 1967
Klein, Naomi. *The Shock Doctrine: The Rise of Disaster Capitalism*. London: Penguin, 2007.
Klein, Naomi. *This Changes Everything: Capitalism vs. the Climate*. New York: Simon & Schuster, 2014.
Klingaman, William K. and Nicholas P. Klingaman. *The Year without Summer: 1816 and the Volcano that Darkened the World and Changed History*. New York: St. Martin's Press, 2013.
Kolbert, Elizabeth. *The Sixth Extinction: An Unnatural History*. New York: Henry Holt, 2014.
Krause, Bernie. *The Great Animal Orchestra: Finding the Origins of Music in the World*. New York: Little, Brown, 2012.
Krech, Shepard, III, John R. McNeill, and Carolyn Merchant, eds. *Encyclopedia of World Environmental History*. 3 vols. New York: Routledge, 2004.
Kurlansky, Mark. *Cod: A Biography of the Fish That Changed the World*. London: Jonathan Cape, 1998.
Lambert, Yves. *La naissance des religions: De la préhistoire aux religions universalistes*. Paris: Armand Colin, 2007.
Landes, David S. *The Wealth and Poverty of Nations: Why Some Are So Rich and Some Are So Poor*. New York: W. W. Norton, 1998.
Latouche, Serge. *Bon pour la casse: Les déraisons de l'obsolescence programmée*. Paris: Les Liens qui Libèrent, 2012.
Latouche, Serge, Pierre Jouventin, and Thierry Paquot. *Pour une écologie du vivant*. Paris, Editions Libre et Solidaire, 2019.
Latour, Bruno. *Face à Gaia*. Paris: La Découverte, 2015.
Lavigne, Franck, Jean-Christophe Komorowski, Vincent Robert, Clive Oppenheimer, Céline M. Vidal, Indyo Pratomo, Irka Hajdas, et al. "Source of the Great A.D. 1257 Mystery Eruption Unveiled, Samalas Volcano, Rinjani Volcanic Complex, Indonesia." *Proceedings of the National Academy of Sciences* 110, no. 42 (October 15, 2015): 16742–47. https://www.pnas.org/content/110/42/16742.

Lebeau, André. *L'enfermement planétaire*, Paris: Gallimard, 2008.
Lebeau, André. *Les horizons terrestres: Réflexions sur la survie de l'humanité.* Paris: Gallimard, 2011.
Leclerc, Georges-Louis, comte de Buffon. *The Epochs of Nature.* Translated and edited by Jan Zalasiewicz. Chicago: Chicago University Press, 2018.
Lefeburé, Antoine. *L'affaire Snowden: Comment les États-Unis espionnent le monde.* Paris: La Découverte, 2014.
Lefèvre, Denis. *Des racines et des gènes: Une histoire mondiale de l'agriculture*, 2 vols. Paris: Rue de l'échiquier, 2018.
Legendre, Pierre. *Dominium mundi: L'empire du management.* Paris: Fayard; Mille et Une Nuits, 2007.
Leiber, Fritz. *The Creature from Cleveland Depths.* New York: Galaxy, 1962.
Leiber, Fritz. *The Night of the Wolf.* New York: Ballantine Books, 1966.
Liebmann, Joshua Farella, Christopher I. Roos, Adam Stack, Sarah Martini, and Thomas M. Swetnam. "Native American Depopulation, Reforestation, and Fire Regimes in the Southwest United States, 1492–1900 CE." PNAS Direct Submission, January 25, 2016. http://www.pnas.org/content/113/6/E696.full.pdf.
Lelieved, J., Y. Proestos, P. Hadjinicolaou, M. Tanarhte, E. Tyrlis, and G. Zittis. "Strongly Increasing Heat Extremes in the Middle East and North Africa (MENA) in the 21st Century." *Climatic Change* 137 (July 2016): 245–60. https://doi.org/10.1007/s10584-016-1665-6.
Leroi-Gourhan, André. *Les religions de la préhistoire.* Paris: PUF, 2015.
Le Roy Ladurie, Emmanuel. *Histoire humaine et comparée du climat.* 3 vols. Paris: Fayard, 2004–2009.
Lewis, Simon L., and Mark A. Maslin. "Defining the Anthropocene." *Nature* 519 (2015): 171–80. https://www.nature.com/articles/nature14258.
"L'homme de Néandertal et l'invention de la culture: Bijoux, rituels, peintures..." Special issue, *Dossier pour la science* 76 (July–September, 2012).
Lieberman, Daniel E. *The Story of the Human Body: Evolution, Health, and Disease.* London: Penguin; New York: Pantheon Books, 2013.
Lindqvist, Sven. *Exterminate All the Brutes! One Man's Odyssey into the Heart of Darkness and the Origins of European Genocide.* New York: New Press, 2007.
Lindqvist, Sven. *Terra Nullius: A Journey through No One's Land.* London: Granta Books, 2007
Lloyd's and Atmospheric and Environmental Research. *Solar Storm Risk to the North American Electric Grid.* London: Lloyd's, 2013. https://www.lloyds.com/news-and-risk-insight/risk-reports/library/natural-environment/solar-storm.
Lorius, Claude, and Laurent Carpentier. *Voyage dans l'anthropocène: Cette nouvelle ère dont nous sommes les héros.* Paris: Actes Sud, 2011.
Loubet, Jean-Louis. *Une autre histoire de l'automobile.* Rennes: Presses Universitaires de Rennes, 2017.

Lovelock, James. *The Revenge of Gaia: Earth's Climate in Crisis and the Fate of Humanity*. New York: Basic Books, 2006.

Lucretius. *On the Nature of Things/De rerum natura*. Translated by Rev. John Selby Watson. London: Henry G. Bohn, 1851.

Maddison, Angus. *The World Economy: Historical Statistics*. Paris: Organisation for Economic Co-operation and Development, 2003.

Maddison, Angus. *The World Economy: A Millennial Perspective*. Paris: Organisation for Economic Co-operation and Development, 2001.

Magni, Catarina. *Les Olmèques: La genèse de l'écriture en Méso-Amérique*. Paris: Errance, 2005.

The Mahabharata. Translated by Bibek Debroy. 10 vols. Gurgaon, Haryana, India: Penguin Books, 2015.

Malhia, Yadvinder, Christopher E. Doughty, Mauro Galetti, Felisa A. Smith, Jens-Christian Svenning, and John W. Terborgh. "Megafauna and Ecosystem Function from the Pleistocene to the Anthropocene." *Proceedings of the National Academy of Sciences* 113, no.4 (January 26, 2016),https://www.pnas.org/content/113/4/838.

Malm, Andreas. *Fossil Capital: The Rise of Steam Power and the Roots of Global Warming*. London: Verso, 2016.

Manier, Bénédicte. *Un million de révolutions tranquilles: Comment les citoyens changent le monde*. Paris: Les Liens qui Libèrent, 2016.

Mann, Charles C. *1491: New Revelations of the Americas before Columbus*. New York: Knopf, 2006.

Mann, Charles C. *1493: Uncovering the New World Columbus Created*. New York: Alfred A. Knopf, 2011.

Margolin, Jean-Louis, and Claude Markovits. *Les Indes et l'Europe: Histoires connectées XVe-XXIe siècle*. Paris: Gallimard, 2015.

Markovits, Claude, ed. *Histoire de l'Inde moderne: 1480–1950*. Paris: Fayard, 1994.

Marks, Robert B. *The Origins of the Modern World: A Global and Ecological Narrative from the Fifteenth to the Twenty-First Century*. Lanham, MD: Rowman & Littlefield, 2015.

Marshall, George. *Don't Even Think about It: Why Our Brains Are Wired to Ignore Climate Change*. New York: Bloomsbury, 2015.

Martin, Paul S. *Twilight of the Mammoths: Ice Age Extinctions and the Rewilding of America*. Berkeley: University of California Press, 2005.

Martinez-Gros, Gabriel. *Brève histoire des empires: Comment ils surgissent, comment ils s'effondrent*. Paris: Le Seuil, 2014.

Mattelart, Armand. *Histoire de la société de l'information*. Paris: La Découverte, 2009.

May, Timothy. *The Mongol Conquests in World History*. London: Reaktion Books, 2012,

Mazoyer, Marcel, and Laurence Roudart. *Histoire des agricultures du monde: Du néolithique a la crise contemporaine*. Paris: Le Seuil, 2002.

McAnany, Patricia A., and Norman Yoffee. *Questioning Collapse: Human Resilience, Ecological Vulnerability, and the Aftermath of Empire*. Cambridge: Cambridge University Press, 2009.
McNeill, John R. *Mosquito Empires: Ecology and the War in the Greater Caribbean, 1620–1914*. Cambridge: Cambridge University Press, 2010.
McNeill, John R. *Something New Under the Sun: An Environmental History of the Twentieth-Century World*. New York: W. W. Norton, 2000.
McNeill, John R. and Erin Stewart Mauldin, eds. *A Companion to Environmental History*. Chichester: Wiley-Blackwell, 2012.
McNeill, John R., and William H. McNeill. *The Human Web: A Bird's-Eye View of World History*. New York: W. W. Norton, 2003.
McNeill, William H. *Plagues and Peoples*. New York: Anchor Press, 1977.
McNeill, William H. *The Pursuit of Power: Technology, Armed Force, and Society since A.D. 1000*. Chicago: University of Chicago Press, 1982.
McNeill, William H. *The Rise of the West: A History of the Human Community*. Chicago: University of Chicago Press, 1963. Reprinted with a retrospective essay. Chicago: University of Chicago Press, 1991.
Meadows, Donatella H., Dennis L. Meadows, Jørgen Randers, and William W. Behrens III. *Dynamics of Growth in a Finite World*. Cambridge, MA: Wright-Allen Press, 1974.
Méheust, Bertrand. *La nostalgie de l'occupation: Peut-on encore se rebeller contre les nouvelles formes d'asservissement?* Paris: La Découverte, 2012.
Méheust, Bertrand. *La politique de l'oxymore: Comment ceux qui nous gouvernent nous masquent la réalité du monde*. Paris: La Découverte, 2009.
Métraux, A. *Easter Island: A Stone-Age Civilization of the Pacific*. London: Andre Deutsch, 1957.
Meyer, Eric Paul. *Une histoire de l'Inde: Les Indiens face a leur passé*. Paris: Albin Michel, 2007.
Mintz, Sidney W. *Sweetness and Power: The Place of Sugar in Modern History*. London: Penguin Books, 1985.
Mitchell, Timothy. *Carbon Democracy: Political Power in the Age of Oil*. London: Verso, 2011.
Mithen, Steven. *After the Ice: A Global Human History, 20,000–5000 B.C.* Cambridge, MA: Harvard University Press, 2004.
Momcilovic, Jerome. *Prodiges d'Arnold Schwarzenegger*. Capricci: Nantes, 2016.
Mora, Camilo, Bénédict Dousset, Iain R. Caldwell, Farrah E. Powell, Rollan C. Geronimo, Coral R. Bielecki, Chelsie W. W. Counsell, et al. "Global Risk of Deadly Heat." *Nature Climate Change* 7 (June 19, 2017): 501–6. https://www.nature.com/articles/nclimate3322.epdf.
Morris, Desmond. *The Naked Ape: A Zoologist's Study of the Human Animal*. London: Jonathan Cape, 1967.
Morris, Ian. *Why the West Rules—For Now*. London: Profile Books, 2010.

Mouhot, Jean-François. *Des esclaves énergétiques: Réflexions sur le changement climatique.* Seyssel: Champ Vallon, 2011.
Moulier-Boutang, Yann. *Cognitive Capitalism.* Translated by Ed Emery. Cambridge: Polity, 2011.
Mouthon, Fabrice. *Le sourire de Prométhée: L'homme et la nature au moyen age.* Paris: La Découverte, 2017.
Musset, Alain. *Le syndrome de Babylone: Géofictions de l'apocalypse.* Paris: Armand Collin, 2012.
Naphy, William, and Andrew Spicer. *The Black Death.* Gloucestershire: Tempus, 2000.
Nicolino, Fabrice. *Un empoisonnement universel: Comment les produits chimiques ont envahi la planète.* Paris: Les Liens qui libèrent, 2014.
Norel, Philippe. *L'histoire économique globale.* Paris: Le Seuil, 2009.
Norel, Philippe. "Techniques chinoises et révolution industrielle britannique," Histoire Globale (blog), March 8, 2010. http://blogs.histoireglobale.com/techniques-chinoises-et-revolution-industrielle-britannique_214.
Olds, Tim. "Survival of the Fittest, The Changing Shapes and Sizes of Olympic Athletes." *Conversation*, August 4, 2016. https://theconversation.com/survival-of-the-fittest-the-changing-shapes-and-sizes-of-olympic-athletes-63184.
Olivieri, Guido. *L'outil périlleux: L'homme, du succès à l'excès.* Geneva: Slatkine, 2007.
Oreskes, Naomi, and Erik M. Conway. *The Collapse of Western Civilization: A View from the Future.* New York: Columbia University Press, 2014.
Oreskes, Naomi, and Erik M. Conway, *Merchants of Doubt: How a Handful of Scientists Obscured the Truth on Issues from Tobacco Smoke to Global Warming.* London: Bloomsbury, 2012.
Orliac, Catherine, and Michel Orliac. *Des dieux regardent les étoiles: Les derniers secrets de L'Île de Pâques.* Paris: Gallimard, 2000.
Orliac, Catherine, and Michel Orliac. *The Silent Gods: Mysteries of Easter Island.* London: Thames and Hudson, 1995.
Orsenna, Erik, and Isabelle de Saint-Aubin, *Géopolitique du moustique: Petit précis de mondialisation IV.* Paris: Fayard, 2017.
Otte, Marcel. *A l'aube spirituelle de l'humanité: Une nouvelle approche de la préhistoire.* Paris: Odile Jacob, 2012.
Overton, Mark. *Agricultural Revolution in England: The Transformation of the Agrarian Economy 1500–1850.* Cambridge: Cambridge University Press, 1996.
Pääbo, Svante. *Neanderthal Man: In search of Lost Genome.* New York: Basic Books, 2014.
Parker, Geoffrey. *Global Crisis: War, Climate Change, and Catastrophe in the 17th Century.* New Haven, CT: Yale University Press, 2013.
Parker, Geoffrey. *The Military Revolution: Military Innovation and the Rise of the West, 1500–1800.* Cambridge: Cambridge University Press, 2016.

Patou-Matthis, Marylène. *Mangeurs de viande: De la préhistoire à nos jours*. Paris: Perrin, 2009.
Pétré-Grenouilleau, Olivier. *Les traites négrières: Essai d'histoire globale*. Paris: Gallimard, 2004.
Picq, Pascal. *De Darwin a Levi-Strauss: L'homme et la diversité en danger*. Paris: Odile Jacob, 2013.
Picq, Pascal. *Il était une fois la paléoanthropologie: Quelques millions d'années et trente ans plus tard*. Paris: Odile Jacob, 2010.
Picq, Pascal. *L'homme est-il un grand singe politique? Essai de primatologie politique et de pataphysique*. Paris: Odile Jacob, 2011.
Picq, Pascal. *Premiers Hommes*. Paris: Flammarion, 2016.
Pliny the Elder, *Natural History: A Selection*. Translated and with an introduction by John F. Healey. London: Penguin Classics, 2004.
Poirier, Jean-Louis. *Cave canem: Hommes et bêtes dans l'Antiquité*. Paris: Les Belles Lettres, 2016.
Pomeranz, Kenneth. *The Great Divergence: China, Europe, and the Making of the Modern World Economy*. Princeton, NJ: Princeton University Press, 2000.
Ponting, Clive. *A New Green History of the World: The Environment and the Collapse of Great Civilizations*. London: Sinclair-Stevenson, 2007.
Procopius. *History of the Wars*. Vol. 1, bk. 2. Translated by H. B. Dewing. Loeb Library of the Greek and Roman Classics. Cambridge, MA: Harvard University Press, 1914.
Pyne, Stephen J. *World Fire: The Culture of Fire on Earth*. New York: Henry Holt, 1995.
Quilliet, Bernard. *La tradition humaniste*. Paris: Fayard, 2002.
Ramade, François. *Le grand massacre: L'avenir des espèces vivantes*. Vanves: Hachette, 1999.
Reséndez, Andrés. *The Other Slavery: The Uncovered Story of Indian Enslavement in America*. Boston: Houghton Mifflin Harcourt, 2016.
Revelle, Roger, and Hans E. Suess. "Carbon Dioxide Exchange between the Atmosphere and Ocean and the Question of an Increase in Atmospheric CO_2 during the Past Decades." *Tellus* 9, no. 1 (February 1957): 19–20.
Ribot, Jean-Christophe, dir. *Et l'homme créa la vache*. Paris: Bonobo Productions; Arte France, 2016. 55 min.
Robin, Marie-Monique. *Les moissons du futur. Comment l'agro-écologie peut nourrir le monde*. Paris: La Découverte, 2012.
Robins, Nicholas A. *Mercury, Mining, and Empire: The Human and Ecological Cost of Colonial Silver Mining in the Andes*. Bloomington: Indiana University Press, 2011.
Roebroeks, Wil, Mark J. Sier, Trine Kellberg Nielsen, Dimitri De Loecker, Josep Maria Parés, Charles E. S. Arps, and Herman J. Mücher. "Use of Red Ochre by Early Neanderthals." *Proceedings of the National Academy of Sciences* 109, no. 6 (January 2012): 1889–1894. https://www.pnas.org/content/109/6/1889.

Rosa, Hartmut. *Alienation and Acceleration: Towards a Critical Theory of Late-Modern Temporality*. Mälmo: Arhus University Press, 2014.
Rostain, Stephen. *Amazonie: Un jardin sauvage ou une forêt domestiquée? Essai d'écologie historique*. Paris: Actes Sud; Errance, 2016.
Ruddiman, William F. *Plows, Plagues, and Petroleum: How Humans Took, Control of Climate*. 2nd ed. Princeton, NJ: Princeton University Press, 2016.
Sadin, Eric. *La silicolonisation du monde: L'irrésistible expansion du libéralisme numérique*. Montreuil: L'Échappée, 2016.
Sadin, Eric. *L'humanité augmentée: L'administration numérique du monde*. Montreuil: L'Échappée, 2013.
Sallmann, Jean-Michel. *Le grand désenclavement du monde, 1200–1600*. Paris: Payot, 2011.
Santolaria, Nicolas. *Dis Siri: Enquête sur le génie à l'intérieur du smartphone*. Paris: Anamosa, 2016.
Saporta, Isabelle. *Le livre noir de l'agriculture: Comment on assassine nos paysans, notre santé et l'environnement*. Paris: Fayard, 2011.
Sarmant, Thierry. *1715: La France et le monde*. Paris: Perrin, 2014.
Schiavone, Aldo. *The End of the Past: Ancient Rome and the Modern West*. Translated by Margery J. Schneider. Cambridge, MA: Harvard University Press, 2002.
Schmidt, Klaus. *Göbekli Tepe: A Stone Age Sanctuary in South-Eastern Anatolia*. Translated by Mirko Wittar. Berlin: Ex Oriente, 2012.
Schneider, Pierre. *Les eléphants de guerre dans l'Antiquité (IVe–Ier siècle av. J.-C.)*. Clermont-Ferrand: Lemme, 2015.
Schwärgel, Christian. *Menschenzeit: Zerstören oder Gestalten? Die Entscheidende Epoche unseres Planeten*. Munich: Riemann, 2010.
Scott, James C. *Against the Grain: A Deep History of the Earliest States*. New Haven, CT: Yale University Press, 2017.
Scott, James C. *The Art of Not Being Governed: An Anarchist History of Upland Southeast Asia*. New Haven, CT: Yale University Press, 2009.
Semal, Luc. *Face à l'effondrement: Militer à l'ombre des catastrophes*. Paris: PUF, 2019.
Senut, Brigitte, and Michel Devillers. *Et le singe se mit debout: Aventures africaines d'une paléontologue*. Paris: Albin Michel, 2008.
Serres, Michel. *Hominescence*. Paris: Le Pommier, 2001.
Serres, Michel. *L'incandescent*. Paris: Le Pommier, 2003.
Serres, Michel. *Petite poucette*. Paris: Le Pommier, 2012.
Serres, Michel. *Rameaux*. Paris: Le Pommier, 2004.
Serres, Michel. *Récits d'humanisme*. Paris: Le Pommier, 2006.
Sigaut, Francois. *Comment Homo devint faber*. Paris: CNRS Éditions, 2012.
Simmons, I. G. *Global Environmental History: 10,000 BC to AD 2000*. Edinburgh: Edinburgh University Press, 2008.
Simone, Raffaele. *Presi nella rete: La mente ai tempi del web*. Milan: Garzanti, 2012.

Sloterdijk, Peter. *L'heure du crime et le temps de l'œuvre d'art*. Paris: Pluriel, 2001.
Snyder, Timothy. *Black Earth: The Holocaust as History and Warning*. London: Bodley Head, 2015.
Snyder, Timothy. *Bloodlands: Europe between Hitler and Stalin*. New York: Basic Books, 2012.
Spinney, Laura. *Pale Rider: The Spanish Flu of 1918 and How It Changed the World*. Jonathan Cape, 2017.
Spinrad, Norman. *Bug Jack Barron*. Garden City, NY: Doubleday, 1969.
Stanziani, Alessandro. *Bâtisseurs d'empires: Russie, Chine et Inde à la croisée des mondes, XVe-XIXe siècle*. Paris: Raisons d'agir, 2012.
Steffen, Will, Wendy Broadgate, Lisa Deutsch, Owen Gaffney, and Cornelia Ludwig. "The Trajectory of the Anthropocene: The Great Acceleration." *Anthropocene Review* 2, no. 1 (2015). http://anr.sagepub.com/content/early/2015/01/08/2053019614564785.full.pdf.
Stiegler, Bernard. *The Age of Disruption: Technology and Madness in Computational Capitalism*. Translated by Daniel Ross. London: Polity, 2019.
Streusand, Douglas E. *Islamic Gunpowder Empires: Ottomans, Safavids, and Mughals*. Boulder, CO: Westview, 2011.
Subrahmanyam, Sanjay. *Courtly Encounters: Translating Courtliness and Violence in Early Modern Eurasia*. Cambridge, MA: Harvard University Press, 2012.
Swan, Shanna H., Katharina M. Main, Fan Liu, Sara L. Stewart, Robin L. Kruse, Antonia M. Calafat, Catherine S. Mao, et al. "Decrease in Anogenital Distance among Male Infants with Prenatal Phthalate Exposure." *Environmental Health Perspectives* 118, no. 8 (August 1, 2005). https://doi.org/10.1289/ehp.8100.
Swynghedauw, Bernard. *L'homme malade de lui-même: Réchauffement du climat, pollutions, modification de la biodiversité . . . ces nouveaux risques qui menacent notre santé*. Paris: Belin, 2015.
Tainter, Joseph A. *The Collapse of Complex Societies*. Cambridge: Cambridge University Press, 1988.
Talty, Stephan. *The Illustrious Dead: The Terrifying Story of How Typhus Killed Napoleon's Greatest Army*. New York, Three Rivers, 2009.
Tardieu, Christine. *Comment nous sommes devenus bipèdes: Le mythe des enfants-loups*. Paris: Odile Jacob, 2012.
Teilhard de Chardin, Pierre. *The Vision of the Past*. London: Collins, 1966.
Testot, Laurent. *Homo canis: Une histoire des chiens et de l'humanité*. Paris: Payot, 2018.
Testot, Laurent. *La nouvelle histoire du monde*. Auxerre: Sciences Humaines Editions, 2019.
Testot, Laurent, ed. *L'histoire globale: Un nouveau regard sur le monde*. Paris: Sciences Humaines, 2008.
Theofanopoulou, Constantina, Simone Gastaldon, Thomas O'Rourke, Bridget

D. Samuels, Angela Messner, Pedro Tiago Martins, Francesco Delogu, Saleh Alamri, and Cedric Boeckx. "Self-Domestication in *Homo sapiens*: Insights from Comparative Genomics." *PLoS One*, October 18, 2017. https://journals.plos.org/plosone/article?id=10.1371/journal.pone.0185306.

Thiébault, Stéphanie. *Archéologie environnementale de la France*. Paris: La Découverte, 2010.

Thierry, François. *La ruine du Qin: Ascension, triomphe et mort du premier empereur de Chine*. Paris: Vuibert, 2013.

Thompson, Edward P. *Whigs and Hunters: The Origin of the Black Act*. London: Breviary Stuff, 2013.

Trautmann, Thomas R. *Elephants and Kings: An Environmental History*. Chicago: University of Chicago Press, 2015.

Truong, Jean-Michel. *Totalement inhumaine*. Paris: Le No Man's Land, 2015.

Tudge, Colin. *Neanderthals, Bandits and Farmers: The Origins of Agriculture*. London: Weidenfeld & Nicolson, 1998.

Uekoetter, Frank, ed. *The Turning Points of Environmental History*. Pittsburgh, PA: University of Pittsburg Press, 2010.

United Nations, Department of Economic and Social Affairs, Population Division. *World Population Prospect: The 2017 Revision, Key Findings and Advance Tables*. Working Paper no. ESA/P/WP/248. New York: United Nations, 2017. https://population.un.org/wpp/Publications/Files/WPP2017_KeyFindings.pdf.

United Nations, Department of Economic and Social Affairs, Population Division. "World Population Prospects 2019: Highlights." https://population.un.org/wpp/Publications/Files/WPP2019_10KeyFindings.pdf.

Vadrot, Claude-Marie. *Guerres et environnement: Panorama des paysages et écosystèmes boulversés*. Paris: Delachaux and Niestlé, 2005.

Vargas, Fred. *L'humanité en peril: Virons de bord, toute!* Paris: Flammarion, 2019.

Verley, Patrick. *L'echelle du monde*. Paris: Gallimard, 1997.

Veyne, Paul. *Quand notre monde est devenu chrétien (312-394)*. Paris: Albin Michel, 2007.

Victoria, Brian. *Zen at War*. Lanham, MD: Rowman & Littlefield, 1997.

Vigne, Jean-Denis. *Les débuts de l'élevage*. Paris: Le Pommier; Universcience, 2012.

Vitalis, André. "La 'révolution numérique': Une révolution technicienne entre liberté et contrôle." *Communiquer* 13 (2015): 44–54. https://journals.openedition.org/communiquer/1494.

Vitaux, Jean. *Histoire de la peste*. Paris: PUF, 2010.

Vogel, Franck. *Bishnoïs: Écologistes depuis le XVe siècle*. N.p.: Blurb, 2013.

Vogel, Franck. *Rajasthan, l'âme d'un prophète*. Aired 2011 on France5; Paris: Gédéon Programmes, 2014 (DVD), 52 min.

Wagner, Peter. *Progress: A Reconstruction*. Cambridge: Polity, 2015.

Walter, Francois. *Hiver: Histoire d'une saison*. Paris: Payot, 2014.

Ward-Perkins, Bryan. *The Fall of Rome and the End of Civilization*. Oxford: Oxford University Press, 2005.
Watson, Andrew M. "The Arab Agricultural Revolution and Its Diffusion, 700–1100." *Journal of Economic History* 34, no. 1 (1974): 8–35. https://doi.org/10.1017/S0022050700079602.
Weatherford, Jack. *Genghis Khan and the Making of the Modern World*. New York: Broadway Books, 2005.
Weaver, John C. *The Great Land Rush and the Making of the Modern World, 1650–1900*. Montreal: McGill-Queen's University Press, 2003.
Weber, Max. *The Protestant Ethic and the Spirit of Capitalism*. London: Charles Scribner's Sons, 1958.
Werdelin, Lars, and Margaret E. Lewis. "Temporal Change in Functional Richness and Evenness in the Eastern African Plio-Pleistocene Carnivoran Guild." *PLoS One*, March 6, 2013). https://doi.org/10.1371/journal.pone.0057944.
White, Sam. *The Climate of Rebellion in the Early Modern Ottoman Empire*. Cambridge: Cambridge University Press, 2011.
Wills, John E., Jr. *1688: A Global History*. New York: W. W. Norton, 2002.
Wilson, Edward O. *The Social Conquest of Earth*. New York: Liveright, 2012.
Winchester, Simon. *Krakatoa: The Day the World Exploded; 27 August 1883*. London: Penguin, 2003.
Wittfogel, Karl A. *Oriental Despotism: A Comparative Study of Total Power*. New Haven, CT: Yale University Press, 1957.
Wolfe, Maryanne. *Proust and the Squid: The Story and Science of the Reading Brain*. New York: Harper Collins, 2007.
Zask, Joëlle. *La démocratie aux champs: Du jardin d'Eden aux jardins partagés, comment l'agriculture cultive les valeurs démocratiques*. Paris: Les Empêcheurs de penser en rond, 2016.
Zhang, David D., Jane Zhang, Harry F. Lee, and Yuan-qing He. "Climate Change and War Frequency in Eastern China over the Last Millennium." *Human Ecology* 35, no.4, June (2007): 403–14. https://www.jstor.org/stable/27654205.
Zhang, Ling. "Traditional Chinese and the Environment." In *Demystifying China: New Understandings of Chinese History*, edited by Naomi Standen, 79–89. Lanham, MD: Rowman & Littlefield, 2013.

INDEX

Aamjiwnaang First Nation community, 362, 366
Abbasid Empire, 149
African Eve, theory of, 38
Agricultural Revolution, xiii, xvii, 54–56, 59–60, 67, 380; Arab, 148–52; "little," 266; Third, 325, 382
agriculture, 81–82, 136, 146, 162, 190, 227; as global invention, 58–62; industrial, 57, 308, 319–20, 325, 366, 384, 390; effect of, on forests, 244, 303–4; intensive, 33, 130, 314–15, 322; in Mesoamerica, 62–65; physical toll of, on people, 74–78; monoculture in, 320; and resources, 312, 355; selection in, 57; vs. horticulture, 69. *See also* Agricultural Revolution; *specific crop names*
AIDS epidemic, 184–85
air pollution. *See under* pollution
Akhenaten, 107–8
Amazon Basin, 164–67
American bison, 326
amorality, temptation of, 350–56
amphibians, 326–27
animals, domestication of, 69–74
anogenital (AG) distance, 363–64
Anthropocene period, 83, 169, 290, 311, 326, 348, 359, 421n5
Antonine Plague, 138
Apache Nation, 194
Arab Agricultural Revolution, 148–52
Ardipithecus, 7–8
Aristotle, 160
Arrhenius, Svante, 304
artificial intelligence, 356
artillery, 220–25
Ashoka, emperor of India, 133
Asian Brown Cloud, 370–71

aurochs, 72–73
Australopithecus, 12, 14–15, 17, 21, 374; brain size of, 31
automobiles, 269
Averroes, 161–62
Axial Age, 99, 115
Aztec civilization, 167–68

Bacon, Francis, 217
Baghdad, 151–52, 153
battleships, 267
Bayly, Christopher A., 203
beavers: Canadian, in Patagonia, 301–2; skins of, trading in, 217
Becquerel, Antoine, 270
Benedict, St., 143
Ben-Ncer, Abdelouahed, 38–39
Bhagavad-Gita (Song of God), 96–99
Big Data, 340
biodiversity, 83, 122, 153, 215, 324, 327, 388; agriculture and, 66–67, 200, DDT and, 319; and extinction, 35, 46, 320; in forests, 166, 245, 327; passenger pigeons and, 323–24; protecting, 242, 380
biological empires, 191–93
biotechnology, 352–53
bipedalism, 16
Bishnoi Hindu community, deforestation and, 106–7, 241–43
bison, American, 326
bisphenol A, 361–62, 363
Black Deal, 178–80
blast furnaces, 158–60; invention of, 131
Blaut, James M., xvii
Bogomils, 143
Bolsonaro, Jair, 372
bonobos, 6

445

books, 338
Bosch, Carl, 313
Brahmanism, 104
brain, use of upright position and, 22
Braudel, Fernand, 56, 267
breastfeeding, 77
Brock, Timothy, 162
Bronze Age, 82, 85; in China, 86; in Egypt, 85–86; in Indus Valley, 86–87; in Mesopotamia, 87
Brook, Timothy, 228, 231
Buddha (Shakyamuni), 105
Buddhism, 104–6
buffalos, destruction of, 323–28
Bug Jack Barron (Spinrad), 348–50
Burton, Richard Francis, 251
Byron, Lord, 286, 289–90

Callendar, Guy Stewart, 306
cannons, 221
capitalism, 270; cognitive, theory of, 344–45; democracy and, 274; development of, 272–73; globalized, 344; as myth, 379–80; printing and, 338; rise of, and protectionism, 273
Carrington, Richard, 291–92
cars, 269
Carson, Rachel, 319, 361
cast iron, 89–90
Cathars, 143
Catlin, George, 304
cats, 71–72, 362
cattle. *See* cows
central Asia, empires of, 134
CFCs (chlorofluorocarbons), 328–29
Chang'an, 151–52
Chardin, Pierre Teilhard de, 336
Chaunu, Pierre, 162–63
chickens, 73
child labor, 276–77
child mortality, 265
chimpanzees, 6, 10; tools and, 19
China: Bronze Age in, 86; coal consumption in, 370; Columbian exchange in, 196; ecological issues in, 370–71; green energy in, 371; Han dynasty, 131–32; humanism as explanation for West overtaking, 276; income inequality in, 368–69; invention of agriculture in, 58–59; liberalism in, 368; Mao's great leap forward, 316–17; as metaphor for accelerated future, 368–73; Ming dynasty, 232–34; Monkey and, xi; Northern Song dynasty, 254; one-child policy in, 371; public servants in, 129–32; Qing dynasty, 197, 240; Song dynasty, 155–58; southern, 61; Tang dynasty, 154–55; Warring States period, 101–4; Yuan dynasty, 232
chlorofluorocarbons (CFCs), 328–29
cholera, 287–88
Christianity, 109–11
Cistercians, 143–44
civilization, 390
Civilization and Climate (Huntington), xiv
climate change, 373–78; political conflict and, 236–39. *See also* global warming
Cline, Eric H., 87, 89
Cluny, Order of, 143
coal, 255, 263, 279–280, 381; consumption of, in China, 370
cognitive capitalism, theory of, 344–45
Cognitive Revolution, xii–xiii, xvii, 38, 39, 337–38, 380
cognitive science, 353
Colborn, Theodora, 360–61
Colombia, war on coca in, 317–18
Columbian exchange, 189–91; in China, 196
Columbian Exchange, The (Crosby), xiv
Columbus, Christopher, 171–74
Comanche Empire, 193–96
communism, 371, 377, 382
compass, 130, 217
computers, 337, 339–40, 345, 392
concentration camps, 325
conflict, political, and climate change, 236–39
Confucius/Confucianism, 101–4
copper, 79–80, 82
Copper Age, 82, 83
Córdoba, Spain, 151
corn. *See* maize
Cortés, Herman, 168, 169–70, 177
cotton, 218–19, 250–51
cows, 72–73, 74, 326, 366–67
Crookes, William, 312
crop rotation, 136
Crosby, Alfred W., Jr., xiv, 189, 191
Crutzen, Paul J., 311
cryonics, 356–57
Cuvier, Georges, 25, 27–28, 34
cybernetics, 338–39

Darwin, Charles, 188
data collection, 343–44
Davis, Mike, 297

Deathworld (Harrison), 383
Deckard, Rick, 323
de Fontenay, Elisabeth, 111
deforestation, 135, 246–47, 392; Bishnoi Hindu community and, 106–7, 241–43; in China, 252; in Enlightenment Europe, 302–3; *Epic of Gilgamesh and*: 302; in France, 303; in sub-Saharan Africa, 303; in tropical islands, 303–4. *See also* forest cycles
Delft, the Netherlands, 216–17
democracy, 338; capitalism and, 274; society's energy and, 278–79
dererum natura, 113
Descartes, René, 274, 367
Devi, Amrita, 241–43
Diamond, Jared, 231, 295–96
Dick, Philip K., 323–24, 339, 388–89
digital memory, 339
Digital Revolution, xiii, xviii, 333–37; analytic frameworks for development of, 344–45
disease. *See* illness
dogs, 70–71, 106, 169, 175
Dominican order, 143
Doughty, Christopher, 45
Drexler, Kim Eric, 354
Dunbar, Claude, 29
Dust Bowl, 320–21
Dutch East India Company, 216, 247–49
Dutch Republic. *See* Netherlands, the

earthworms, 186–89
Easter Island (Rapa Nui), 293–98
Eastern Roman Empire, 144
East India Company (EIC), 247–50; opium and, 251–52
Ebola epidemic, 184
ecological amnesia, 389
Ecological Imperialism (Crosby), 191
economic revolution, 253–54
Egan, Timothy, 321
Egypt, Bronze Age in, 85–86, 87
Ehrlich, Anne, 309
Ehrlich, Paul, 309
EIC. *See* East India Company (EIC)
electricity, 270, 292
elephants, xv–xvi, 43–44, 116–18, 389–90; plight of, 119–22
Elvin, Mark, 162, 228, 231
empathy, 29
endocrine disrupters, 360–61, 362
Energy Revolution, xiii, xviii
energy slaves, 283–84

engines: invention of, 254–55, 264; mobile, 267; steam, 267
England: enclosures in, 214–15; wool trade in, 215–16
environmental crises: gold rules for, 330
Environmental history, xiv–xv
Epic of Gilgamesh, 302
Epicureanism, 112–14
epidemics, 136
Eric the Red, 230
erosion, 135–36
European thought, 111–14
Evelyn, John, 259–61, 329
evolution, xvi
evolutionist humanism, 275–76
Evolutive Revolution, xiii, xviii, 335, 347
exapting, 335–36
externalized memory, 337–39

Fagan, Brian, 228
Faraday, Michael, 270
fertility, 357; mutation and, 358; preventive genetic medicine and, 357–58
fertilizers, 312–13
Feynman, Richard P., 353–54
financial speculation, 346–47
fire, emergence of, 31–33
Fleming, Alexander, 184
Fleming, John, 34
flu, Spanish, 183–84
Flynn, James R., 364
Flynn effect, 364
Ford, Henry, 281
Fordism, 281
forest cycles: in Enlightenment Europe, 246; in France, 244–45; in Levant, 245–46. *See also* deforestation
forests. *See* deforestation
fossil fuels, 279
Fourier, Joseph, 304
France, 253; deforestation in, 303; forest cycles in, 244–45
Franciscan order, 144
free trade, 269, 273, 379
Freon gas, 328–29

GAFAM companies (Google, Apple, Facebook, Amazon, and Microsoft), 334, 341
genetically modified (GM) organisms, 352–53
genetic diversity, 37–38
genetic medicine, preventive, 357–58
Genghis Khan, 158–60

Germany, 254
germs, 164, 168, 174–77, 184, 189, 191, 391. *See also* illness
GHLCA (gorilla-human last common ancestor), 6
Gibbon, Edward, 272
Glacken, Clarence, J., xiv
glass, industrial production of, 268
global connections, 381
global history, defined, xiii
globalization, 190, 193; archaic, 203; biological, 190–91; economic, 269, 344; European, 216
global warming, 56, 84, 185, 247, 306–7, 373–78; humanity's ability to fight, 330–31; omens of, 302–7; scenarios of, 240, 387–88
Golden Rule, 99–101
gorillas, 6, 10
Great Acceleration, the, 337–41
Great American Interchange, 12–13
Great Britain, 253–54
Great Convergence, 368, 369
Great Divergence, 252–53, 368
Great Rift, 15
Great Smog of London, 329
green energy, in China, 371
greenhouse effect/greenhouse gases, 290, 304, 306, 319, 341, 373–74, 388
Greenland, 229–31
Green Revolution (Third Agriculture Revolution), 325, 382
Grotius, Hugo, 213
Guericke, Otto von, 264
gunpowder, 130, 217, 220–25

Haber, Fritz, 311–14, 320
Han dynasty, 131–32
Han Empire, 136–37
Hansen, James, 306–7
Harari, Yuval Noah, 28, 275
Hardt, Michael, 344
Harmand, Sonia, 20
Harrison, Harry, 383
Hartwell, Robert, 156
Hebraism, 108
Hedin, Sven, 305
Heinrich, Bernd, 5
herring trade, 216–17
Hesiod, 84–85
hierarchical religion, 67–68
Hinduism, 104–5
Hnsen, Victor D., 225
Holocaust, 314–15

Hominidae/hominids, 6, 8, 227
Homo, 13; appearance of, 15–18; species of, 29–31
Homo sapiens, 19; birthplace of, 39–40. *See also* humans
Hong Liangji, 270–71
horses, 73–74, 193–96
Hublin, Jean-Jacques, 38–39
humanism, 274–76, 355; evolutionist, 26–276; as explanation for West overtaking China, 276; liberal, 275–76; socialist, 275–76. *See also* transhumanism
humans, 6, 10; genome of, 352; modern, competing theories explaining emergence of, 30–31; momentum and, 9–109. See also *Homo*; *Homo sapiens*
Humboldt, Alexander von, 257
Hume, David, 273, 276–77
Huntington, Ellsworth, xiv, 305
Huntington, Samuel, 252
hybridization, 57, 325–26, 410n5

illness, 74, 76, 136, 137–38, 144–45, 168, 174, 176–78, 183–85, 190, 191, 229, 287–88. *See also* germs
India: Brahmanism, 104–5; Buddhism, 104–6; economy of, in early modern times, 219–20; empires of, 132–34; Hinduism, 104–5; Jainism, 104; as mercantile power, 218–19; Monkey and, xi; Mughal Empire, 218–19, 240; Vedism, 104
Indigenous people of Americas, destruction of, 168–69
industrial capitalism, 344
industrial revolution (general), 253
Industrial Revolution, the, 265–66, 381–82; greenhouse gases and, 290; mobilization of capital and, 272; stages of early, 267–70
Indus Valley, Bronze Age in, 86–87
information, horizontal circulation of, 345
international law, founding of, 213
internet, 340, 342–43
invisible hand, 273
Irish potato famine, 297–98
irrigation systems, 136
Islam, 110, 147–48
Italy, 254

Jainism, 104
Japan, 239–40, 254; invention of China in, 59
Jaspers, Karl, 99
Jefferson, Thomas, 26

Jenner, Edward, 183
Jesuit order, 255–58
Jones, Eric L., 252
Judaism, 108–9; Rabbinic, 110
Justinian, reign of, 144–45
Justinian plague, 176, 177–78

Keeling, Charles D., 306, 374
Kenyanthropus, 20
Keyes, David, 146
Khaldun, Ibn, 139–41, 160–61
Klein, Naomi, 375, 383–84
Krakatoa volcano eruption, 289
Krause, Bernie, 327
Kublai Khan, 160
Kurzweil, Ray, 356

laissez-faire economics, 273
language, 29, 337–38
Lao Tzu, 102
Late Antique Little Ice Age, 146–47
Leclerc, Georges Louis, comte de Buffon, 26, 303
legalism, 102–3
Leiber, Fritz, 335
Le Moyne, Charles III, second Baron of Longueuil, 25–26
Lepanto, battle of, 222
Le Roy Laduriein, Emmanuel, 238–39
Levant, forests and, 245–46
Lewis, Simon, 229
liberal humanism, 275–76
liberalism, 272; in China, 368
Lieberman, Daniel E., 21
life expectancies, 265–66
Little Ice Age, 76, 162, 169, 187, 216, 226–29, 288–89; in East Asia, 231–32
loess, use of, in China, 60
Lucretius, 113–14
Luddites, 278
Lyell, Charles, 34
Lyssenko, Trofim Denissovitch, 317

Mackenzie, George, 289
Maddison, Angus, 322
Mahabharata, 95–99
Maislin, Mark A., 229
maize, 62–65, 190; in China, 196
malaria, 181–82, 255–56
Malthus, Thomas, 270–71
Malthusian check, 254
mammoths, 43

Man and Nature (Marsh), xiv
Manhattan Project, 316
manioc, 62
Man's Role in Changing the Face of the Earth (Thomas), xiv
Mao Zedong, 103, 316–17
market capitalism, 268
Marsh, George Perkins, xiv
marsupial mammals, 13
Marx, Karl, 272
mass media, 345
mastodons, 43
Maunder, Annie Russell, 2228
Maunder, Edward Walter, 228
Maunder minimum, 228–29
Maurya dynasty, 133
Mazdayasna (Zoroastrianism), 108
McNeill, John R., 180, 309, 310, 322, 330
McNeill, William H., xiv–xv, 252
Meadows report (1972), 309–10
Medieval Warm Period, 229
memory, digital, 339
memory, externalized, 337–39
mercantilism, 214
mercury, silver mining and, 208–12, 392
Mesoamerica, agriculture in, 62–65
Mesopotamia, Bronze Age in, 87
metalwork, 80
microbes. *See* germs
middle classes, 369–70, 382; expansion of, and pollution, 372
Midgley, Thomas, 328
military revolution, 253
Minamata disease, 362
Ming dynasty, 232–34
Mintz, Sidney W., 198
Mitchell, Timothy, 278–81
mitochondria Eve, theory of, 38
mobile engines, 267
Modi, Narendra, 372
Mohism, 103–4
momentum, humans and, 9–10
money, 122–26
Monkey, x–xiii, xvi; China and, xi; India and, xi; seven revolutions in saga of, xii–xiii
monocultures, 320
monotheism, 107–11
Moral Revolution, xiii, xviii, 115, 334, 380
Muhammed, 147
multitude, concept of, 344
mummies, powdered, eating of, 203–4
Mungo Lady and Mungo Man, 45

Musk, Elon, 332
mutation: in germs, 175; in humans, 83, 260, 358; unwitting, 359–67; in wheat, 51

nanotechnologies, 351–52, 354
Napoleon Bonaparte, 182, 183, 239, 267–68, 285–86
Nash, Roderick F., xiv
NBIC technologies (nanotechnoloogy, biotechnology, information technology, cognitive science), xviii, 350–55, 356
Negri, Antonio, 344
neoliberal ideology, 356, 382
Netherlands, the, 253–54; windmills in, 262–63
Newcomen, Thomas, 264
new empire, theory of, 344
New Zealand, 195
nitrate fertilizers, 312–13
noosphere, 336
North Atlantic Drift, 54
Northern Song dynasty, 254
Nuygens, Christiaan, 264

obesity, 367
ocean liners, 267
oil, 280–81
Olduvai Gorge (Tanzania), 22–23
omnivores, 22
one-child policy (China), ramifications of, 371–72
opium, East India Company and, 251–52
Orrorin, 7–8
Ottoman Empire, 234–36, 240
Ötzi, 80–82
Out of Africa 2 hypothesis, 40, 41
owls, 323
ozone layer, 328, 329

Pääbo, Svante, 41
pachyderms. *See* elephants
Palliser's Triangle (Canada), 321
Panama, Isthmus of, formation of, 11–12
paper, 130, 217
Paranthropus, 11, 13, 14, 17
Parker, Geoffrey, 228, 236, 238
Patagonia, 300–302
pesticides, 314, 318, 320, 362
philosophy, 111–14; Chinese, 101–4
Physiological Revolution, xii, xvii, 15, 21–22, 380
pigs, 72

Pires, Tomé, 166
Pizarro, Francisco, 177
placental mammals, 12–13
Plagues and Peoples (McNeil), xiv–xv
Pleistocene era, 33–37
plow, 266; invention of, 130–31
political conflict, climate change and, 236–39
political revolution, 253
pollution: air, 259–62, 329–30, 370, 392; chemical, 359–65; expansion of middle class and, 372; industrial waste, 341; mercury, 392
Pommeranz, Kenneth, 252, 254
postmedia era, theory, 345
potatoes, 62, 196
pottery, discovery of, 57–58; in China, 59; in Japan, 59
printing, 268, 338; capitalism and, 338
proboscideans, 43–44
Prometheus, xii, xvi
protectionism, 266; capitalism and, 273
Pyne, Stephen J., 32

Qing dynasty, 197
Qin Shi Huang, 103, 211
quinine, 255–58

rabbits, 175–76
reductions (in missions), 256–57
Reformation, 338
refrigeration, 328–29
religion, 390–91; hierarchical, 67–68
Revelle, Roger, 306
revolutions: economic, 253–54; industrial, 253; military, 253; political, 253; scientific, 253
rice, cultivation of, 61–62
robots, 346–47
Rolfe, John, 186
Roman Empire, fall of, 137–38, 142–43
Rome, 126–29
Royal Society, 305

San people, 389
Sargon the Great, 85–86
Schumpeter, Joseph, 267
science, experimental, 338
scientific reasoning, 275
scientific revolution, 253
Scott, James C., 69
Scott, Ridley, 323
sea cows, 298–99

Seldon, John, 213
Shakyamuni (Buddha), 105
Shang Yang, 102-3
Shankara, 105
Shelley, Mary, 289-91
Shelley, Percy, 289-90
silence, conquering the planet, 327-28
silk roads, 131-32
silver, mining and refining of, 208-12
slavery, 276-77, 282-84; American Indian tribes and, 195-96; moral abolition and, 283
slaves, energy, 283-84
slave trade, 180-81, 190; sugar and, 198-203
Sloterdijk, Peter, 204
Smith, Adam, 272-73, 276
Smith, Robert, 26
smog, 329-30, 392
socialist humanism, 275-76
Song dynasty, 155-56; decline of, 156-58
sound, absence of, 327-28
Spain, 151, 254
Spanish flu, 183-84
Spinrad, Norman, 348
state, the, 391
steam engines, 254-55, 259, 263, 267, 311, 329, 418n6
stoics, 112
Suess, Hans E., 306
sugarcane, 197-203, 255
Sunnism, 160
sweet potatoes, in China, 196
synthetic biology, 352-53

Talas, Battle of, 149
Tambora volcanic eruption, 285-86; climate effects of, 286-88; reasons for impact of, 288-89
Tang dynasty, 154-55
Taoism, 102
tea, 249, 255
telegraph, 267, 292-93
television, 346
Tenochtitlan, 167, 170-72
terraces, building of, 136
textile industry, British, 250-51, 265
Theodosius, emperor, 141-42
Third Agricultural Revolution (Green Revolution), 325, 382
Thirty Years War (1618-1648), 236-39
This Changes Everything (Klein), 375, 383-84
Thomas, William M., xiv

Thoreau, Henry David, 304
time: acceleration of, 337-41; centralization of, 269-70
tin, 79-80, 85
tobacco, 186-89
Toba hypothesis, 38
tools, 19-21
Tordesillas, Treaty of, 213
Toumaï (*Sahelanthropus tchadensis*), 6-7, 8
Traces on the Rhodian Share (Glacken), xiv
trade unions, 278
trains, steam, 261, 267-69
transhumanism, 355-56. *See also* humanism
Trump, Donald, 372
tunnel history, xvi-xvii
Turing, Alan, 339-40
Turner, William, 286
typhus, 288

Umayyad Empire, 148
unions, 278
United Provinces. *See* Netherlands, the
urbanization, 260, 269, 279

Vedas/Vedism, 104-5
Vikings, 229-31
Vitalis, Andre, 344
volcanoes, 38, 146, 161-62, 227, 228, 260, 285-89, 392

Waldensians, 143
Wang Mang, 136-37
war, 67, 313-18, 379, 383
Ward-Perkins, Bryan, 142
water mills, 252; invention of, 131
Watson, Andrew M., 149-50
Watt, James, 264, 311
weapons, 23, 37, 67, 87, 125, 221, 223-24; biological, 175-76, 185
Werdelin, Lars, 23, 28
whale hunting, 299-300
wheat, 50-52, 390; civilization of, 56-58
White, Sam, 228
Wilderness and the American Mind (Nash), xiv
Wilson, Alan, 38
windmills, 252, 270
Wittfogel, Karl August, 66
wool trade, English, 215-16
workers' rights, 277-80
worms. *See* earthworms
writing, 78, 82, 90-92, 294, 338, 345

yellow fever, 181–82
Younger Dryas, 54, 227, 330
Yuan dynasty, 232
Yuanzhen slough, 233

Zhang, David D., 238
Zomia zone, 68–69
Zoroastrianism (Mazdayasna), 108